Partial Differential Equations with Variable Exponents

Variational Methods and Qualitative Analysis

MONOGRAPHS AND RESEARCH NOTES IN MATHEMATICS

Published Titles

Application of Fuzzy Logic to Social Choice Theory, John N. Mordeson, Davender S. Malik and Terry D. Clark

Blow-up Patterns for Higher-Order: Nonlinear Parabolic, Hyperbolic Dispersion and Schrödinger Equations, Victor A. Galaktionov, Enzo L. Mitidieri, and Stanislav Pohozaev

Difference Equations: Theory, Applications and Advanced Topics, Third Edition, Ronald E. Mickens

Dictionary of Inequalities, Second Edition, Peter Bullen

Iterative Optimization in Inverse Problems, Charles L. Byrne

Modeling and Inverse Problems in the Presence of Uncertainty, H. T. Banks, Shuhua Hu, and W. Clayton Thompson

Monomial Algebras, Second Edition, Rafael H. Villarreal

Partial Differential Equations with Variable Exponents: Variational Methods and Qualitative Analysis, Vicenţiu D. Rădulescu and Dušan D. Repovš

Signal Processing: A Mathematical Approach, Second Edition, Charles L. Byrne

Sinusoids: Theory and Technological Applications, Prem K. Kythe

Special Integrals of Gradshetyn and Ryzhik: the Proofs – Volume I, Victor H. Moll

Forthcoming Titles

Actions and Invariants of Algebraic Groups, Second Edition, Walter Ferrer Santos and Alvaro Rittatore

Analytical Methods for Kolmogorov Equations, Second Edition, Luca Lorenzi

Complex Analysis: Conformal Inequalities and the Bierbach Conjecture, Prem K. Kythe

Computational Aspects of Polynomial Identities: Volume l, Kemer's Theorems, 2nd Edition Belov Alexey, Yaakov Karasik, Louis Halle Rowen

Cremona Groups and Icosahedron, Ivan Cheltsov and Constantin Shramov

Geometric Modeling and Mesh Generation from Scanned Images, Yongjie Zhang

Groups, Designs, and Linear Algebra, Donald L. Kreher

Handbook of the Tutte Polynomial, Joanna Anthony Ellis-Monaghan and Iain Moffat

Lineability: The Search for Linearity in Mathematics, Juan B. Seoane Sepulveda, Richard W. Aron, Luis Bernal-Gonzalez, and Daniel M. Pellegrinao

Forthcoming Titles (continued)

Line Integral Methods and Their Applications, Luigi Brugnano and Felice Iaverno

Microlocal Analysis on R^n and on NonCompact Manifolds, Sandro Coriasco

Nonlinear Functional Analysis in Banach Spaces and Banach Algebras: Fixed Point Theory Under Weak Topology for Nonlinear Operators and Block Operators with Applications Aref Jeribi and Bilel Krichen

Practical Guide to Geometric Regulation for Distributed Parameter Systems, Eugenio Aulisa and David S. Gilliam

Reconstructions from the Data of Integrals, Victor Palamodov

Special Integrals of Gradshetyn and Ryzhik: the Proofs – Volume II, Victor H. Moll

Stochastic Cauchy Problems in Infinite Dimensions: Generalized and Regularized Solutions, Irina V. Melnikova and Alexei Filinkov

Symmetry and Quantum Mechanics, Scott Corry

MONOGRAPHS AND RESEARCH NOTES IN MATHEMATICS

Partial Differential Equations with Variable Exponents

Variational Methods and Qualitative Analysis

Vicenţiu D. Rădulescu

King Abdulaziz University

Jeddah, Saudi Arabia

Dušan D. Repovš

University of Ljubljana

Ljubljana, Slovenia

CRC Press
Taylor & Francis Group
Boca Raton London New York

CRC Press is an imprint of the
Taylor & Francis Group, an **informa** business
A CHAPMAN & HALL BOOK

CRC Press
Taylor & Francis Group
6000 Broken Sound Parkway NW, Suite 300
Boca Raton, FL 33487-2742

Version Date: 20150526

ISBN-13: 978-1-4987-0341-3 (hbk)

Visit the Taylor & Francis Web site at
http://www.taylorandfrancis.com

and the CRC Press Web site at
http://www.crcpress.com

Vicenţiu D. Rădulescu dedicates this book with deep gratitude to the memory of his beloved parents, Professor Dumitru Rădulescu (1914–1982) and Ana Rădulescu (1923–2011)

Contents

Preface xiii

List of Figures xv

List of Tables xvii

Symbol Description xix

About the Authors xxi

I Isotropic and Anisotropic Function Spaces 1

1 Lebesgue and Sobolev Spaces with Variable Exponents 3

1.1 History of function spaces with variable exponents 4
1.2 Lebesgue spaces with variable exponents 7
1.3 Sobolev spaces with variable exponents 10
1.4 Dirichlet energies and Euler-Lagrange equations 11
1.5 Lavrentiev phenomenon 12
1.6 Anisotropic function spaces 14
1.7 Orlicz spaces 16
1.8 Glossary 20

II Variational Analysis of Problems with Variable Exponents 23

2 Nonlinear Degenerate Problems in Non-Newtonian Fluids 25

2.1 Physical motivation 26
2.2 A boundary value problem with nonhomogeneous differential operator 27
2.2.1 Properties of the energy functional 30
2.2.2 A mountain pass-type critical point 37
2.3 Nonlinear eigenvalue problems with two variable exponents . 42
2.3.1 Ekeland variational principle versus the mountain pass geometry 43

2.4 A sublinear perturbation of the eigenvalue problem associated to the Laplace operator . . . 48
2.5 Variable exponents versus Morse theory and local linking . . 54
2.5.1 Verification of the compactness condition 57
2.5.2 Computation of critical groups 61
2.6 The Caffarelli-Kohn-Nirenberg inequality with variable exponents . . . 66
2.6.1 A compact embedding 69
2.6.2 Application to a nonlinear problem with three variable exponents . . . 71
2.7 Glossary 73

3 Spectral Theory for Differential Operators with Variable Exponents 75

3.1 Continuous spectrum for differential operators with two variable exponents . . . 76
3.2 A nonlinear eigenvalue problem with three variable exponents and lack of compactness . . . 82
3.3 Concentration phenomena: the case of several variable exponents and indefinite potential . . . 92
3.3.1 Concentration of the spectrum 95
3.3.2 Optimization of the eigenvalues 102
3.3.3 Further properties on closed balls 103
3.4 Anisotropic problems with lack of compactness and nonlinear boundary condition . . . 107
3.5 Glossary 115

4 Nonlinear Problems in Orlicz-Sobolev Spaces 117

4.1 Existence and multiplicity of solutions 118
4.1.1 Some properties of Orlicz-Sobolev spaces 122
4.1.2 Absorption case: existence of infinitely many solutions 125
4.1.3 Desorption case: existence of solutions for large parameters . . . 132
4.2 A continuous spectrum for nonhomogeneous operators 134
4.2.1 Case of small positive parameters 136
4.2.2 Existence of low-energy solutions 140
4.3 Nonlinear eigenvalue problems with indefinite potential . . . 145
4.3.1 Concentration properties of the spectrum 149
4.3.2 Optimization property for the Rayleigh quotient . . . 156
4.3.3 More about the Rayleigh quotient on spheres 158
4.4 Multiple solutions in Orlicz-Sobolev spaces 161
4.4.1 More about Orlicz-Sobolev spaces 163
4.4.2 Multiplicity versus the three critical points theorem . 166

4.5 Neumann problems in Orlicz-Sobolev spaces 171
4.5.1 More about generalized Orlicz-Sobolev spaces 174
4.5.2 Conditions for concentration of the spectrum 179
4.6 Glossary 191

III Anisotropic Problems: Continuous and Discrete 193

5 Anisotropic Problems 195

5.1 Eigenvalue problems for anisotropic elliptic equations 196
5.1.1 Existence of weak solutions 199
5.1.2 Concentration properties of the spectrum 203
5.2 Combined effects in anisotropic elliptic equations 207
5.2.1 Existence of infinitely many high-energy solutions . . 212
5.2.2 Case of small positive parameters 216
5.3 Anisotropic problems with no-flux boundary condition 220
5.3.1 The first domination case 223
5.3.2 The second domination case 229
5.4 Bifurcation for a singular problem modelling the equilibrium of anisotropic continuous media 233
5.4.1 Existence of low-energy solutions 238
5.4.2 Additional properties and examples 243
5.5 Glossary 247

6 Difference Equations with Variable Exponents 249

6.1 Eigenvalue problems associated to anisotropic difference operators 250
6.1.1 Location properties for the spectrum 253
6.2 Homoclinic solutions of difference equations with variable exponents 258
6.3 Low-energy solutions for discrete anisotropic equations . . . 268
6.4 Glossary 278

Appendix A. Ekeland Variational Principle 279

Appendix B. Mountain Pass Theorem 283

Bibliography 289

Index 297

Preface

> A wide variety of physically significant problems arising in very different areas (such as electricity, hydrodynamics, heat, magnetism, optics, elasticity, etc.) have a family resemblance and should be treated by common methods.
>
> Henri Poincaré, [80]

Partial differential equations are a precise, elegant, rich, and captivating subject, which is quite old, and its history is broad and deep. Elliptic partial differential equations are startling due to their elegance and clarity. One progresses very rapidly from the basics of the *linear* Laplace, Poisson, and Helmholtz equations to profound results concerning *nonlinear* elliptic problems and the qualitative analysis of their solutions. The study of physical models has remained up to the present one of the fundamental concerns of the development of partial differential equations. The idea of using analytic methods to study partial differential equations has its foundations in the ideas of Poincaré. This modern approach provides a new way to view the analysis of linear or nonlinear partial differential equations. As stated by Brezis and Browder [11], "Poincaré emphasized that a wide variety of physically significant problems arising in very different areas (such as electricity, hydrodynamics, heat, magnetism, optics, elasticity, etc.) have a family resemblance–un *air de famille* in Poincaré's words–and should be treated by common methods."

Presented from the point of view of modern research work in the field, this monograph is intended for researchers and graduate students who wish to obtain a thorough introduction to the theory of nonlinear partial differential equations with *variable exponents*, in particular, those of elliptic type. Our main purpose is to introduce the reader to the most important variational methods in the case of elliptic partial differential equations described by nonhomogeneous differential operators and containing one or more power-type nonlinearities with variable exponents. We give a systematic treatment of the basic mathematical theory and constructive methods for these classes of nonlinear elliptic equations, as well as their applications to various processes arising in applied sciences.

The analysis developed in this book is based on the notion of a *generalized* or *weak solution*. This concept is one of the most elegant and most powerful in partial differential equations and the introduction of this notion represents a central methodological turning point in the development of this field of mathematics. Weak solutions are in strong relationship with *Sobolev spaces*, which occur in a wide range of questions, in both pure and applied mathematics. These function spaces turned out to be one of the most powerful tools in analysis created in the 20th century.

Our approach leads not only to the basic results of existence and multiplicity of weak solutions, but also to several qualitative properties, including spectral analysis, bifurcation, or asymptotic analysis. The authors skillfully weave these varied topics to reveal a number of surprising interactions. Perhaps more importantly, the topics are presented with an understanding and explanation of their interrelations with other important parts of mathematics, including topology, differential geometry, mathematical physics, and potential theory. Even though the book examines elliptic equations with variable exponent from many different points of view, it uses the calculus of variations as its unifying theme.

The excellent books by Brezis [10] and Ciarlet [18] present some of the most appropriate material at the interplay between functional analysis and linear or nonlinear partial differential equations.

It is a pleasure to thank our editor Bob Stern for encouraging us to write this book, and for making the process as smooth and carefree as possible.

The work on this monograph was partially supported by Slovenian Research Agency grant No. P1-0292-0101.

Vicenţiu D. Rădulescu and Dušan D. Repovš
January 2015

List of Figures

1.1 Graph of the function defined in Example 1 for $p = 4$ 18
1.2 Graph of the function defined in Example 2 for $p = s = 2$ 18
1.3 Graph of the function defined in Example 3 for $p = 3$ 19

2.1 Graph of the function defined in (2.3) for $\lambda = 1$, $\gamma = 4$, $\beta = 3$ 29
2.2 Graph of the function defined in Example 6(i) 49
2.3 Graph of the function defined in Example 6(ii) 50
2.4 A special case of the function defined in Example 6(iii) 50

4.1 Graph of the function $g(t) = 3t|t| + 2t \sin(\sin(t)) + t^2 \cos(t) \cos(\sin(t))$, $t \in [-25, 25]$ 180
4.2 Graph of the function $\Phi(t)$, $t \in [-8, 8]$ 190

A.1 Geometric illustration of the Ekeland variational principle 281
B.1 Mountain pass landscape between "villages" e_0 and e_1 285

List of Tables

1.1 Dirichlet energy integral and the corresponding Euler-Lagrange equation 12
1.2 Admissible potentials 20

2.1 Variable exponent potentials and the corresponding differential operators 29
2.2 Admissible nonlinearity not satisfying the Ambrosetti-Rabinowitz condition 55

3.1 Existence of solutions depending on the values of the parameter 109

4.1 Admissible potentials 121
4.2 Homeomorphisms and distinguished values 146

Symbol Description

Symbol	Description
$L^{p(\cdot)}$	Lebesgue space with variable exponent
$W^{1,p(\cdot)}$	Sobolev space with variable exponent
$W_0^{1,\vec{p}(\cdot)}$	Anisotropic Sobolev space
L_Φ	Orlicz space
W^1L_Φ	Orlicz-Sobolev space
$W_0^1L_\Phi$	Orlicz-Sobolev space of functions vanishing on the boundary
$\rho_{p(\cdot)}$	Modular
$\Delta_{p(\cdot)}$	$p(\cdot)$-Laplace operator
$\Delta_{\vec{p}(\cdot)}$	Anisotropic Laplace operator
$*$	Convolution product
$\langle\cdot\rangle$	Duality product
Du	Symmetric part of ∇u
$\Delta u(k)$	Forward difference operator
$\Delta^2_{p(k)}$	$p(k)$-Laplace difference operator
$\rightharpoonup$	Weak convergence
$C^{0,\frac{1}{\lvert\log t\rvert}}$	Space of logarithmic Hölder continuous functions
∇	Gradient of a scalar field
div	Divergence of a vector field
lim inf	Inferior limit
lim sup	Superior limit
χ	Characteristic function of a set
$p^*(\cdot)$	Critical Sobolev exponent
K_Φ	Orlicz class
$\mathbf{AC}[0,1]$	Set of absolutely continuous trajectories
$\mathbf{Lip}[0,1]$	Set of Lipschitz trajectories
$\lvert u\rvert_{p(\cdot)}$	Luxemburg norm
G_σ	Gaussian filter
$\mathbb{Z}[a,b]$	Discrete interval
$\frac{\partial}{\partial\nu}$	Neumann derivative
Δ	Laplace operator
$\partial\Omega$	Boundary of open set Ω
$\Phi^\star$	Complementary function of Φ
$\Phi_\star^{-1}$	Orlicz-Sobolev conjugate
$\lvert\Omega\rvert$	Lebesgue measure of Ω
$L^p(\Omega;w)$	Weighted Lebesgue space
$\mathcal{D}_a^{1,2}(\Omega)$	Weighted Sobolev space
$C_q(\Phi,u)$	Critical group of Φ at u
$C_q(\Phi,\infty)$	Critical group of Φ at infinity
$\mathcal{P}$	Family of continuous paths

About the Authors

Vicenţiu D. Rădulescu received his Ph.D. at the Université Pierre et Marie Curie (Paris 6) in 1995 under the supervision of Haim Brezis. In 2003 he defended his Habilitation *Mémoire* at the same university. Rădulescu is Distinguished Adjunct Professor at the King Abdulaziz University of Jeddah, Professorial Fellow at the "Simion Stoilow" Mathematics Institute of the Romanian Academy, and Professor of Mathematics at the University of Craiova. He is the author of more than 200 research papers in nonlinear analysis and several books, including *Variational and Nonvariational Methods in Nonlinear Analysis and Boundary Value Problems* (Kluwer, 2003), *Singular Elliptic Problems: Bifurcation and Asymptotic Analysis* (Oxford University Press, 2008), *Problems in Real Analysis: Advanced Calculus on the Real Axis* (Springer, 2009), *Variational Principles in Mathematical Physics, Geometry and Economics: Qualitative Analysis of Nonlinear Equations and Unilateral Problems* (Cambridge University Press, 2010), *Nonlinear PDEs: Mathematical Models in Biology, Chemistry and Population Genetics* (Springer, 2012), *Variational Methods for Nonlocal Fractional Problems* (Cambridge University Press, 2015). He was a Highly Cited Researcher (2014). He was elected to the Accademia Peloritana dei Pericolanti (2014).

Dušan D. Repovš received his Ph.D. at the Florida State University in 1983. Repovš is Professor of Mathematics at the University of Ljubljana and Head of the Topology, Geometry and Nonlinear Analysis Group at the Institute of Mathematics, Physics and Mechanics in Ljubljana. He is the author of more than 300 research papers in mathematics and the book *Continuous Selections of Multivaled Mappings* (Kluwer, 1998). He has given more than 400 invited talks at conferences and universities around the world. He has directed numerous national and international research grants. He was awarded the Ambassador for Science of the Republic of Slovenia Award (1995), the Republic of Slovenia Science Award (1997), the Bogolyubov Memorial Medal (2009) and the Honorary Doctorate of the University of Craiova (2014). He was elected to the Engineering Academy of Slovenia (1996), the New York Academy of Sciences (1998) and the European Academy of Sciences (2004). He is a member of the American Mathematical Society, the European Mathematical Society, the London Mathematical Society, the Mathematical Society of Japan, the Moscow Mathematical Society, Société Mathématique de France, Société Mathématique de Suisse, Unione Matematica Italiana, and Österreichische Mathematische Gesellschaft.

Part I

Isotropic and Anisotropic Function Spaces

Chapter 1

Lebesgue and Sobolev Spaces with Variable Exponents

1.1 History of function spaces with variable exponents 4
1.2 Lebesgue spaces with variable exponents 7
1.3 Sobolev spaces with variable exponents 10
1.4 Dirichlet energies and Euler-Lagrange equations 11
1.5 Lavrentiev phenomenon 12
1.6 Anisotropic function spaces 14
1.7 Orlicz spaces 16
1.8 Glossary 20

> The principal aim of mathematical education is to develop certain faculties of the mind, and among these, intuition is not the least precious. It is through it that the mathematical world remains in touch with the real world.
>
> Henri Poincaré (1854–1912),
> Science et Méthode, Paris, 1908

One of the reasons for the huge development of the theory of classical Lebesgue and Sobolev spaces L^p and $W^{1,p}$ (where $1 \leq p \leq \infty$) is the description of many phenomena arising in applied sciences. For instance, many materials can be modeled with sufficient accuracy using the function spaces L^p and $W^{1,p}$, where p is a fixed constant. For some nonhomogeneous materials, for instance electrorheological fluids (sometimes referred to as "smart fluids"), this approach is not adequate, but rather the exponent p should be allowed to vary. This leads us to the study of variable exponent Lebesgue and Sobolev spaces, $L^{p(x)}$ and $W^{1,p(x)}$, where p is a real–valued function.

1.1 History of function spaces with variable exponents

Variable exponent Lebesgue spaces appeared in the literature in 1931 in the paper by Orlicz [76]. He was interested in the study of function spaces that contain all measurable functions $u : \Omega \to \mathbb{R}$ such that

$$\rho(\lambda u) = \int_\Omega \varphi(\lambda |u(x)|)dx,$$

for some $\lambda > 0$ and φ satisfying some natural assumptions, where Ω is an open set in $\mathbb{R}^N$. This space is denoted by L^φ and it is now called *Orlicz space.* However, we point out that in [76] the case $|u(x)|^{p(x)}$ corresponding to variable exponents was not included. In the 1950's these problems were systematically studied by Nakano [75], who developed the theory of *modular function spaces.* Nakano explicitly mentioned variable exponent Lebesgue spaces as an example of more general spaces he considered, see Nakano [75, p. 284]. Later, Polish mathematicians investigated the modular function spaces, see Musielak [74]. Variable exponent Lebesgue spaces on the real line have been independently developed by Russian researchers. In that context, we refer to the work of Tsenov [103] and Sharapudinov [97]. They were interested in the minimization of functionals like

$$\int_a^b |u(x) - v(x)|^{p(x)} dx,$$

where u is a given function and v varies over a finite dimensional subspace of $L^{p(x)}[a, b]$. Zhikov [106] started a new direction of investigation, which created the relationship between spaces with variable exponent and variational integrals with nonstandard growth conditions. We also point out the contributions of Marcellini [48], who studied minimization problems with (p, q)-growth, namely

$$\inf \int_\Omega F(x, |\nabla u|)dx,$$

where $t^p \leq F(x, t) \leq t^q + 1$ for all $t \geq 0$. The case corresponding to the variable exponent corresponds to $F(x, t) = t^{p(x)}$, where $p : \Omega \to (1, \infty)$ is a bounded function.

In 1991, Kovacik and Rakosnik [42] established several basic properties of spaces $L^{p(x)}$ and $W^{1,p(x)}$ with variable exponents. Their results were extended by Fan and Zhao [30] in the framework of Sobolev spaces $W^{m,p(x)}$. Pioneering regularity results for functionals with nonstandard growth are due to Acerbi and Mingione [1]. Density of smooth functions in $W^{k,p(x)}(\Omega)$ and related Sobolev embedding properties are due to Edmunds and Rakosnik [23, 24]. We also point out the important contributions of the Finnish research group on variable exponent spaces and image processing, whose main goal was to study nonlinear potential theory in variable exponent Sobolev spaces. The abstract theory of Lebesgue and Sobolev spaces with variable exponents was developed in the monograph by Diening, Harjulehto, Hästö, and Ruzicka [21].

The study of differential equations and variational problems involving $p(x)$-growth conditions is a consequence of their applications. In 1920 Bingham was surprised to discover that some paints do not run like honey. He studied such a behavior and described a strange phenomenon. There are fluids that first flow, then stop spontaneously (Bingham fluids). Inside them, the forces that create the flows reach a threshold. As this threshold is not reached, the fluid flow deforms as a solid. Invented in the 17th century, the "Flemish medium" makes painting oil thixotropic: it flows under pressure of the brush, but freezes as soon as you leave it to rest. While the exact composition of the Flemish medium remains unknown, it is known that the bonds form gradually between its components, which is why the picture freezes in a few minutes. Thanks to this wonderful medium, Rubens was able to paint *La Kermesse* in only 24 hours.

Recent systematic study of partial differential equations with variable exponents was motivated by the description of several relevant models in electrorheological and thermorheological fluids, image processing, or robotics. In what follows, we give two relevant examples that justify the mathematical study of models involving variable exponents. The first example is due to Chen, Levine, Rao [17] and it concerns applications to image restoration. Let us consider an input I that corresponds to shades of gray in a domain $\Omega \subset \mathbb{R}^2$. We assume that I is made up of the true image corrupted by the noise. Suppose that the noise is additive, that is, $I = T + \eta$ where T is the true image and η is a random variable with zero mean. Thus, the effect of the noise can be eliminated by smoothing the input, since this will cause the effect of the zero-mean random variables at nearby locations to cancel. Smoothing corresponds to minimizing the energy

$$\mathcal{E}_1(u) = \int_\Omega (|\nabla u(x)|^2 + |u(x) - I(x)|^2)dx\,.$$

Unfortunately, smoothing destroys the small details of the image, so this procedure is not useful. A better approach is the *total variation smoothing.* Since an edge in the image gives rise to a very large gradient, the level sets around the edge are very distinct, so this method does a good job of preserving edges. Total variation smoothing corresponds to minimizing the energy

$$\mathcal{E}_2(u) = \int_\Omega (|\nabla u(x)| + |u(x) - I(x)|^2)dx\,.$$

Unfortunately, total variation smoothing not only preserves edges, but it also creates edges where there were none in the original image. This is called the *staircase effect.*

Looking at $\mathcal{E}_1$ and $\mathcal{E}_2$, Chen, Levine and Rao suggested that an appropriate energy is

$$\mathcal{E}(u) = \int_\Omega (|\nabla u(x)|^{p(x)} + |u(x) - I(x)|^2) dx \,,$$

where $1 \leq p(x) \leq 2$.

This function should be close to 1 where there are likely to be no edges, and close to 2 where there are likely to be no edges. The approximate location of the edges can be determined by just smoothing the input data and looking where the gradient is large.

The diffusion resulting from the model proposed by Chen, Levine and Rao is a combination of the Gaussian smoothing and regularization based on the total variation. More exactly, the following adaptive model was proposed

$$\min_{I=u+v,\ u\in \mathrm{BV}\cap L^2(\Omega)} \int_\Omega \varphi(x, \nabla u)\, dx + \lambda \cdot \|u\|^2_{L^2(\Omega)} \,, \tag{1.1}$$

where $\Omega \subset \mathbb{R}^2$ is an open domain,

$$\varphi(x,r) = \begin{cases} \frac{1}{p(x)} |r|^{p(x)}, & \text{for} \quad |r| \leq \beta \\ |r| - \frac{\beta \cdot p(x) - \beta^{p(x)}}{p(x)}, & \text{for} \quad |r| > \beta \,, \end{cases}$$

where $\beta > 0$ is fixed and $1 < \alpha \leq p(x) \leq 2$. The function $p(x)$ involved here depends on the location of x in the model. For instance, it can be

$$p(x) = 1 + \frac{1}{1 + k|\nabla G_\sigma * I|^2} \,,$$

where $G_\sigma(x) = \frac{1}{\sigma} \exp(-|x|^2/(4\sigma^2))$ is the Gaussian filter and $k > 0$ and $\sigma > 0$ are fixed parameters. For problem (1.1), Chen, Levine and Rao established the existence and uniqueness of the solution and the long-time behavior of the associated flow of the proposed model. The effectiveness of the model in image restoration is illustrated by some experimental results included in their paper.

The next example is related to electrorheological fluids. The constitutive equation for the motion of an electrorheological fluid is

$$u_t + \operatorname{div} S(u) + (u \cdot \nabla) u + \nabla \pi = f, \tag{1.2}$$

where $u : \mathbb{R}^{3,1} \to \mathbb{R}^3$ is the velocity of the fluid at a point in space-time, $\pi : \mathbb{R}^{3,1} \to \mathbb{R}$ is the pressure, $f : \mathbb{R}^{3,1} \to \mathbb{R}^3$ represents external forces, and the stress tensor $S : W^{1,1}_{loc} \to \mathbb{R}^{3,3}$ is of the form

$$S(u)(x) = \mu(x)[1 + |Du(x)|^2]^{(p(x)-2)/p(x)} Du(x),$$

where $Du = (\nabla u + \nabla u^T)/2$ is the symmetric part of the gradient of u.

We observe that the highest order differential term in (1.2) is

$$\operatorname{div}\left((1+|Du(x)|^2)^{(p(x)-2)/p(x)}Du(x)\right).$$

The degenerate case corresponds to the Laplace operator with variable exponent.

For more applications, we refer to the pioneering work by Halsey [36] and the monograph by Ruzicka [95]. A survey of recent contributions to the study of partial differential equations with variable exponents is contained in the paper by Rădulescu [85].

Our goal in this book is to present recent results concerning the qualitative analysis of nonlinear elliptic partial differential equations with variable exponents and to point out new phenomena with respect to related problems described by homogeneous differential operators. For this reason, we do not focus on the theory of associated function spaces but we refer instead to the book by Diening, Harjulehto, Hästö and Ruzicka [21] for details.

1.2 Lebesgue spaces with variable exponents

With the emergence of nonlinear problems in applied sciences, standard Lebesgue and Sobolev spaces demonstrated their limitations in applications. The class of nonlinear problems with variable exponent growth is a new research field and it reflects a new kind of physical phenomena.

In this section we introduce the Lebesgue spaces with variable exponent and we list their main properties.

Let Ω be a bounded domain in $\mathbb{R}^N$. For any continuous function $p : \Omega \to (1, \infty)$, we denote

$$p^- = \text{ess inf}_{x\in\Omega}p(x) \quad \text{and} \quad p^+ = \text{ess sup}_{x\in\Omega}p(x).$$

Usually it is assumed that $p^+ < +\infty$, since this condition is known to imply many desirable features of the associated variable exponent Lebesgue space $L^{p(x)}(\Omega)$. This function space is defined by

$$L^{p(x)}(\Omega) = \left\{u;\ u \text{ is measurable and } \int_\Omega |u(x)|^{p(x)}\,dx < \infty\right\}.$$

On this space we define the *Luxemburg norm* by the formula

$$|u|_{p(x)} = \inf \left\{ \mu > 0;\ \int_\Omega \left| \frac{u(x)}{\mu} \right|^{p(x)} dx \leq 1 \right\}.$$

Equipped with this norm, $L^{p(x)}(\Omega)$ becomes a Banach space. If $1 < p^- \leq p^+ < \infty$ then $L^{p(x)}(\Omega)$ is a reflexive uniformly convex Banach space. Moreover, for **any** measurable bounded exponent p, the space $L^{p(x)}(\Omega)$ is separable.

For a constant function p, the variable exponent Lebesgue space coincides with the standard Lebesgue space.

Warning

If $1 < p^- \leq p^+ < \infty$ and $p : \overline{\Omega} \to [1, \infty)$ is smooth, then the formula

$$\int_\Omega |u(x)|^p dx = p \int_0^\infty t^{p-1} \left| \{x \in \Omega;\ |u(x)| > t\} \right| dt$$

has **no** variable exponent analogue.

If $0 < |\Omega| < \infty$ and p_1, p_2 are variable exponents such that $p_1(x) \leq p_2(x)$ almost everywhere in Ω, there exists a continuous embedding $L^{p_2(x)}(\Omega) \hookrightarrow L^{p_1(x)}(\Omega)$.

We denote by $L^{p'(x)}(\Omega)$ the conjugate space of $L^{p(x)}(\Omega)$, where $1/p(x) + 1/p'(x) = 1$. For any $u \in L^{p(x)}(\Omega)$ and $v \in L^{p'(x)}(\Omega)$ the Hölder type inequality holds.

$$\left| \int_\Omega uv\, dx \right| \leq \left(\frac{1}{p^-} + \frac{1}{p'^-} \right) |u|_{p(x)} |v|_{p'(x)} \tag{1.3}$$

Warning

Variable exponent Lebesgue spaces do **not** have the *mean continuity property*: if p is continuous and nonconstant in an open ball B, there exists a function $u \in L^{p(x)}(B)$ such that

$u(x+h) \notin L^{p(x)}(B)$ for all $h \in \mathbb{R}^N$ with arbitrary small norm.

Most of the problems in the development of the theory of $L^{p(x)}$ spaces arise from the fact that these spaces are virtually never translation invariant. The use of convolution is also limited: the Young inequality holds if and only if p is constant.

$$\|f * g\|_{L^{p(x)}} \leq C \|f\|_{L^{p(x)}} \|g\|_{L^1}$$

An important role in manipulating the generalized Lebesgue and Sobolev spaces is played by the *modular* of the $L^{p(x)}(\Omega)$ space, which is the mapping $\rho_{p(x)} : L^{p(x)}(\Omega) \to \mathbb{R}$ defined by

$$\rho_{p(x)}(u) = \int_\Omega |u(x)|^{p(x)}\, dx.$$

Assuming that p is unbounded, we have $p^+ = +\infty$. In this case, we denote $\Omega_\infty = \{x \in \Omega;\ p(x) = +\infty\}$ and define the modular by setting

$$\rho_{p(x)}(u) = \int_{\Omega \setminus \Omega_\infty} |u(x)|^{p(x)}\, dx + \text{ess sup}_{x \in \Omega_\infty} |p(x)|\,.$$

If (u_n), $u \in L^{p(x)}(\Omega)$ then the following relations hold:

$$|u|_{p(x)} > 1 \quad \Rightarrow \quad |u|_{p(x)}^{p^-} \leq \rho_{p(x)}(u) \leq |u|_{p(x)}^{p^+}\,, \tag{1.4}$$

$$|u|_{p(x)} < 1 \quad \Rightarrow \quad |u|_{p(x)}^{p^+} \leq \rho_{p(x)}(u) \leq |u|_{p(x)}^{p^-}\,, \tag{1.5}$$

$$|u_n - u|_{p(x)} \to 0 \quad \Leftrightarrow \quad \rho_{p(x)}(u_n - u) \to 0\,. \tag{1.6}$$

1.3 Sobolev spaces with variable exponents

We define the variable exponent Sobolev space by

$$W^{1,p(x)}(\Omega) = \{u \in L^{p(x)}(\Omega) : |\nabla u| \in L^{p(x)}(\Omega)\}.$$

On $W^{1,p(x)}(\Omega)$ we can consider one of the following equivalent norms

$$\|u\|_{p(x)} = |u|_{p(x)} + |\nabla u|_{p(x)}$$

or

$$\|u\| = \inf\left\{\mu > 0;\ \int_\Omega \left(\left|\frac{\nabla u(x)}{\mu}\right|^{p(x)} + \left|\frac{u(x)}{\mu}\right|^{p(x)}\right) dx \leq 1\right\}.$$

We define $W_0^{1,p(x)}(\Omega)$ as the closure of the set of compactly supported $W^{1,p(x)}$-functions with respect to the norm $\|u\|_{p(x)}$. When smooth functions are dense, we can also use the closure of $C_0^\infty(\Omega)$ in $W^{1,p(x)}(\Omega)$.

When $p^- > 1$, the function spaces $W^{1,p(x)}(\Omega)$ and $W_0^{1,p(x)}(\Omega)$ are reflexive uniformly convex Banach spaces. Moreover, for **any** measurable bounded exponent p, the spaces $W^{1,p(x)}(\Omega)$ and $W_0^{1,p(x)}(\Omega)$ are separable.

Set

$$\varrho_{p(x)}(u) = \int_\Omega (|\nabla u(x)|^{p(x)} + |u(x)|^{p(x)})\, dx.$$

Then for all (u_n), $u \in W_0^{1,p(x)}(\Omega)$ the following holds

$$\|u\| > 1 \Rightarrow \|u\|^{p^-} \leq \varrho_{p(x)}(u) \leq \|u\|^{p^+}, \tag{1.7}$$

$$\|u\| < 1 \Rightarrow \|u\|^{p^+} \leq \varrho_{p(x)}(u) \leq \|u\|^{p^-}, \tag{1.8}$$

$$\|u_n - u\| \to 0 \quad \Leftrightarrow \quad \varrho_{p(x)}(u_n - u) \to 0. \tag{1.9}$$

Next, we recall some embedding results regarding variable exponent Lebesgue–Sobolev spaces. If $p,\ \theta : \Omega \to (1,\infty)$ are Lipschitz continuous and $p^+ < N$ and $p(x) \le \theta(x) \le p^\star(x)$ for any $x \in \Omega$, where $p^\star(x) = Np(x)/(N-p(x))$, then there exists a continuous embedding $W_0^{1,p(x)}(\Omega) \hookrightarrow L^{\theta(x)}(\Omega)$. Furthermore, assuming that Ω_0 is a bounded subset of Ω, the embedding $W_0^{1,p(x)}(\Omega_0) \hookrightarrow L^{\theta(x)}(\Omega_0)$ is continuous and compact.

1.4 Dirichlet energies and Euler-Lagrange equations

> By the *Dirichlet principle* we understand a method for solving boundary value problems via minimum problems for variational integrals. This principle goes back to Gauss (1839), Lord Kelvin and Dirichlet. Weierstarss (1870) was the first to underline the shortcomings of this principle.
>
> David Hilbert (1862–1943)

In many situations arising in applications, solving a linear or a nonlinear partial differential equation is equivalent to minimizing the associated energy functional. When such questions arise in mechanics, physics or other applied sciences, they often have a natural interpretation, in relationship with the least action principle or minimization of the energy. In the language of the calculus of variations, we say that the *Euler-Lagrange equation* is associated with the *minimization* of the associated *energy* functional. Among other numerous applications, this allows us to solve the Dirichlet problem for the Poisson equation by means of the Dirichlet principle.

In Table 1.1 we describe the energy functional and the Euler-Lagrange equation for the cases where p is a constant (resp., variable) exponent. We observe that the latter case gives rise to distinct energies and related associated Euler-Lagrange equations.

We provide the following example related to minimizers of the one-dimensional Dirichlet energy with variable exponents (see [37] for more details). We say that a function $u \in W^{1,p(x)}(0,1)$ is a minimizer with boundary values 0 and $a > 0$ if $u(0) = 0$, $u(1) = a$, and

$$\int_0^1 |u'(y)|^{p(y)}dy \le \int_0^1 |v'(y)|^{p(y)}dy,$$

for all $v \in W^{1,p(x)}(0,1)$ with $v(0) = 0$, $v(1) = a$.

TABLE 1.1: Dirichlet energy integral and the corresponding Euler-Lagrange equation

Energy functional	Euler-Lagrange equation
$\int_\Omega \lvert\nabla u\rvert^p dx$	$\operatorname{div}(\lvert\nabla u\rvert^{p-2}\nabla u) = 0$
$\int_\Omega \lvert\nabla u\rvert^{p(x)} dx$	$\operatorname{div}(p(x)\,\lvert\nabla u\rvert^{p(x)-2}\nabla u) = 0$
$\int_\Omega \frac{1}{p(x)} \lvert\nabla u\rvert^{p(x)} dx$	$\operatorname{div}(\lvert\nabla u\rvert^{p(x)-2}\nabla u) = 0$

If p is constant, then the minimizer is linear, namely $u(x) = ax$. Let us assume that $p(x) = 3\chi_{(0,1/2)} + 2\chi_{(1/2,1)}$. Assume that u is a minimizer and denote $u(1/2) = b$. Then $u|_{(0,1/2)}$ is the solution of the classical energy integral problem with boundary values 0 and b, and $u|_{(1/2,1)}$ is the solution with values b and a. Thus, these functions are linear. Furthermore u has Dirichlet energy $4b^3 + 2(a-b)^2$. The function $b \longmapsto 2b^3 + (a-b)^2$ has a minimum at $b = (\sqrt{12a+1} - 1)/6$, which determines the minimizer of the variable exponent problem. A computation shows that the minimizer is convex if $a > 2/3$, concave if $a < 2/3$ and linear if $a = 2/3$.

1.5 Lavrentiev phenomenon

Zhikov [106] showed that smooth functions are in general *not* dense in $W^{1,p(x)}(\Omega)$. This property is related to the *Lavrentiev phenomenon*, which asserts that there are variational problems for which the infimum over the smooth functions is strictly greater than the infimum over all functions that satisfy the same boundary conditions. An equivalent formulation asserts that a Lagrangian L exhibits the Lavrentiev phenomenon if the infimum taken over the set of absolutely continuous trajectories $\mathbf{AC}[0,1]$ is strictly lower than the infimum taken over the set of Lipschitzian trajectories $\mathbf{Lip}[0,1]$, with fixed boundary conditions. The first example of such a phenomenon is due to Lavrentiev [45], who proved that

$$\inf_{u\in W^{1,1}(0,1),u(0)=0,u(1)=1} \int_0^1 (x-u^3)^2 |u'(x)|^6 dx = 0,$$

while

$$\inf_{u\in W^{1,\infty}(0,1),u(0)=0,u(1)=1} \int_0^1 (x-u^3)^2 |u'(x)|^6 dx > 0.$$

A similar conclusion can be obtained by using the Lagrangian $L(t, x, x') = (x^3 - t^2)x'^6$ on the interval $[0, 1]$.

If p is logarithmic Hölder continuous (notation: $p \in C^{0, \frac{1}{|\log t|}}(\overline{\Omega})$), that is,

$$|p(x) - p(y)| \leq \frac{C}{|\log|x - y||} \quad \forall x, y \in \overline{\Omega}, \ |x - y| \leq 1/2,$$

then the smooth functions are dense in $W^{1,p(x)}(\Omega)$ and so the Sobolev space $W_0^{1,p(x)}(\Omega)$ is the closure of $C_0^\infty(\Omega)$ under the norm $\|\cdot\|$. Edmunds and Rakosnik [23] derived the same conclusion under a local monotonicity condition on p.

Since Ω is bounded and $p \in C_+(\overline{\Omega})$ is logarithmic Hölder continuous, we have

$$|u|_{p(x)} \leq C\,|\nabla u|_{p(x)} \quad \forall\, u \in W_0^{1,p(x)}(\Omega) \ \ \text{[Poincaré inequality]},$$

where $C = C(p, |\Omega|, \operatorname{diam}(\Omega), N)$. The Poincaré inequality holds under a much weaker assumption on p than the Sobolev inequality and embedding, namely that the exponent p is *not too discontinuous.*

If Ω is *bounded* then the following embeddings exist:

$$C^{0,1}(\overline{\Omega}) \subset W^{1,q}(\Omega) \text{ (if } q > N) \subset C^{0, \frac{1}{|\log t|}}(\overline{\Omega}).$$

If Ω is *unbounded,* then p is said to be *logarithmic Hölder continuous* if

$$|p(x) - p(y)| \leq \frac{C}{|\log|x - y||} \quad \forall x, y \in \overline{\Omega}, \ |x - y| \leq 1/2$$

and

$$|p(x) - p(y)| \leq \frac{C}{\log(e + |x|)|} \quad \forall x, y \in \Omega, \ |y| \geq |x|.$$

In this case we cannot require $p \in W^{1,q}(\Omega)$, since $\int_\Omega |p(x)|^q dx = \infty$. Let

$$W^{1,(\infty,q(\cdot))}(\Omega) := \{u \in L^\infty(\Omega);\ |\nabla u| \in L^{q(\cdot)}(\Omega)\},$$

where $N < q_- \leq q_+ < \infty$.

In the case of unbounded domains, we deduce the following property:

If Ω is *unbounded* then the hypotheses
(i) $p \in C^{0,1}(\overline{\Omega})$;
(ii) $p \in W^{1,(\infty,q(\cdot))}(\Omega)$ with $N < q_- \leq q_+ < \infty$;
(iii) $p \in C^{0,\frac{1}{|\log t|}}(\overline{\Omega})$
are *independent* of each other.

1.6 Anisotropic function spaces

In this section, we introduce a natural generalization of the *isotropic* variable exponent Sobolev space $W_0^{1,p(x)}(\Omega)$.

Let $\overrightarrow{p} : \overline{\Omega} \to \mathbb{R}^N$ denote the vector function $\overrightarrow{p} = (p_1, \ldots, p_N)$. We define the anisotropic variable exponent Sobolev space $W_0^{1,\overrightarrow{p}(\cdot)}(\Omega)$, as the closure of $C_0^\infty(\Omega)$ with respect to the norm

$$\|u\|_{\overrightarrow{p}(\cdot)} = \sum_{i=1}^{N} |\partial_{x_i} u|_{p_i(\cdot)} .$$

In the case when $p_i \in C_+(\overline{\Omega})$ are constant functions for any $i \in \{1, .., N\}$, the resulting anisotropic Sobolev space is denoted by $W_0^{1,\overrightarrow{p}}(\Omega)$, where $\overrightarrow{p}$ is the constant vector $(p_1, ..., p_N)$.

The function space $W_0^{1,\overrightarrow{p}}(\Omega)$ is a reflexive Banach space for any $\overrightarrow{p} \in \mathbb{R}^N$ with $p_i > 1$ for all $i \in \{1, .., N\}$. This result can be easily extended to $W_0^{1,\overrightarrow{p}(\cdot)}(\Omega)$. Indeed, denoting by $X = L^{p_1(\cdot)}(\Omega) \times \ldots \times L^{p_N(\cdot)}(\Omega)$ and considering the operator $T : W_0^{1,\overrightarrow{p}(\cdot)}(\Omega) \to X$, defined by $T(u) = \nabla u$, it is clear that $W_0^{1,\overrightarrow{p}(\cdot)}(\Omega)$ and X are isometric by T, since $\|Tu\|_X = \sum_{i=1}^N |\partial_{x_i} u|_{p_i(\cdot)} = \|u\|_{\overrightarrow{p}(\cdot)}$. Thus, $T(W_0^{1,\overrightarrow{p}(\cdot)}(\Omega))$ is a closed subspace of X, which is a reflexive Banach space. It follows that $T(W_0^{1,\overrightarrow{p}(\cdot)}(\Omega))$ is reflexive and consequently, $W_0^{1,\overrightarrow{p}(\cdot)}(\Omega)$ is also a reflexive Banach space.

In order to facilitate working with the space $W_0^{1,\overrightarrow{p}(\cdot)}(\Omega)$ we introduce $\overrightarrow{P}_+$, $\overrightarrow{P}_- \in \mathbb{R}^N$ as

$$\overrightarrow{P}_+ = (p_1^+, ..., p_N^+), \quad \overrightarrow{P}_- = (p_1^-, ..., p_N^-).$$

Set P_+^+, P_-^+, $P_-^- \in \mathbb{R}^+$ defined as

$$P_+^+ = \max\{p_1^+, ..., p_N^+\}, \quad P_-^+ = \max\{p_1^-, ..., p_N^-\}, \quad P_-^- = \min\{p_1^-, ..., p_N^-\}.$$

A fundamental assumption in the theory of anisotropic Sobolev spaces is

$$\sum_{i=1}^{N} \frac{1}{p_i^-} > 1. \tag{1.10}$$

Define $P_-^\star \in \mathbb{R}^+$ and $P_{-,\infty} \in \mathbb{R}^+$ by

$$P_-^\star = \frac{N}{\sum_{i=1}^N 1/p_i^- - 1}, \qquad P_{-,\infty} = \max\{P_-^+, P_-^\star\}.$$

The next result establishes a useful compactness property in the framework of anisotropic Sobolev spaces.

Theorem 1 *Assume that $\Omega \subset \mathbb{R}^N$ ($N \geq 3$) is a bounded domain with smooth boundary. Assume that relation* (1.10) *is fulfilled and $q \in C(\overline{\Omega})$ satisfies*

$$1 < q(x) < P_{-,\infty} \quad \textit{for all } x \in \overline{\Omega}. \tag{1.11}$$

Then the embedding

$$W_0^{1,\vec{p}(\cdot)}(\Omega) \hookrightarrow L^{q(\cdot)}(\Omega)$$

is compact.

Proof. Clearly $L^{p_i(\cdot)}(\Omega)$ is continuously embedded in $L^{p_i^-}(\Omega)$ for any $i \in \{1, ..., N\}$, since $p_i^- \leq p_i(x)$ for all $x \in \overline{\Omega}$. Thus, for each $i \in \{1, ..., N\}$ there exists a positive constant $C_i > 0$ such that

$$|\phi|_{p_i^-} \leq C_i |\phi|_{p_i(\cdot)} \quad \text{for all} \quad \phi \in L^{p_i(\cdot)}(\Omega).$$

If $u \in W_0^{1,\vec{p}(\cdot)}(\Omega)$ then $\partial_{x_i} u \in L^{p_i(\cdot)}(\Omega)$ for each $i \in \{1, ..., N\}$. The above inequalities imply

$$\|u\|_{\vec{P}_-} = \sum_{i=1}^{N} |\partial_{x_i} u|_{p_i^-} \leq C \sum_{i=1}^{N} |\partial_{x_i} u|_{p_i(\cdot)} = C\|u\|_{\vec{p}(\cdot)},$$

where $C = \max\{C_1, ..., C_N\}$. Thus, we can deduce that $W_0^{1,\vec{p}(\cdot)}(\Omega)$ is continuously embedded in $W_0^{1,\vec{P}_-}(\Omega)$. On the other hand, since relation (1.11) holds we infer that $q^+ < P_{-,\infty}$. This fact implies that $W_0^{1,\vec{P}_-}(\Omega)$ is compactly embedded in $L^{q^+}(\Omega)$. Finally, since $q(x) \leq q^+$ for each $x \in \overline{\Omega}$ we can deduce that $L^{q^+}(\Omega)$ is continuously embedded in $L^{q(\cdot)}(\Omega)$. The above piece of information yields the conclusion that $W_0^{1,\vec{p}(\cdot)}(\Omega)$ is compactly embedded in $L^{q(\cdot)}(\Omega)$. □

1.7 Orlicz spaces

Let $\phi_i : \mathbb{R} \to \mathbb{R}$ $(i = 1, 2)$ be odd increasing homeomorphisms from $\mathbb{R}$ onto $\mathbb{R}$. Set

$$\Phi_i(t) = \int_0^t \phi_i(s)\, ds, \quad (\Phi_i)^\star(t) = \int_0^t (\phi_i)^{-1}(s)\, ds, \qquad \text{for all } t \in \mathbb{R},\ i = 1, 2\,.$$

We first observe that Φ_i, $i = 1, 2$, are *Young functions*, that is, $\Phi_i(0) = 0$, Φ_i are convex, and $\lim_{x\to\infty} \Phi_i(x) = +\infty$. We also have $\Phi_i(x) = 0$ if and only if $x = 0$, $\lim_{x\to 0} \Phi_i(x)/x = 0$, and $\lim_{x\to\infty} \Phi_i(x)/x = +\infty$. The functions Φ_i are called *N–functions*. The functions $(\Phi_i)^\star$, $i = 1, 2$, are called the *complementary* functions of Φ_i, $i = 1, 2$, and they satisfy

$$(\Phi_i)^\star(t) = \sup\{st - \Phi_i(s);\ s \geq 0\}, \qquad \text{for all } t \geq 0\,.$$

We observe that $(\Phi_i)^\star$, $i = 1, 2$, are also N–functions and that the Young inequality holds:

$$st \leq \Phi_i(s) + (\Phi_i)^\star(t), \qquad \text{for all } s, t \geq 0\,.$$

The Orlicz spaces $L_{\Phi_i}(\Omega)$, $i = 1, 2$, defined by the N–functions Φ_i, are the spaces of measurable functions $u : \Omega \to \mathbb{R}$ such that

$$\|u\|_{L_{\Phi_i}} := \sup\left\{\int_\Omega uv\, dx;\ \int_\Omega (\Phi_i)^\star(|g|)\, dx \leq 1\right\} < \infty\,.$$

Then $(L_{\Phi_i}(\Omega), \|\cdot\|_{L_{\Phi_i}})$, $i = 1, 2$, are Banach spaces whose norm is equivalent to the Luxemburg norm

$$\|u\|_{\Phi_i} := \inf\left\{k > 0;\ \int_\Omega \Phi_i\left(\frac{u(x)}{k}\right)\, dx \leq 1\right\}.$$

For Orlicz spaces, the Hölder inequality reads as follows:

$$\int_\Omega uvdx \leq 2\,\|u\|_{L_{\Phi_i}}\,\|v\|_{L_{(\Phi_i)^\star}} \qquad \text{for all } u \in L_{\Phi_i}(\Omega) \text{ and } v \in L_{(\Phi_i)^\star}(\Omega),\ i = 1, 2\,.$$

Next, we introduce the Orlicz-Sobolev spaces. We denote by $W^1L_{\Phi_i}(\Omega)$, $i = 1, 2$, the Orlicz-Sobolev spaces defined by

$$W^1L_{\Phi_i}(\Omega) := \left\{u \in L_{\Phi_i}(\Omega);\ \frac{\partial u}{\partial x_i} \in L_{\Phi_i}(\Omega),\ i = 1, ..., N\right\}.$$

These are Banach spaces with respect to the norms

$$\|u\|_{1,\Phi_i} := \|u\|_{\Phi_i} + \||\nabla u|\|_{\Phi_i},\ i = 1, 2\,.$$

We also define the Orlicz-Sobolev spaces $W_0^1 L_{\Phi_i}(\Omega)$, $i = 1, 2$, as the closure of $C_0^\infty(\Omega)$ in $W^1 L_{\Phi_i}(\Omega)$. On $W_0^1 L_{\Phi_i}(\Omega)$ we can consider the equivalent norm

$$\|u\|_i := \||\nabla u|\|_{\Phi_i}.$$

For an easier manipulation of the spaces defined above, we define

$$(\phi_i)_0 := \inf_{t>0} \frac{t\phi_i(t)}{\Phi_i(t)} \quad \text{and} \quad (\phi_i)^0 := \sup_{t>0} \frac{t\phi_i(t)}{\Phi_i(t)},\ i \in \{1, 2\}\,.$$

A natural assumption is that for each $i \in \{1, 2\}$ we have

$$1 < (\phi_i)_0 \leq \frac{t\phi_i(t)}{\Phi_i(t)} \leq (\phi_i)^0 < \infty, \quad \forall\, t \geq 0\,. \tag{1.12}$$

The above relation implies that each Φ_i, $i \in \{1, 2\}$, satisfies the Δ_2-condition, that is,

$$\Phi_i(2t) \leq K\Phi_i(t), \quad \forall\, t \geq 0\,, \tag{1.13}$$

where K is a positive constant.

On the other hand, the following relations hold:

$$\|u\|_i^{(\phi_i)^0} \leq \int_\Omega \Phi_i(|\nabla u|)\, dx \leq \|u\|_i^{(\phi_i)_0}, \quad \forall\, u \in W_0^1 L_{\Phi_i}(\Omega),\ \|u\|_i < 1\,, \tag{1.14}$$

$$\|u\|_i^{(\phi_i)_0} \leq \int_\Omega \Phi_i(|\nabla u|)\, dx \leq \|u\|_i^{(\phi_i)^0}, \quad \forall\, u \in W_0^1 L_{\Phi_i}(\Omega),\ \|u\|_i > 1\,. \tag{1.15}$$

We assume that for each $i \in \{1, 2\}$ the function Φ_i satisfies the following condition

$$\text{the function } [0, \infty) \ni t \longmapsto \Phi_i(\sqrt{t}) \text{ is convex}\,. \tag{1.16}$$

Conditions (1.13) and (1.16) assure that for all $i \in \{1, 2\}$ the Orlicz spaces $L_{\Phi_i}(\Omega)$ are uniformly convex spaces and thus, they are reflexive Banach spaces. This fact also implies that the Orlicz-Sobolev spaces $W_0^1 L_{\Phi_i}(\Omega)$, $i \in \{1, 2\}$, are reflexive Banach spaces.

We provide examples of functions $\phi : \mathbb{R} \to \mathbb{R}$ which are odd, increasing homeomorphisms from $\mathbb{R}$ onto $\mathbb{R}$ and satisfy conditions (1.12) and (1.16).

Example 1 *Let*

$$\phi(t) = p|t|^{p-2}t, \quad \forall\, t \in \mathbb{R}\,,$$

with $p > 1$. *For this function we have*

$$(\phi)_0 = (\phi)^0 = p\,.$$

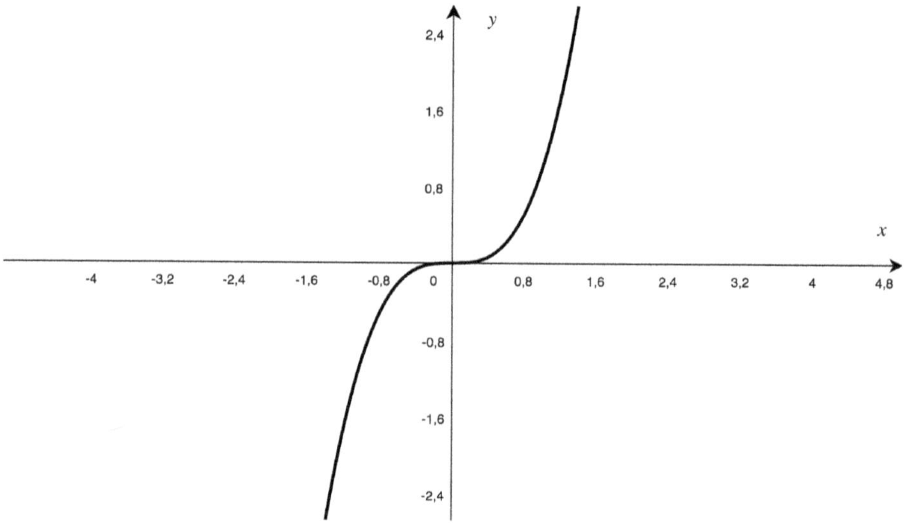

FIGURE 1.1: Graph of the function defined in Example 1 for $p = 4$

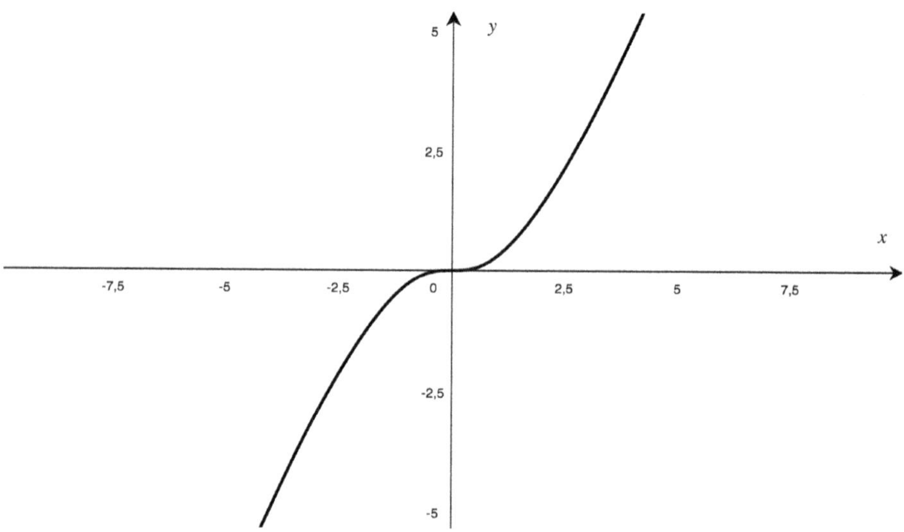

FIGURE 1.2: Graph of the function defined in Example 2 for $p = s = 2$

In this particular case, the corresponding Orlicz space $L_\Phi(\Omega)$ is the classical Lebesgue space $L^p(\Omega)$ while the Orlicz-Sobolev space $W_0^1 L_\Phi(\Omega)$ is the classical Sobolev space $W_0^{1,p}(\Omega)$.

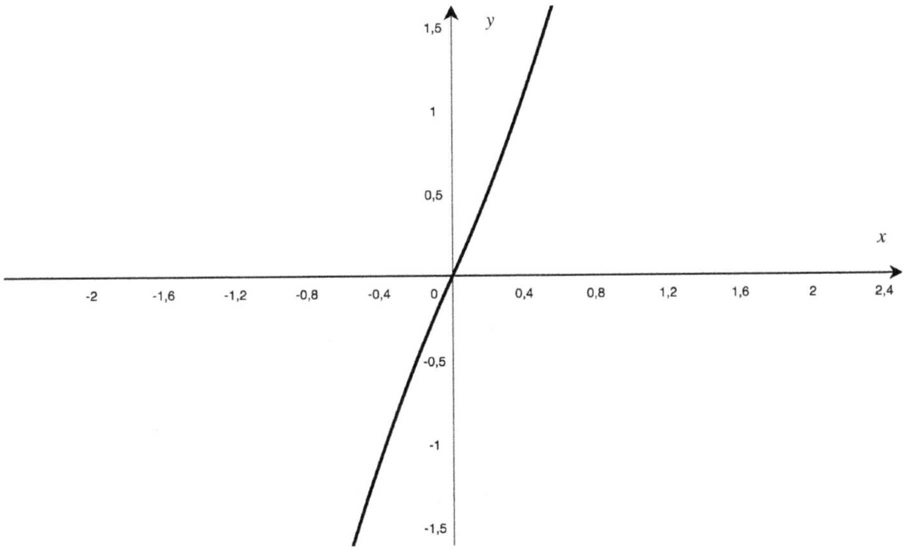

FIGURE 1.3: Graph of the function defined in Example 3 for $p = 3$

Example 2 *Consider*

$$\phi(t) = \log(1 + |t|^s)|t|^{p-2}t, \quad \forall\, t \in \mathbb{R},$$

with $p,\ s > 1$. *We deduce that*

$$(\phi)_0 = p, \qquad (\phi)^0 = p + s\,.$$

Example 3 *Let*

$$\phi(t) = \frac{|t|^{p-2}t}{\log(1 + |t|)}, \quad \text{if } t \neq 0, \quad \phi(0) = 0\,,$$

with $p > 2$. *In this case we have*

$$(\phi)_0 = p - 1, \qquad (\phi)^0 = p\,.$$

We summarize these examples in the following table.

TABLE 1.2: Admissible potentials

Potential	ϕ_0	ϕ^0
$\phi(t) = p\|t\|^{p-2}t$	p	p
$\phi(t) = \log(1+\|t\|^s)\|t\|^{p-2}t$	p	$p+s$
$\phi(t) = \dfrac{\|t\|^{p-2}t}{\log(1+\|t\|)}$	$p-1$	p

Features of Orlicz-Sobolev spaces

The spaces $L_\Phi(\Omega)$ and $W^1L_\Phi(\Omega)$ generalize the usual spaces $L^p(\Omega)$ and $W^{1,p}(\Omega)$, in which the role played by the convex mapping $t \longmapsto |t|^p$ is assumed by a more general convex function $\Phi(t)$. One of the main features of Orlicz-Sobolev spaces is that they fill a gap in the classical theory of Sobolev embeddings. Indeed, if $kp = N$ and $p > 1$ then $W^{k,p}(\Omega)$ is continuously embedded into $L^q(\Omega)$ for any $p \leq q < \infty$, but there is no *smallest* target L^q space for these embeddings, in the sense that $W^{k,p}(\Omega) \not\subseteq L^\infty(\Omega)$. However, if the class of target spaces is enlarged to contain Orlicz spaces then, as showed by Trudinger [102], the best such target space is $L_\Phi(\Omega)$, where $\Phi(t) = \exp(|t|^{p/(p-1)}) - 1$.

1.8 Glossary

Bingham Fluids: Fluids that flow and stop spontaneously. They were named after the American scientist Eugene C. Bingham (1878–1945).

Dirichlet Principle: A method for solving boundary value problems via minimum problems for variational integrals. Dirichlet principle was named after the German mathematician Peter Gustav Lejeune Dirichlet (1805–1859).

Embedding Theorem: A result concerning the existence of embeddings of certain Sobolev and Lebesgue spaces.

Hölder Inequality: A fundamental inequality between integrals and an indispensable tool for the study of Lebesgue spaces. The result was named after the German mathematician Otto Hölder (1859–1937).

Lavrentiev Phenomenon: A Lagrangian exhibits the Lavrentiev phenomenon if the infimum taken over the set of absolutely continuous trajectories is strictly lower than the infimum taken over the set of Lipschitzian trajectories, with fixed boundary conditions. The phenomenon was named after the Russian mathematician Mikhail Lavrentiev (1900–1980).

Lebesgue Space: The L^p space is a function space defined using a natural generalization of the p-norm for finite-dimensional vector spaces ($1 \leq p \leq \infty$). Lebesgue spaces were named after the French mathematician Henri Lebesgue (1875–1941).

Poincaré Inequality: An important inequality that allows to obtain bounds for a function using bounds of its derivatives and the geometry of its domain of definition. The inequality was named after the French mathematician Henri Poincaré (1854–1912).

Smart Fluid: A fluid whose properties (for example the viscosity) can be changed by applying electric or magnetic field.

Sobolev Critical Exponent: If $\Omega \subset \mathbb{R}^N$ is an open set and $p : \Omega \to [1, \infty)$, then $p^*(x) = Np(x)/(N - p(x))$ if $p(x) < N$ and $p^*(x) = +\infty$ if $p(x) \geq N$.

Sobolev Space: A vector space of functions equipped with a norm that is a combination of L^p–norms of the function itself as well as its derivatives up to a given order. Sobolev spaces were named after the Russian mathematician Sergei L. Sobolev (1908–1989).

Winslow Effect: Describes the behavior of certain fluids that become solids or quasi-solids when subjected to an electric field. The result was named after the American engineer Willis M. Winslow (1904–2000).

Part II

Variational Analysis of Problems with Variable Exponents

Chapter 2

Nonlinear Degenerate Problems in Non-Newtonian Fluids

2.1	Physical motivation	26
2.2	A boundary value problem with nonhomogeneous differential operator	27
	2.2.1 Properties of the energy functional	30
	2.2.2 A mountain pass-type critical point	37
2.3	Nonlinear eigenvalue problems with two variable exponents	42
	2.3.1 Ekeland variational principle versus the mountain pass geometry	43
2.4	A sublinear perturbation of the eigenvalue problem associated to the Laplace operator	48
2.5	Variable exponents versus Morse theory and local linking	54
	2.5.1 Verification of the compactness condition	57
	2.5.2 Computation of critical groups	61
2.6	The Caffarelli-Kohn-Nirenberg inequality with variable exponents	66
	2.6.1 A compact embedding	69
	2.6.2 Application to a nonlinear problem with three variable exponents	71
2.7	Glossary	73

Mediocrity knows nothing higher than itself, but talent instantly recognizes genius.

Sir Arthur Conan Doyle (1859–1930)

In this chapter we consider three classes of nonlinear Dirichlet problems with variable exponents and we establish qualitative and spectral properties of solutions. All problems studied in this chapter involve *sublinear* terms with respect to the differential operator. The study of sublinear elliptic problems was initiated by Brezis *et al.* in their pioneering papers [12, 13].

In the first part of this chapter we develop a mathematical theory based on variational arguments, in order to establish the existence and multiplicity of solutions for problems with nonstandard growth and a general nonhomogeneous differential operator. Next, we consider a nonlinear eigenvalue problem with variable exponent and sublinear growth and we establish the existence of a continuous spectrum in a neighborhood of the origin. This study will be extended in the next chapters to problems involving differential operators with *several variable exponents*, indefinite potential, and various nonlinearities. Finally, we establish a striking result for a nonlinear eigenvalue problem with a sublinear term. In this case, we prove that the spectrum is continuous and concentrates at infinity. This result can be extended to other classes of differential operators, for instance to the poly-harmonic operators. Throughout, we will be interested in the study of combined effects of the nonlinear terms, differential operators, parameters, and various perturbations. The methods and techniques developed in this chapter can be applied to several classes of problems with variable exponents, including elliptic systems of gradient type with $p(x)$ growth. This chapter is based on the results established in [85, 86, 49, 51, 57, 59, 82].

2.1 Physical motivation

The study of differential equations and variational problems involving $p(x)$-growth conditions was motivated by their applications. Materials requiring such more advanced theory have been studied experimentally since the middle of the 20th century. We mainly refer to electrorheological fluids, which are fluids whose viscosity depends on the electric field in the fluid. In such fluids (for instance lithium polymetachrylate), viscosity is inversely proportional to the strength of the electric field. The string-like formations in the fluid can raise the viscosity by as much as five orders of magnitude. Electrorheological fluids have been used in robotics and space technology. The experimental research has been done mainly in the United States, for instance in NASA laboratories.

In what follows, we describe a model that takes into account the delicate interaction between the electromagnetic fields and the moving fluids. In particular, in the context of continuum mechanics, these fluids are treated as non-Newtonian fluids. The system modeling the phenomenon arising from this study is

$$\begin{cases} \text{div}\,\mathbf{E} = 0 \qquad \text{curl}\,\mathbf{E} = 0 \\ \dfrac{\partial v}{\partial t} - \text{div} S(x, \mathbf{E}, \mathcal{E}(v)) + [\nabla v]v + \nabla \pi = g(x, \mathrm{E}) \\ \text{div}\, v = 0 \end{cases} \tag{2.1}$$

where $\mathbf{E}(x)$ is the electromagnetic field, $v : \Omega(\subset \mathbb{R}^3) \to \mathbb{R}^3$ is the velocity of the field, $\mathcal{E}(v)$ is the symmetric part of the gradient, S is the extra stress tensor and π is the pressure.

The constitutive relation for the extra stress tensor S is

$$S(x, \mathbf{E}, z) = \nu(\mathbf{E})(1 + |z|^2)^{(p-2)/2} z + \text{ terms of the same growth}$$

for all symmetric 3×3 matrices z and where $p = p(|\mathbf{E}|^2)$. The structure of the system allows to determine $\mathbf{E}$ so that it depends on x and thus $p = p(x)$.

The extra stress tensor S is chosen so that it is a monotone vector field satisfying the ellipticity condition

$$D_z S(x, \mathbf{E}, z)\lambda \otimes \lambda \geq \nu(\mathbf{E})(1 + |z|^2)^{(p-2)/2} |\lambda|^2$$

where $\nu(\mathbf{E}) \geq \overline{\nu} > \mathbf{0}$, for any 3×3 symmetric matrices z, λ with null trace.

For the system described above, Rajagopal and Ruzicka [90] established an existence theory, which works particularly in the stationary case

$$-\text{div} S(x, \mathcal{E}(v)) + [\nabla v]v + \nabla \pi = g(x).$$

2.2 A boundary value problem with nonhomogeneous differential operator

Let $\Omega \subset \mathbb{R}^N$ $(N \geq 3)$ be a bounded domain with smooth boundary and assume that $p : \overline{\Omega} \to (1, \infty)$ is a continuous function.

In this chapter we are concerned with the qualitative analysis of solutions of the nonlinear boundary value problems of the type

$$\begin{cases} -\text{div}(a(x, \nabla u)) = f(x, u), & \text{for } \ x \in \Omega \\ u = 0, & \text{for } \ x \in \partial\Omega\,. \end{cases} \tag{2.2}$$

The interest for studying such problems is due to the presence of the nonhomogeneous elliptic operator div $(a(x, \nabla u))$. A basic example of such operator is the $p(x)$-Laplace operator, which is defined by

$$\Delta_{p(x)} u = \text{div}\,(|\nabla u|^{p(x)-2} \nabla u).$$

Throughout this chapter we assume that $a(x, \xi) : \overline{\Omega} \times \mathbb{R}^N \to \mathbb{R}^N$ is the continuous derivative with respect to ξ of the mapping $A : \overline{\Omega} \times \mathbb{R}^N \to \mathbb{R}$, $A = A(x, \xi)$, that is, $a(x, \xi) = \nabla_\xi A(x, \xi)$. Suppose that a and A satisfy the following hypotheses:

(A1) We have $A(x, 0) = 0$, for all $x \in \overline{\Omega}$.

(A2) There exists a positive constant c_1 such that

$$|a(x, \xi)| \leq c_1(1 + |\xi|^{p(x)-1}),$$

for all $x \in \overline{\Omega}$ and $\xi \in \mathbb{R}^N$.

(A3) The following inequality holds

$$0 \leq (a(x,\xi) - a(x,\psi)) \cdot (\xi - \psi),$$

for all $x \in \overline{\Omega}$ and $\xi, \psi \in \mathbb{R}^N$, with equality if and only if $\xi = \psi$.

(A4) There exists $k > 0$ such that

$$A\left(x, \frac{\xi+\psi}{2}\right) \leq \frac{1}{2}A(x,\xi) + \frac{1}{2}A(x,\psi) - k|\xi-\psi|^{p(x)}$$

for all $x \in \overline{\Omega}$ and $\xi, \psi \in \mathbb{R}^N$.

(A5) We have

$$|\xi|^{p(x)} \leq a(x,\xi) \cdot \xi \leq p(x)\, A(x,\xi),$$

for all $x \in \overline{\Omega}$ and $\xi \in \mathbb{R}^N$.

The following important examples illustrate special cases of such mappings.

Example 4 *Set* $A(x,\xi) = \frac{1}{p(x)}|\xi|^{p(x)}$, $a(x,\xi) = |\xi|^{p(x)-2}\xi$. *Then the above hypotheses are fulfilled and the corresponding differential operator is the* $p(x)$*-Laplace operator*

$$\operatorname{div}\,(|\nabla u|^{p(x)-2}\nabla u).$$

Example 5 *Set* $A(x,\xi) = \frac{1}{p(x)}[(1+|\xi|^2)^{p(x)/2} - 1]$, $a(x,\xi) = (1+|\xi|^2)^{(p(x)-2)/2}\xi$. *Then we obtain the generalized mean curvature operator*

$$\operatorname{div}((1+|\nabla u|^2)^{(p(x)-2)/2}\nabla u).$$

Example 5 is related to the *Plateau problem* for *minimal surfaces*, which consists of finding a surface with least area spanning a given smooth simple closed curve in $\mathbb{R}^3$. In nonparametric form, such a surface is described by a function u satisfying the *minimal surface equation*

$$\operatorname{div}\left(\frac{\nabla u}{\sqrt{1+|\nabla u|^2}}\right) = 0,$$

which is the Euler-Lagrange equation for the area functional $\int (1+|\nabla u|^2)^{1/2}$.

We summarize the properties of the above examples in Table 2.1.

In this chapter we study problem (2.2) in the special case

$$f(x,t) = \lambda(t^{\gamma-1} - t^{\beta-1}) \tag{2.3}$$

with $1 < \beta < \gamma < \inf_{x\in\overline{\Omega}} p(x)$ and $t \geq 0$. More precisely, we consider the degenerate boundary value problem

$$\begin{cases} -\operatorname{div}(a(x,\nabla u)) = \lambda(u^{\gamma-1} - u^{\beta-1}), & \text{for} \quad x \in \Omega \\ u = 0, & \text{for} \quad x \in \partial\Omega \\ u \geq 0, & \text{for} \quad x \in \Omega. \end{cases} \tag{2.4}$$

TABLE 2.1: Variable exponent potentials and the corresponding differential operators

$A(x,\xi)$	$a(x,\xi)$	**Differential operator**
$\frac{1}{p(x)}\|\xi\|^{p(x)}$	$\|\xi\|^{p(x)-2}\xi$	$\text{div}\,(\|\nabla u\|^{p(x)-2}\nabla u)$
$\frac{1}{p(x)}[(1+\|\xi\|^2)^{p(x)/2}-1]$	$(1+\|\xi\|^2)^{(p(x)-2)/2}\xi$	$\text{div}((1+\|\nabla u\|^2)^{(p(x)-2)/2}\nabla u)$

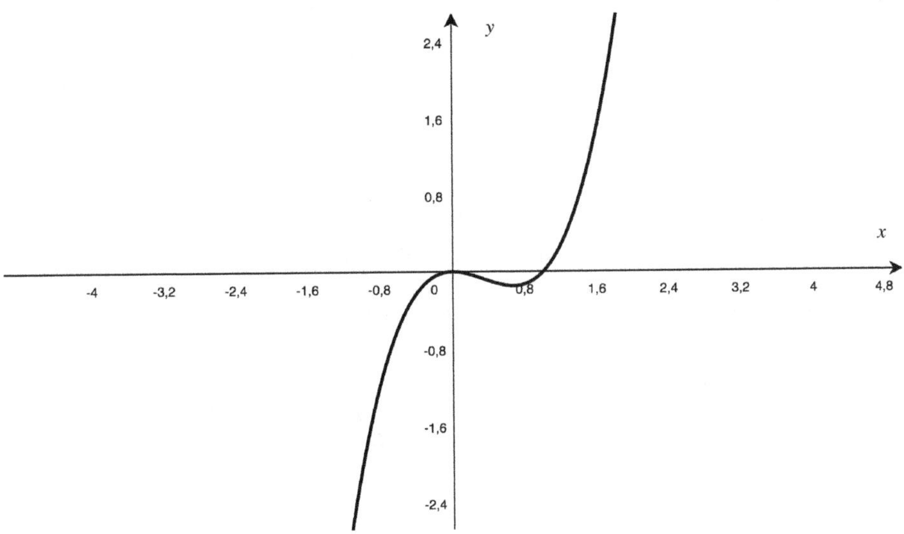

FIGURE 2.1: Graph of the function defined in (2.3) for $\lambda = 1$, $\gamma = 4$, $\beta = 3$

We say that $u \in W_0^{1,p(x)}(\Omega)$ is a *weak solution* of problem (2.4) if $u \geq 0$ a. e. in Ω and

$$\int_\Omega a(x,\nabla u)\cdot\nabla\phi\,dx - \lambda\int_\Omega u^{\gamma-1}\phi\,dx + \lambda\int_\Omega u^{\beta-1}\phi\,dx = 0$$

for all $\phi \in W_0^{1,p(x)}(\Omega)$.

The purpose of this chapter is to study problem (2.4) when its right-hand side is, roughly speaking, the difference of two sublinear terms. The combined effects of these two nonlinearities in relationship with the nonhomogeneous differential operator, change considerably the structure of the solution set. The next multiplicity result asserts that problem (2.4) has at least two nontrivial weak solutions provided that $\lambda > 0$ is *large enough* and the operators A and a satisfy conditions (A1)-(A5) stated earlier. We refer to the pioneering paper by Ambrosetti, Brezis and Cerami [3], who were the first to study combined effects of concave and convex nonlinearities in some elliptic problems.

Theorem 2 *Assume that hypotheses (A1)-(A5) are fulfilled. Then there exists $\lambda^\star > 0$ such that for all $\lambda > \lambda^\star$ problem* (2.4) *has at least two distinct nonnegative, nontrivial weak solutions, provided that $p^+ < \min\{N, Np^-/(N-p^-)\}$.*

2.2.1 Properties of the energy functional

Let E denote the generalized Sobolev space $W_0^{1,p(x)}(\Omega)$. We associate to the nonlinear problem (2.4) the energy functional $I : E \to \mathbb{R}$ defined by

$$I(u) = \int_\Omega A(x, \nabla u)\, dx - \frac{\lambda}{\gamma} \int_\Omega u_+^\gamma\, dx + \frac{\lambda}{\beta} \int_\Omega u_+^\beta\, dx\,,$$

where $u_+(x) = \max\{u(x), 0\}$.

In this section we establish several basic properties of I.

Proposition 1 *The functional I is well-defined on E and $I \in C^1(E, \mathbb{R})$ with the derivative given by*

$$\langle I'(u), \phi\rangle = \int_\Omega a(x, \nabla u) \cdot \nabla\phi\, dx - \lambda \int_\Omega u_+^{\gamma-1}\phi\, dx + \lambda \int_\Omega u_+^{\beta-1}\phi\, dx,$$

for all $u,\ \phi \in E$.

In order to prove Proposition 1, we define the functional $\Lambda : E \to \mathbb{R}$ by

$$\Lambda(u) = \int_\Omega A(x, \nabla u)\, dx, \quad \forall u \in E.$$

The next result shows that Λ has convenient qualitative properties.

Lemma 1 *(i) The functional Λ is well-defined on E.*
(ii) The functional Λ is of class $C^1(E, \mathbb{R})$ and

$$\langle \Lambda'(u), \phi\rangle = \int_\Omega a(x, \nabla u) \cdot \nabla\phi\, dx,$$

for all $u, \phi \in E$.

Proof. (i) For any $x \in \Omega$ and $\xi \in \mathbb{R}^N$ we have

$$A(x, \xi) = \int_0^1 \frac{d}{dt} A(x, t\xi)\, dt = \int_0^1 a(x, t\xi) \cdot \xi\, dt.$$

Using hypotheses (A2) we obtain

$$\begin{aligned} A(x,\xi) &\leq c_1 \int_0^1 (1+|\xi|^{p(x)-1}t^{p(x)-1})|\xi|\,dt \\ &\leq c_1|\xi| + \frac{c_1}{p(x)}|\xi|^{p(x)} \\ &\leq c_1|\xi| + \frac{c_1}{p^-}|\xi|^{p(x)}, \quad \forall x \in \overline{\Omega},\ \xi \in \mathbb{R}^N. \end{aligned} \tag{2.5}$$

The above inequality and (A5) imply

$$0 \leq \int_\Omega A(x,\nabla u)\,dx \leq c_1 \int_\Omega |\nabla u|\,dx + \frac{c_1}{p^-}\int_\Omega |\nabla u|^{p(x)}\,dx, \quad \forall u \in E.$$

Using the Hölder inequality and relations (1.4) and (1.5) we deduce that Λ is well defined on E.

(ii) **Existence of the Gâteaux derivative.** Let u, $\phi \in E$. Fix $x \in \Omega$ and $0 < |r| < 1$. Then by the mean value theorem, there exists $\nu \in [0,1]$ such that

$$|A(x,\nabla u(x)+r\nabla\phi(x)) - A(x,\nabla u)|/|r| = |a(x,\nabla u(x)+\nu\, r\, \nabla\phi(x))||\nabla\phi(x)|.$$

Using condition (A2) we obtain

$$\begin{aligned} &|A(x,\nabla u(x)+r\nabla\phi(x)) - A(x,\nabla u)|/|r| \leq \\ &[c_1 + c_1(|\nabla u(u)| + |\nabla\phi(x)|)^{p(x)-1}]|\nabla\phi(x)| \leq \\ &[c_1 + c_1 2^{p^+}(|\nabla u(x)|^{p(x)-1} + |\nabla\phi(x)|^{p(x)-1})]|\nabla\phi(x)|. \end{aligned}$$

Next, by the Hölder inequality, we have

$$\int_\Omega c_1|\nabla\phi|\,dx \leq |c_1|_{\frac{p(x)}{p(x)-1}} \cdot |\nabla\phi|_{p(x)}$$

and

$$\int_\Omega |\nabla u|^{p(x)-1}|\nabla\phi|\,dx \leq ||\nabla u|^{p(x)-1}|_{\frac{p(x)}{p(x)-1}} \cdot |\nabla\phi|_{p(x)}.$$

The above inequalities imply

$$c_1[1 + 2^{p^+}(|\nabla u(x)|^{p(x)-1} + |\nabla\phi(x)|^{p(x)-1})]|\nabla\phi(x)| \in L^1(\Omega).$$

It follows from the Lebesgue theorem that

$$\langle\Lambda'(u),\phi\rangle = \int_\Omega a(x,\nabla u)\cdot\nabla\phi\,dx.$$

Continuity of the Gâteaux derivative. Assume that $u_n \to u$ in E. Let us define $\theta(x,u) = a(x,\nabla u)$. Using hypothesis (A2) we deduce that $\theta(x,u_n) \to \theta(x,u)$ in $(L^{q(x)}(\Omega))^N$, where $q(x) = \frac{p(x)}{p(x)-1}$. By the Hölder inequality we obtain

$$|\langle\Lambda'(u_n) - \Lambda'(u),\phi\rangle| \leq |\theta(x,u_n) - \theta(x,u)|_{q(x)}|\nabla\phi|_{p(x)}$$

and so

$$\|\Lambda'(u_n) - \Lambda'(u)\| \leq |\theta(x, u_n) - \theta(x, u)|_{q(x)} \to 0, \quad \text{as } n \to \infty.$$

The proof of Lemma 1 is thus complete. □

The following result extends a property initially established by Stampacchia [99] in the framework of standard Sobolev spaces, to spaces with variable exponents.

Lemma 2 *If $u \in E$ then u_+, $u_- \in E$ and*

$$\nabla u_+ = \begin{cases} 0, & \text{if } [u \leq 0] \\ \nabla u, & \text{if } [u > 0], \end{cases} \qquad \nabla u_- = \begin{cases} 0, & \text{if } [u \geq 0] \\ \nabla u, & \text{if } [u < 0] \end{cases}$$

where $u_\pm = \max\{\pm u(x), 0\}$ for all $x \in \Omega$.

Proof. Let $u \in E$ be fixed. Then there exists a sequence $(\phi_n) \in C_0^\infty(\Omega)$ such that

$$|\nabla(\phi_n - u)|_{p(x)} \to 0.$$

Since $1 < p^- \leq p(x)$ for all $x \in \Omega$, it follows that $L^{p(x)}$ is continuously embedded in $L^{p^-}(\Omega)$ and thus

$$|\nabla(\phi_n - u)|_{p^-} \to 0.$$

Hence $u \in W_0^{1,p^-}(\Omega)$. We obtain

$$u_+, u_- \in W_0^{1,p^-}(\Omega) \subset W_0^{1,1}(\Omega). \tag{2.6}$$

On the other hand, the Stampacchia theorem [99] (see also Theorem 7.6 of Gilbarg and Trudinger [35]) implies

$$\nabla u_+ = \begin{cases} 0, & \text{if } [u \leq 0] \\ \nabla u, & \text{if } [u > 0] \end{cases} \quad \nabla u_- = \begin{cases} 0, & \text{if } [u \geq 0] \\ \nabla u, & \text{if } [u < 0]. \end{cases}$$

From the above equalities we deduce that

$$|u_+(x)|^{p(x)} \leq |u(x)|^{p(x)}, \quad |\nabla u_+(x)|^{p(x)} \leq |\nabla u|^{p(x)}, \quad \text{a.e. } x \in \Omega \tag{2.7}$$

and

$$|u_-(x)|^{p(x)} \leq |u(x)|^{p(x)}, \quad |\nabla u_-(x)|^{p(x)} \leq |\nabla u|^{p(x)}, \quad \text{a.e. } x \in \Omega. \tag{2.8}$$

Since $u \in E$ we have

$$|u(x)|^{p(x)}, \ |\nabla u(x)|^{p(x)} \in L^1(\Omega). \tag{2.9}$$

By equations (2.7)–(2.9) and the Lebesgue dominated convergence theorem, we obtain that u_+, $u_- \in L^{p(x)}(\Omega)$ and $\rho_{p(x)}(|\nabla u_+|) < \infty$, $\rho_{p(x)}(|\nabla u_-|) < \infty$. It follows that

$$u_+, u_- \in W^{1,p(x)}(\Omega). \tag{2.10}$$

By equations (2.6) and (2.10) we conclude that

$$u_+, u_- \in W^{1,p(x)}(\Omega) \cap W_0^{1,1}(\Omega).$$

Since $p \in C^{0,\alpha}(\overline{\Omega})$, Theorem 2.6 and Remark 2.9 of Fan and Zhao [30] show that $E = W^{1,p(x)}(\Omega) \cap W_0^{1,1}(\Omega)$. Thus, u_+, $u_- \in E$ and the proof of Lemma 2 is complete. □

From Lemmas 1 and 2 we can deduce that Proposition 1 holds.

Remark 1 *If u is a critical point of I then using Lemma 2 and condition (A5) we obtain the following*

$$\begin{aligned}
0 &= \langle I'(u), u_- \rangle \\
&= \int_\Omega a(x, \nabla u) \cdot \nabla u_- \, dx - \lambda \int_\Omega (u_+)^{\gamma-1} u_- \, dx + \lambda \int_\Omega (u_+)^{\beta-1} u_- \, dx \\
&= \int_\Omega a(x, \nabla u) \cdot \nabla u_- \, dx = \int_\Omega a(x, \nabla u_-) \cdot \nabla u_- \, dx \geq \int_\Omega |\nabla u_-|^{p(x)} \, dx.
\end{aligned}$$

Thus, we can deduce that $u \geq 0$. It follows that the nontrivial critical points of I are nonnegative solutions of problem (2.4).

The above remark shows that we can prove Theorem 2 using the critical points theory. More exactly, we first show that for large enough $\lambda > 0$, the functional I has a global minimizer $u_1 \geq 0$ such that $I(u_1) < 0$. Next, by means of the mountain pass theorem, a second critical point u_2 with $I(u_2) > 0$ can be obtained.

Lemma 3 *The functional Λ is weakly lower semi-continuous.*

Proof. Since Λ is convex (by condition (A4)), it follows from Corollary III.8 of Brezis [10] that it is enough to show that Λ is lower semi-continuous. For this purpose, we fix $u \in E$ and $\epsilon > 0$. Using the convexity of Λ, we deduce that for any $v \in E$ the following inequality holds

$$\int_\Omega A(x, \nabla v) \, dx \geq \int_\Omega A(x, \nabla u) \, dx + \int_\Omega a(x, \nabla u) \cdot (\nabla v - \nabla u) \, dx.$$

Using condition (A2) and the Hölder inequality we obtain

$$\begin{aligned}
\int_\Omega A(x,\nabla v)\,dx &\geq \int_\Omega A(x,\nabla u)\,dx - \int_\Omega |a(x,\nabla u)||\nabla v - \nabla u|\,dx \\
&\geq \int_\Omega A(x,\nabla u)\,dx - c_1\int_\Omega |\nabla(v-u)|\,dx \\
&- c_1\int_\Omega |\nabla u|^{p(x)-1}|\nabla(v-u)|\,dx \\
&\geq \int_\Omega A(x,\nabla u)\,dx - c_2|1|_{q(x)}|\nabla(v-u)|_{p(x)} \\
&- c_3||\nabla u|^{p(x)-1}|_{q(x)}|\nabla(v-u)|_{p(x)} \\
&\geq \int_\Omega A(x,\nabla u)\,dx - c_4\|v-u\| \\
&\geq \int_\Omega A(x,\nabla u)\,dx - \epsilon
\end{aligned}$$

for all $v \in E$ with $\|v-u\| < \delta = \epsilon/c_4$, where c_2, c_3, c_4 are positive constants, and $q(x) = \frac{p(x)}{p(x)-1}$. We conclude that Λ is weakly lower semi-continuous. The proof of Lemma 3 is complete. □

The next result establishes that the infimum of a related *Rayleigh quotient* is positive.

Lemma 4 *There exists $\lambda_1 > 0$ such that*

$$\lambda_1 = \inf_{u\in E,\ \|u\|>1} \frac{\displaystyle\int_\Omega \frac{1}{p(x)}|\nabla u|^{p(x)}\,dx}{\displaystyle\int_\Omega |u|^{p^-}\,dx}.$$

Proof. We know that E is continuously embedded in $L^{p^-}(\Omega)$. It follows that there exists $C > 0$ such that

$$\|u\| \geq C|u|_{p^-}, \quad \forall u \in E.$$

On the other hand, by (1.4) we have

$$\int_\Omega |\nabla u|^{p(x)}\,dx \geq \|u\|^{p^-}, \quad \forall u \in E \text{ with } \|u\| > 1.$$

Combining the above inequalities we obtain

$$\int_\Omega \frac{1}{p(x)}|\nabla u|^{p(x)}\,dx \geq \frac{C^{p^-}}{p^+}\int_\Omega |u|^{p^-}\,dx, \quad \forall u \in E \text{ with } \|u\| > 1.$$

The proof of Lemma 4 is complete. □

The simplest method in the calculus of variations is the *direct method*: one looks for a *minimum* of a functional, which is to be obtained as the limit (in some appropriate sense) of a minimizing sequence. In what follows, we study the energy functional I in relationship with the direct method in the calculus of variations.

Proposition 2 *(i) The functional I is bounded from below and is coercive. (ii) The functional I is weakly lower semi-continuous.*

Proof. (i) Since $1 < \beta < \gamma < p^-$ we have

$$\lim_{t\to\infty} \frac{\frac{1}{\gamma}t^\gamma - \frac{1}{\beta}t^\beta}{t^{p^-}} = 0.$$

Then for any $\lambda > 0$ there exists $C_\lambda > 0$ such that

$$\lambda\left(\frac{1}{\gamma}t^\gamma - \frac{1}{\beta}t^\beta\right) \le \frac{\lambda_1}{2}t^{p^-} + C_\lambda, \quad \forall t \ge 0,$$

where λ_1 is as defined in Lemma 4.

Condition (A5) and the above inequality show that for any $u \in E$ with $\|u\| > 1$ we have

$$\begin{aligned} I(u) &\ge \int_\Omega \frac{1}{p(x)}|\nabla u|^{p(x)}\,dx - \frac{\lambda_1}{2}\int_\Omega |u|^{p^-}\,dx - C_\lambda\mu(\Omega) \\ &\ge \frac{1}{2}\int_\Omega \frac{1}{p(x)}|\nabla u|^{p(x)}\,dx - C_\lambda\mu(\Omega) \\ &\ge \frac{1}{2p^+}\|u\|^{p^-} - C_\lambda\mu(\Omega). \end{aligned}$$

This shows that I is bounded from below and also coercive.

(ii) Using Lemma 3 we deduce that Λ is weakly lower semi-continuous. We show that I is also weakly lower semi-continuous. Let $(u_n) \subset E$ be a sequence which weakly converges to u in E. Since Λ is weakly lower semi-continuous we have

$$\Lambda(u) \le \liminf_{n\to\infty} \Lambda(u_n). \tag{2.11}$$

On the other hand, since E is compactly embedded in both $L^\gamma(\Omega)$ and $L^\beta(\Omega)$ it follows that (u_{n+}) strongly converges to u_+ in both $L^\gamma(\Omega)$ and $L^\beta(\Omega)$. This fact together with relation (2.11) imply

$$I(u) \le \liminf_{n\to\infty} I(u_n).$$

Therefore I is weakly lower semi-continuous. The proof of Proposition 2 is complete. □

Proposition 2 implies that there exists a global minimizer $u_1 \in E$ of I. The following result establishes an additional property of this minimizer, namely $u_1 \ne 0$, provided that λ is sufficiently large.

Proposition 3 *There exists $\lambda^\star > 0$ such that $\inf_E I < 0$ for all $\lambda > \lambda^\star$.*

Proof. Let $\Omega_1 \subset \Omega$ be a large enough compact subset and let $u_0 \in E$ be such that $u_0(x) = t_0$ in Ω_1 and $0 \le u_0(x) \le t_0$ in $\Omega \setminus \Omega_1$, where $t_0 > 1$ is chosen so that

$$\frac{1}{\gamma} t_0^\gamma - \frac{1}{\beta} t_0^\beta > 0.$$

We have

$$\begin{aligned}\frac{1}{\gamma}\int_\Omega u_0^\gamma \, dx - \frac{1}{\beta}\int_\Omega u_0^\beta \, dx &\ge \frac{1}{\gamma}\int_{\Omega_1} u_0^\gamma \, dx - \frac{1}{\beta}\int_{\Omega_1} u_0^\beta \, dx - \frac{1}{\beta}\int_{\Omega\setminus\Omega_1} u_0^\beta \, dx \\ &\ge \frac{1}{\gamma}\int_{\Omega_1} u_0^\gamma \, dx - \frac{1}{\beta}\int_{\Omega_1} u_0^\beta \, dx \\ &- \frac{1}{\beta} t_0^\beta \, \mu(\Omega \setminus \Omega_1) > 0.\end{aligned}$$

We deduce that $I(u_0) < 0$ for large enough $\lambda > 0$. The proof of Proposition 3 is complete. □

Since Proposition 3 holds, it follows that $u_1 \in E$ is a *nontrivial* weak solution of problem (2.4).

Fix $\lambda \ge \lambda^\star$. Set

$$g(x,t) = \begin{cases} 0, & \text{for} \quad t < 0 \\ t^{\gamma-1} - t^{\beta-1}, & \text{for} \quad 0 \le t \le u_1(x) \\ u_1(x)^{\gamma-1} - u_1(x)^{\beta-1}, & \text{for} \quad t > u_1(x) \end{cases}$$

and

$$G(x,t) = \int_0^t g(x,s) \, ds.$$

Define the truncated functional $J : E \to \mathbb{R}$ by

$$J(u) = \int_\Omega A(x, \nabla u) \, dx - \lambda \int_\Omega G(x,u) \, dx.$$

The same arguments as those we used for functional I imply that $J \in C^1(E, \mathbb{R})$ and

$$\langle J'(u), \phi \rangle = \int_\Omega a(x, \nabla u) \cdot \nabla \phi \, dx - \lambda \int_\Omega g(x,u)\phi \, dx,$$

for all u, $\phi \in E$.

On the other hand, we point out that if $u \in E$ is a critical point of J then $u \ge 0$. The proof can be carried out as in the case of functional I.

Next, we prove that the global minimizer of I plays a central role with respect to *all* critical points of the truncated energy J.

Lemma 5 *If u is a critical point of J then $u \le u_1$.*

Proof. We have

$$
\begin{aligned}
0 &= \langle J'(u) - I'(u_1), (u-u_1)_+\rangle \\
&= \int_\Omega (a(x,\nabla u) - a(x,\nabla u_1)) \cdot \nabla (u-u_1)_+ \, dx \\
&- \lambda \int_\Omega [g(x,u) - (u_1^{\gamma-1} - u_1^{\beta-1})](u-u_1)_+ \, dx \\
&= \int_{[u>u_1]} (a(x,\nabla u) - a(x,\nabla u_1)) \cdot \nabla (u-u_1) \, dx.
\end{aligned}
$$

From condition (A3) we deduce that the above equality holds if and only if $\nabla u = \nabla u_1$. It follows that $\nabla u(x) = \nabla u_1(x)$ for all $x \in \omega := \{y \in \Omega;\ u(y) > u_1(y)\}$. Hence

$$\int_\omega |\nabla(u-u_1)|^{p(x)} \, dx = 0$$

and thus

$$\int_\Omega |\nabla(u-u_1)_+|^{p(x)} \, dx = 0.$$

By relation (1.5) we obtain

$$\|(u-u_1)_+\| = 0.$$

Since $u - u_1 \in E$, we have by Lemma 2 that $(u-u_1)_+ \in E$. Thus we obtain by that $(u-u_1)_+ = 0$ in Ω, that is, $u \le u_1$ in Ω. The proof of Lemma 5 is complete. □

2.2.2 A mountain pass-type critical point

We first determine a critical point $u_2 \in E$ of J such that $J(u_2) > 0$ via the mountain pass theorem. Next, Lemma 5 guarantees that $0 \le u_2 \le u_1$ in Ω. Therefore

$$g(x,u_2) = u_2^{\gamma-1} - u_2^{\beta-1} \quad \text{and} \quad G(x,u_2) = \frac{1}{\gamma} u_2^\gamma - \frac{1}{\beta} u_2^\beta$$

and thus

$$J(u_2) = I(u_2) \quad \text{and} \quad J'(u_2) = I'(u_2).$$

More exactly, we find

$$I(u_2) > 0 = I(0) > I(u_1) \qquad \text{and} \qquad I'(u_2) = 0\,.$$

This shows that u_2 is a weak solution of problem (2.4) such that $0 \le u_2 \le u_1$, $u_2 \ne 0$ and $u_2 \ne u_1$.

In order to find u_2 described above we prove that J satisfies one of the geometric hypotheses of the mountain pass theorem, namely that J has a "mountain" near the origin.

Lemma 6 *There exist $\rho \in (0, \|u_1\|)$ and $a > 0$ such that $J(u) \geq a$, for all $u \in E$ with $\|u\| = \rho$.*

Proof. Let $u \in E$ be fixed and such that $\|u\| < 1$. It is clear that

$$\frac{1}{\gamma}t^\gamma - \frac{1}{\beta}t^\beta \leq 0, \quad \forall t \in [0,1].$$

Define

$$\Omega_u = \{x \in \Omega;\ u(x) > \min\{1, u_1(x)\}\}.$$

If $x \in \Omega \setminus \Omega_u$ then $u(x) < \min\{1, u_1(x)\} \leq u_1(x)$ and we have

$$G(x,u) = \frac{1}{\gamma}u_+^\gamma - \frac{1}{\beta}u_+^\beta \leq 0.$$

If $x \in \Omega_u \cap \{x;\ u_1(x) < u(x) < 1\}$ then

$$G(x,u) = \frac{1}{\gamma}u_1^\gamma - \frac{1}{\beta}u_1^\beta + (u_1^{\gamma-1} - u_1^{\beta-1})(u - u_1) \leq 0.$$

Define

$$\Omega_{u,1} = \Omega_u \setminus \{x;\ u_1(x) < u(x) < 1\}.$$

Thus, provided that $\|u\| < 1$, by invoking condition (A5), the above estimates and relation (1.5), we obtain

$$\begin{aligned} J(u) &\geq \int_\Omega \frac{1}{p(x)}|\nabla u|^{p(x)}\,dx - \lambda \int_{\Omega_{u,1}} G(x,u)\,dx \\ &\geq \frac{1}{p^+}\|u\|^{p^+} - \lambda \int_{\Omega_{u,1}} G(x,u)\,dx. \end{aligned} \tag{2.12}$$

Since $p^+ < \min\{N, Np^-/(N-p^-)\}$ it follows that $p^+ < p^\star(x)$ for all $x \in \Omega$. Then there exists $q \in (p^+, Np^-/(N-p^-))$ such that E is continuously embedded in $L^q(\Omega)$. Thus there exists a positive constant $C > 0$ such that

$$|u|_q \leq C\|u\|, \quad \forall u \in E.$$

Using the definition of G and the above estimate, we obtain

$$\begin{aligned}
\lambda \int_{\Omega_{u,1}} G(x,u)\,dx &= \lambda \int_{\Omega_{u,1}\cap[u<u_1]} \left(\frac{1}{\gamma}u_+^\gamma - \frac{1}{\beta}u_+^\beta\right)dx \\
&+\lambda \int_{\Omega_{u,1}\cap[u>u_1]} \left(\frac{1}{\gamma}u_1^\gamma - \frac{1}{\beta}u_1^\beta\right)dx \\
&+\lambda \int_{\Omega_{u,1}\cap[u>u_1]} (u_1^{\gamma-1} - u_1^{\beta-1})(u-u_1)dx \\
&\le \frac{\lambda}{\gamma}\int_{\Omega_{u,1}\cap[u<u_1]} \frac{1}{\gamma}\,u_+^\gamma\,dx \\
&+\frac{\lambda}{\gamma}\int_{\Omega_{u,1}\cap[u>u_1]} u_1^\gamma dx \\
&+\lambda \int_{\Omega_{u,1}\cap[u>u_1]} u_1^{\gamma-1} u dx \\
&\le \lambda D \int_{\Omega_{u,1}} u_+^\gamma dx \le \lambda D \int_{\Omega_{u,1}} u_+^q dx \\
&\le \lambda D_1 \|u_1\|^q,
\end{aligned} \tag{2.13}$$

where D and D_1 are positive constants.

Combining inequalities (2.12) and (2.13), we find that for a small enough $\rho \in (0, \min\{1, \|u_1\|\})$, we have

$$J(u) \ge \left(\frac{1}{p^+} - \lambda D_1 \|u\|^{q-p^+}\right)\|u\|^{p^+}.$$

Taking into account that $q > p^+$, we can infer that the conclusion of Lemma 6 holds. □

Lemma 7 *The functional J is coercive.*

Proof. For each $u \in E$ with $\|u\| > 1$ by condition (A5), relation (1.4) and the Hölder inequality we have (with similar arguments as in the proof of Lemma 6):

$$\begin{aligned}
J(u) &\ge \int_\Omega \frac{1}{p(x)}|\nabla u|^{p(x)}\,dx - \lambda \int_{[u>u_1]} G(x,u)\,dx - \lambda \int_{[u<u_1]} G(x,u)\,dx \\
&\ge \frac{1}{p^+}\|u\|^{p^-} - C_2 3\|u\|^\gamma - C_2,
\end{aligned}$$

where C_1, C_2 and C_3 are positive constants. Since $\gamma < p^-$ the above inequality implies that $J(u) \to \infty$ as $\|u\| \to \infty$, that is, J is coercive. The proof of Lemma 7 is complete. □

The following result yields a sufficient condition which ensures that a weakly convergent sequence in E converges strongly, too.

Lemma 8 *Assume that the sequence (u_n) weakly converges to u in E and that*

$$\limsup_{n\to\infty} \int_\Omega a(x, \nabla u_n) \cdot (\nabla u_n - \nabla u)\, dx \le 0.$$

Then (u_n) strongly converges to u in E.

Proof. Using relation (2.5) we deduce that there exists a positive constant c_5 such that

$$A(x, \xi) \le c_5(|\xi| + |\xi|^{p(x)}), \quad \forall x \in \overline{\Omega},\ \xi \in \mathbb{R}^N.$$

The above inequality implies

$$A(x, \nabla u_n) \le c_5(|\nabla u_n| + |\nabla u_n|^{p(x)}), \quad \forall x \in \overline{\Omega}. \tag{2.14}$$

The fact that u_n weakly converges to u in E implies that there exists $R > 0$ such that $\|u_n\| \le R$ for all n. By relation (2.14) and inequalities (1.4) and (1.5), we deduce that $\{\int_\Omega A(x, \nabla u_n)\, dx\}$ is bounded. Then, up to a subsequence, we have $\int_\Omega A(x, \nabla u_n)\, dx \to c$. By Lemma 3 we obtain

$$\int_\Omega A(x, \nabla u)\, dx \le \liminf_{n\to\infty} \int_\Omega A(x, \nabla u_n)\, dx = c.$$

On the other hand, since Λ is convex, we have

$$\int_\Omega A(x, \nabla u)\, dx \ge \int_\Omega A(x, \nabla u_n)\, dx + \int_\Omega a(x, \nabla u_n) \cdot (\nabla u - \nabla u_n)\, dx.$$

Next, by hypothesis $\limsup\limits_{n\to\infty} \int_\Omega a(x, \nabla u_n) \cdot (\nabla u_n - \nabla u)\, dx \le 0$, we conclude that $\int_\Omega A(x, \nabla u)\, dx = c$.

Taking into account that $(u_n + u)/2$ weakly converges to u in E and using Lemma 3 we have

$$c = \int_\Omega A(x, \nabla u)\, dx \le \liminf_{n\to\infty} \int_\Omega A\left(x, \nabla \frac{u_n + u}{2}\right) dx. \tag{2.15}$$

Assume to the contrary that u_n does not converge to u in E. Then by (1.6) it follows that there exist $\epsilon > 0$ and a subsequence (u_{n_m}) of (u_n) such that

$$\int_\Omega |\nabla(u_{n_m} - u)|^{p(x)}\, dx \ge \epsilon, \quad \forall m. \tag{2.16}$$

By condition (A4) we have

$$\frac{1}{2}A(x, \nabla u) + \frac{1}{2}A(x, \nabla u_{n_m}) - A\left(x, \nabla \frac{u + u_{n_m}}{2}\right) \ge k|\nabla(u_{n_m} - u)|^{p(x)}. \tag{2.17}$$

Relations (2.16) and (2.17) yield

$$\begin{aligned}\frac{1}{2}\int_\Omega A(x,\nabla u)\,dx \quad &+\frac{1}{2}\int_\Omega A(x,\nabla u_{n_m})\,dx-\int_\Omega A\left(x,\nabla\frac{u+u_{n_m}}{2}\right)\\ &\geq k\int_\Omega |\nabla(u_{n_m}-u)|^{p(x)}\,dx\geq k\epsilon.\end{aligned}$$

Letting $m\to\infty$ in the above inequality we obtain

$$c-k\epsilon\geq \limsup_{m\to\infty}\int_\Omega A\left(x,\nabla\frac{u+u_{n_m}}{2}\right)\,dx$$

and this is in contradiction with (2.15). It follows that u_n strongly converges to u in E and Lemma 8 is proved. □

Now we return to the proof of Theorem 2. Using Lemma 6 and the mountain pass theorem, we obtain a sequence $(u_n)\subset E$ such that

$$J(u_n)\to c>0 \quad\text{and}\quad J'(u_n)\to 0 \tag{2.18}$$

where

$$c=\inf_{\gamma\in\Gamma}\max_{t\in[0,1]} J(\gamma(t))$$

and

$$\Gamma=\{\gamma\in C([0,1],E);\ \gamma(0)=0,\ \gamma(1)=u_1\}.$$

By relation (2.18) and Lemma 7 we obtain that (u_n) is bounded and thus passing eventually to a subsequence, still denoted by (u_n), we may assume that there exists $u_2\in E$ such that u_n weakly converges to u_2. Since E is compactly embedded in $L^i(\Omega)$ for any $i\in[1,p^-]$, it follows that u_n strongly converges to u_2 in $L^i(\Omega)$ for all $i\in[1,p^-]$. Hence

$$\begin{aligned}\langle\Lambda'(u_n)-\Lambda'(u_2),u_n-u_2\rangle \quad &=\langle J'(u_n)-J'(u_2),u_n-u_2\rangle\\ &+\lambda\int_\Omega[g(x,u_n)-g(x,u_2)](u_n-u_2)\,dx=o(1),\end{aligned}$$

as $n\to\infty$. By Lemma 8 we deduce that u_n strongly converges to u_2 in E and using relation (2.18) we find

$$J(u_2)=c>0 \quad\text{and}\quad J'(u_2)=0.$$

Therefore $J(u_2)=c>0$ and $J'(u_2)=0$. By Lemma 5 we deduce that $0\leq u_2\leq u_1$ in Ω. Therefore

$$g(x,u_2)=u_2^{\gamma-1}-u_2^{\beta-1} \quad\text{and}\quad G(x,u_2)=\frac{1}{\gamma}u_2^\gamma-\frac{1}{\beta}u_2^\beta$$

and thus

$$J(u_2)=I(u_2) \quad\text{and}\quad J'(u_2)=I'(u_2).$$

We conclude that u_2 is a critical point of I and thus a solution of problem (2.4). Furthermore, $I(u_2)=c>0$ and $I(u_2)>0>I(u_1)$. Thus u_2 is not trivial and $u_2\neq u_1$. The proof of Theorem 2 is now complete. □

2.3 Nonlinear eigenvalue problems with two variable exponents

A basic result in the elementary theory of linear partial differential equations asserts that the spectrum of the Laplace operator in $H_0^1(\Omega)$ is discrete, where Ω is a bounded open set in $\mathbb{R}^N$ with smooth boundary. More precisely, the problem

$$\begin{cases} -\Delta u = \lambda u & \text{in} \quad \Omega \\ u = 0 & \text{on} \quad \partial\Omega \end{cases}$$

has an unbounded sequence of eigenvalues

$$0 < \lambda_1 < \lambda_2 \leq \ldots \leq \lambda_n \leq \ldots .$$

This celebrated result goes back to the Riesz-Fredholm theory of self-adjoint and compact operators on Hilbert spaces.

The anisotropic case

$$\begin{cases} -\Delta u = \lambda a(x) u & \text{in} \quad \Omega \\ u = 0 & \text{on} \quad \partial\Omega \end{cases}$$

was considered by Bocher [7], Hess and Kato [38] and Minakshisundaram and Pleijel [64]. For instance, Minakshisundaram and Pleijel proved that the above eigenvalue problem has an unbounded sequence of positive eigenvalues if $a \in L^\infty(\Omega)$, $a \geq 0$ in Ω, and $a > 0$ in $\Omega_0 \subset \Omega$, where $|\Omega_0| > 0$.

We are concerned with a related eigenvalue problem but with a particular feature: it contains two variable exponents. We see that our hypotheses imply the existence of a *continuous spectrum* which concentrates around the origin. Unfortunately, we do not have additional properties but we will establish in the next chapter various concentration properties of the spectrum.

Consider the nonlinear eigenvalue problem

$$\begin{cases} -\text{div}(|\nabla u|^{p(x)-2}\nabla u) = \lambda |u|^{q(x)-2}u, & \text{for} \quad x \in \Omega \\ u = 0, & \text{for} \quad x \in \partial\Omega\,, \end{cases} \tag{2.19}$$

where $\Omega \subset \mathbb{R}^N$ $(N \geq 3)$ is a bounded domain with smooth boundary, $\lambda > 0$ is a real number, and p, q are continuous on $\overline{\Omega}$.

The case $p(x) = q(x)$ was considered by Fan, Zhang and Zhao [32]. Using the Ljusternik-Schnirelmann critical point theory, they established the existence of a sequence of eigenvalues. Denoting by Λ the set of all nonnegative eigenvalues, they showed that $\sup \Lambda = +\infty$ and they pointed out that we have $\inf \Lambda > 0$ only under additional assumptions. We remark that for the p-Laplace operator (corresponding to $p(x) \equiv p$) we always have $\inf \Lambda > 0$.

In this section, we study problem (2.19) under the basic assumption

$$1 < \min_{x\in\overline{\Omega}} q(x) < \min_{x\in\overline{\Omega}} p(x) < \max_{x\in\overline{\Omega}} q(x). \tag{2.20}$$

Our main result establishes the existence of a *continuous family of eigenvalues* of problem (2.19) in a neighborhood of the origin. More precisely, we show that there exists $\lambda^\star > 0$ such that *any* $\lambda \in (0, \lambda^\star)$ is an eigenvalue of problem (2.19).

Theorem 3 *Assume that condition* (2.20) *is fulfilled and that* $\max_{x\in\overline{\Omega}} p(x) < N$ *and* $q(x) < p^\star(x)$ *for all* $x \in \overline{\Omega}$. *Then there exists* $\lambda^\star > 0$ *such that any* $\lambda \in (0, \lambda^\star)$ *is an eigenvalue of problem* (2.19).

The above property implies that the infimum of the associated Rayleigh quotient is zero, namely

$$\inf_{u\in W_0^{1,p(x)}(\Omega)\setminus\{0\}} \frac{\displaystyle\int_\Omega |\nabla u|^{p(x)}\, dx}{\displaystyle\int_\Omega |u|^{q(x)}\, dx} = 0.$$

Thus, for any positive constant C, there exists $u_0 \in W_0^{1,p(x)}(\Omega)$ such that

$$C\int_\Omega |u_0|^{q(x)}\, dx \geq \int_\Omega |\nabla u_0|^{p(x)}\, dx.$$

2.3.1 Ekeland variational principle versus the mountain pass geometry

Let E denote the generalized Sobolev space $W_0^{1,p(x)}(\Omega)$.

For any $\lambda > 0$ the energy functional corresponding to problem (2.19) is defined as $J_\lambda : E \to \mathbb{R}$,

$$J_\lambda(u) = \int_\Omega \frac{1}{p(x)}|\nabla u|^{p(x)}\, dx - \lambda\int_\Omega \frac{1}{q(x)}|u|^{q(x)}\, dx.$$

Standard arguments together with those provided in the previous sections imply that $J_\lambda \in C^1(E, \mathbb{R})$ and

$$\langle J_\lambda'(u), v\rangle = \int_\Omega |\nabla u|^{p(x)-2}\nabla u\nabla v\, dx - \lambda\int_\Omega |u|^{q(x)-2}uv\, dx,$$

for all $u,\ v \in E$. Thus, the weak solutions of problem (2.19) coincide with the critical points of J_λ. If such a weak solution exists and is nontrivial then the corresponding λ is an eigenvalue of problem (2.19).

The following property shows that the energy functional J satisfies one of the geometric hypotheses of the mountain pass theorem in the good sense, namely it has a mountain *near the origin.*

Lemma 9 *There exists $\lambda^\star > 0$ such that for any $\lambda \in (0, \lambda^\star)$ there exist ρ, $a > 0$ such that $J_\lambda(u) \geq a > 0$ for any $u \in E$ with $\|u\| = \rho$.*

Proof. Since $q(x) < p^\star(x)$ for all $x \in \overline{\Omega}$ it follows that E is continuously embedded in $L^{q(x)}(\Omega)$. So there exists a positive constant c_1 such that

$$|u|_{q(x)} \leq c_1 \|u\|, \quad \forall\, u \in E. \tag{2.21}$$

We fix $\rho \in (0,1)$ such that $\rho < 1/c_1$. Then relation (2.21) implies

$$|u|_{q(x)} < 1, \quad \forall\, u \in E, \text{ with } \|u\| = \rho.$$

Furthermore, relation (1.5) yields

$$\int_\Omega |u|^{q(x)}\, dx \leq |u|^{q^-}_{q(x)}, \quad \forall\, u \in E, \text{ with } \|u\| = \rho. \tag{2.22}$$

Relations (2.21) and (2.22) imply

$$\int_\Omega |u|^{q(x)}\, dx \leq c_1^{q^-} \|u\|^{q^-}, \quad \forall\, u \in E, \text{ with } \|u\| = \rho. \tag{2.23}$$

Taking into account relations (1.5) and (2.23) we deduce that for any $u \in E$ with $\|u\| = \rho$ the following inequalities hold

$$\begin{aligned} J_\lambda(u) &\geq \frac{1}{p^+}\int_\Omega |\nabla u|^{p(x)}\, dx - \frac{\lambda}{q^-}\int_\Omega |u|^{q(x)}\, dx \\ &\geq \frac{1}{p^+}\|u\|^{p^+} - \frac{\lambda}{q^-} c_1^{q^-} \|u\|^{q^-} \\ &= \frac{1}{p^+}\rho^{p^+} - \frac{\lambda}{q^-} c_1^{q^-} \rho^{q^-} \\ &= \rho^{q^-}\left(\frac{1}{p^+}\rho^{p^+ - q^-} - \frac{\lambda}{q^-} c_1^{q^-}\right). \end{aligned}$$

By the above inequality we note that if we define

$$\lambda^\star = \frac{\rho^{p^+ - q^-}}{2p^+} \cdot \frac{q^-}{c_1^{q^-}} \tag{2.24}$$

then for any $\lambda \in (0, \lambda^\star)$ and any $u \in E$ with $\|u\| = \rho$ there exists $a = \frac{\rho^{p^+}}{2p^+} > 0$ such that

$$J_\lambda(u) \geq a > 0.$$

The proof of Lemma 9 is complete. □

The next result shows that the energy J fulfills the second geometric assumption of the mountain pass theorem in the *wrong sense*, namely it has a valley *near the origin*, hence not *far away*, as required. In particular, this shows that the functional J does not have a mountain pass geometry.

Lemma 10 *There exists $\phi \in E$ such that $\phi \geq 0$, $\varphi \neq 0$ and $J_\lambda(t\phi) < 0$, for small enough $t > 0$.*

Proof. Assumption (2.20) implies that $q^- < p^-$. Let $\epsilon_0 > 0$ be such that $q^- + \epsilon_0 < p^-$. On the other hand, since $q \in C(\overline{\Omega})$ it follows that there exists an open set $\Omega_0 \subset \Omega$ such that $|q(x) - q^-| < \epsilon_0$ for all $x \in \Omega_0$. Thus, we can conclude that $q(x) \leq q^- + \epsilon_0 < p^-$ for all $x \in \Omega_0$.

Let $\phi \in C_0^\infty(\Omega)$ be such that $\text{supp}(\phi) \supset \overline{\Omega}_0$, $\phi(x) = 1$ for all $x \in \overline{\Omega}_0$ and $0 \leq \phi \leq 1$ in Ω. Then using the above information for any $t \in (0,1)$ we obtain

$$\begin{aligned}
J_\lambda(t\phi) &= \int_\Omega \frac{t^{p(x)}}{p(x)} |\nabla\phi|^{p(x)}\,dx - \lambda \int_\Omega \frac{t^{q(x)}}{q(x)} |\phi|^{q(x)}\,dx \\
&\leq \frac{t^{p^-}}{p^-} \int_\Omega |\nabla\phi|^{p(x)}\,dx - \frac{\lambda}{q^+} \int_\Omega t^{q(x)} |\phi|^{q(x)}\,dx \\
&\leq \frac{t^{p^-}}{p^-} \int_\Omega |\nabla\phi|^{p(x)}\,dx - \frac{\lambda}{q^+} \int_{\Omega_0} t^{q(x)} |\phi|^{q(x)}\,dx \\
&\leq \frac{t^{p^-}}{p^-} \int_\Omega |\nabla\phi|^{p(x)}\,dx - \frac{\lambda \cdot t^{q^- + \epsilon_0}}{q^+} \int_{\Omega_0} |\phi|^{q(x)}\,dx.
\end{aligned}$$

Therefore

$$J_\lambda(t\phi) < 0$$

for $t < \delta^{1/(p^- - q^- - \epsilon_0)}$ with

$$0 < \delta < \min\left\{1, \frac{\frac{\lambda \cdot p^-}{q^+} \int_{\Omega_0} |\phi|^{q(x)}\,dx}{\int_\Omega |\nabla\phi|^{p(x)}\,dx}\right\}.$$

Finally, we point out that $\int_\Omega |\nabla\phi|^{p(x)}\,dx > 0$. Indeed, it is clear that

$$\int_{\Omega_0} |\phi|^{q(x)}\,dx \leq \int_\Omega |\phi|^{q(x)}\,dx \leq \int_{\Omega_0} |\phi|^{q^-}\,dx.$$

On the other hand, $W_0^{1,p(x)}(\Omega)$ is continuously embedded in $L^{q^-}(\Omega)$ and thus, there exists a positive constant c_2 such that

$$|\phi|_{q^-} \leq c_2 \|\phi\|.$$

The last two inequalities imply that

$$\|\phi\| > 0$$

and combining this fact with relations (1.4) or (1.5) we deduce that

$$\int_{\Omega} |\nabla \phi|^{p(x)} \, dx > 0.$$

The proof of Lemma 10 is complete. □

Proof of Theorem 3. Let $\lambda^\star > 0$ be defined as in (2.24) and choose any $\lambda \in (0, \lambda^\star)$. It follows by Lemma 9 that on the boundary of the ball centered at the origin and of radius ρ in E, denoted by $B_\rho(0)$, we have

$$\inf_{\partial B_\rho(0)} J_\lambda > 0. \tag{2.25}$$

On the other hand, by Lemma 10, there exists $\phi \in E$ such that $J_\lambda(t\phi) < 0$ for all small enough $t > 0$. Moreover, relations (2.23) and (1.5) imply that for any $u \in B_\rho(0)$ we have

$$J_\lambda(u) \geq \frac{1}{p^+} \|u\|^{p^+} - \frac{\lambda}{q^-} c_1^{q^-} \|u\|^{q^-}.$$

It follows that

$$-\infty < \underline{c} := \inf_{B_\rho(0)} J_\lambda < 0.$$

We now let $0 < \epsilon < \inf_{\partial B_\rho(0)} J_\lambda - \inf_{B_\rho(0)} J_\lambda$. Applying the Ekeland variational principle to the functional $J_\lambda : \overline{B_\rho(0)} \to \mathbb{R}$, we find $u_\epsilon \in \overline{B_\rho(0)}$ such that

$$\begin{aligned} J_\lambda(u_\epsilon) &< \inf_{\overline{B_\rho(0)}} J_\lambda + \epsilon \\ J_\lambda(u_\epsilon) &< J_\lambda(u) + \epsilon \cdot \|u - u_\epsilon\|, \quad u \neq u_\epsilon. \end{aligned}$$

Since

$$J_\lambda(u_\epsilon) \leq \inf_{\overline{B_\rho(0)}} J_\lambda + \epsilon \leq \inf_{B_\rho(0)} J_\lambda + \epsilon < \inf_{\partial B_\rho(0)} J_\lambda ,$$

we deduce that $u_\epsilon \in B_\rho(0)$. Now, we define $I_\lambda : \overline{B_\rho(0)} \to \mathbb{R}$ by $I_\lambda(u) = J_\lambda(u) + \epsilon \cdot \|u - u_\epsilon\|$. It is clear that u_ϵ is a minimum point of I_λ and thus

$$\frac{I_\lambda(u_\epsilon + t \cdot v) - I_\lambda(u_\epsilon)}{t} \geq 0$$

for small enough $t > 0$ and any $v \in B_1(0)$. The above relation yields

$$\frac{J_\lambda(u_\epsilon + t \cdot v) - J_\lambda(u_\epsilon)}{t} + \epsilon \cdot \|v\| \geq 0.$$

Letting $t \to 0$ it follows that $\langle J'_\lambda(u_\epsilon), v\rangle + \epsilon \cdot \|v\| > 0$ and we infer that $\|J'_\lambda(u_\epsilon)\| \leq \epsilon$.

We deduce that there exists a sequence $\{w_n\} \subset B_\rho(0)$ such that

$$J_\lambda(w_n) \to \underline{c} \quad \text{and} \quad J'_\lambda(w_n) \to 0. \tag{2.26}$$

It is clear that the sequence $\{w_n\}$ is bounded in E. Thus, there exists $w \in E$ such that, up to a subsequence, $\{w_n\}$ weakly converges to w in E. Since $q(x) < p^\star(x)$ for all $x \in \overline{\Omega}$ we deduce that E is compactly embedded in $L^{q(x)}(\Omega)$, hence $\{w_n\}$ strongly converges to w in $L^{q(x)}(\Omega)$. So by relations (1.6) and (1.3),

$$\lim_{n\to\infty} \int_\Omega |w_n|^{q(x)-2} w_n (w_n - w)\, dx = 0.$$

On the other hand, relation (2.26) yields

$$\lim_{n\to\infty} \langle J'_\lambda(w_n), w_n - w\rangle = 0.$$

Using the above information we find

$$\lim_{n\to\infty} \int_\Omega |\nabla w_n|^{p(x)-2} \nabla w_n \nabla (w_n - w)\, dx = 0. \tag{2.27}$$

Relation (2.27) and the fact that $\{w_n\}$ weakly converges to w in E enable us to apply Theorem 3.1 of Fan and Zhang [31] and obtain that $\{w_n\}$ strongly converges to w in E. So by (2.26),

$$J_\lambda(w) = \underline{c} < 0 \quad \text{and} \quad J'_\lambda(w) = 0. \tag{2.28}$$

We conclude that w is a nontrivial weak solution of problem (2.19) and thus any $\lambda \in (0, \lambda^\star)$ is an eigenvalue of problem (2.19).

The proof of Theorem 3 is complete. □

Let us now assume that the hypotheses of Theorem 3 are fulfilled and that furthermore,

$$\max_{\overline{\Omega}} p(x) < \max_{\overline{\Omega}} q(x).$$

Then, using similar arguments as in the proof of Lemma 10, we can find some $\psi \in E$ such that

$$\lim_{t\to\infty} J_\lambda(t\psi) = -\infty.$$

This fact combined with Lemma 9 and the mountain pass theorem implies that there exists a sequence $\{u_n\}$ in E such that

$$J_\lambda(u_n) \to \overline{c} > 0 \quad \text{and} \quad J'_\lambda(u_n) \to 0 \text{ in } E^\star. \tag{2.29}$$

However, relation (2.29) is not useful because we cannot show that the sequence $\{u_n\}$ is bounded in E since the functional J_λ does not satisfy a relation of the Ambrosetti-Rabinowitz type. This enables us to affirm that we cannot obtain a critical point for J_λ by using this method.

On the other hand, we point out that we would fail in trying to show that the functional J_λ is coercive since by relation (2.20) we have $q^+ > p^-$. Thus, we cannot apply (as in the homogeneous case) a result as Theorem 1.2 of Struwe [100] in order to obtain a critical point of the functional J_λ.

2.4 A sublinear perturbation of the eigenvalue problem associated to the Laplace operator

Let $\Omega \subset \mathbb{R}^N$ be a bounded domain with smooth boundary. A fundamental result in elementary functional analysis asserts that the spectrum of the Laplace operator is discrete and unbounded. More precisely, the *linear* eigenvalue problem

$$\begin{cases} -\Delta u = \lambda u, & \text{in} \quad \Omega \\ u = 0, & \text{on} \quad \partial\Omega \end{cases} \tag{2.30}$$

has an unbounded sequence of eigenvalues

$$0 < \lambda_1 < \lambda_2 \leq \cdots \leq \lambda_n \leq \cdots .$$

The lowest eigenvalue λ_1 of problem (2.30) can be characterized as the minimum of the Rayleigh quotient, that is,

$$\lambda_1 = \inf_{u \in H_0^1(\Omega) \setminus \{0\}} \frac{\displaystyle\int_\Omega |\nabla u|^2 \, dx}{\displaystyle\int_\Omega u^2 \, dx} . \tag{2.31}$$

Moreover, λ_1 is simple, that is, all associated eigenfunctions are simply multiples of each other (see Gilbarg and Trudinger [35]).

We are concerned with a *sublinear perturbation* of this eigenvalue problem and we prove some new results. More precisely, we consider the problem

$$\begin{cases} -\Delta u = \lambda f(x, u), & \text{in} \quad \Omega \\ u = 0, & \text{on} \quad \partial\Omega , \end{cases} \tag{2.32}$$

where $f : \Omega \times \mathbb{R} \to \mathbb{R}$ is a Carathéodory function and $\lambda \in \mathbb{R}$ is a real number.

We say that λ is an *eigenvalue* of problem (2.32) if there exists $u \in H_0^1(\Omega) \setminus \{0\}$ such that for any $v \in H_0^1(\Omega)$,

$$\int_\Omega \nabla u \nabla v \, dx - \lambda \int_\Omega f(x, u) v \, dx = 0 .$$

Moreover, if λ is an eigenvalue of problem (2.32) then $u \in H_0^1(\Omega) \setminus \{0\}$ given in the above definition is called the *eigenfunction* corresponding to λ.

We study problem (2.32) for the special case when

$$f(x, t) = \begin{cases} h(x, t), & \text{if} \quad t \geq 0 \\ t, & \text{if} \quad t < 0 , \end{cases} \tag{2.33}$$

where $h : \Omega \times [0, \infty) \to \mathbb{R}$ is a Carathéodory function satisfying the following hypotheses

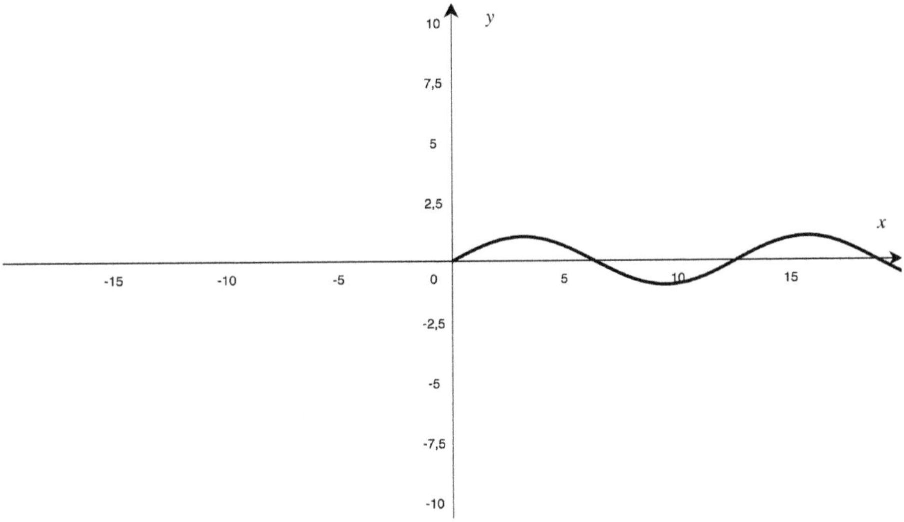

FIGURE 2.2: Graph of the function defined in Example 6(i)

(H1) there exists a positive constant $C \in (0,1)$ such that $|h(x,t)| \leq Ct$ for any $t \geq 0$ and a.e. $x \in \Omega$;

(H2) there exists $t_0 > 0$ such that $H(x,t_0) := \int_0^{t_0} h(x,s)\, ds > 0$, for a.e. $x \in \Omega$;

(H3) $\lim_{t\to\infty} \frac{h(x,t)}{t} = 0$, uniformly with respect to x.

Example 6 *We point out some examples of functions h which satisfy the hypotheses (H1)-(H3):*

(i) $h(x,t) = \sin(t/2)$, for any $t \geq 0$ and any $x \in \Omega$;

(ii) $h(x,t) = k\log(1+t)$, for any $t \geq 0$ and any $x \in \Omega$, where $k \in (0,1)$ is a constant;

(iii) $h(x,t) = g(x)(t^{q(x)-1} - t^{p(x)-1})$, for any $t \geq 0$ and any $x \in \Omega$, where p, $q : \overline{\Omega} \to (1,2)$ are continuous functions satisfying $\max_{\overline{\Omega}} p < \min_{\overline{\Omega}} q$, and $g \in L^\infty(\Omega)$ satisfies $0 < \inf_\Omega g \leq \sup_\Omega g < 1$.

Under these assumptions, we prove that the first eigenvalue of the Laplace operator in $H_0^1(\Omega)$ is an *isolated* eigenvalue of problem (2.32) and that *any* sufficiently large λ is an eigenvalue, whereas the interval $(0, \lambda_1)$ does *not* contain any eigenvalues. This shows that problem (2.32) has both isolated eigenvalues and a *continuous spectrum* in a neighborhood of $+\infty$.

Theorem 4 *Assume that f is given by relation (2.33) and that conditions (H1), (H2) and (H3) are fulfilled. Then λ_1 defined in (2.31) is an isolated*

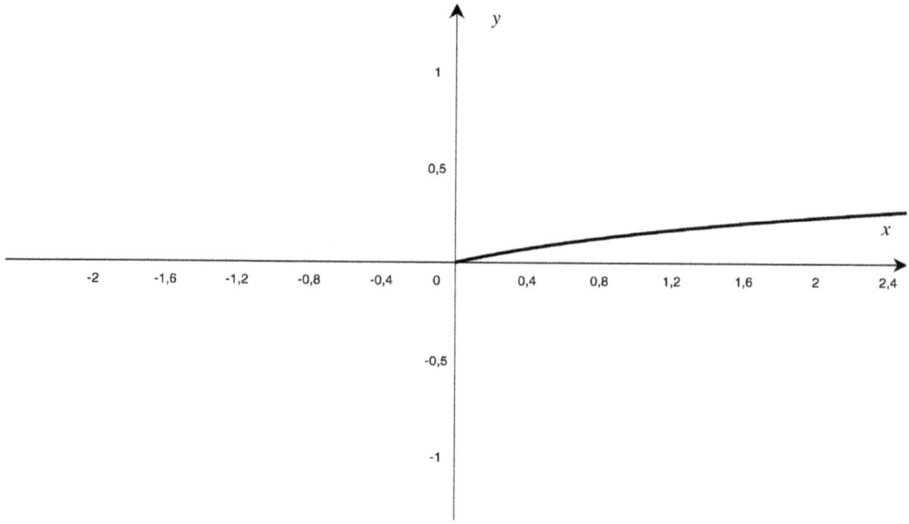

FIGURE 2.3: Graph of the function defined in Example 6(ii)

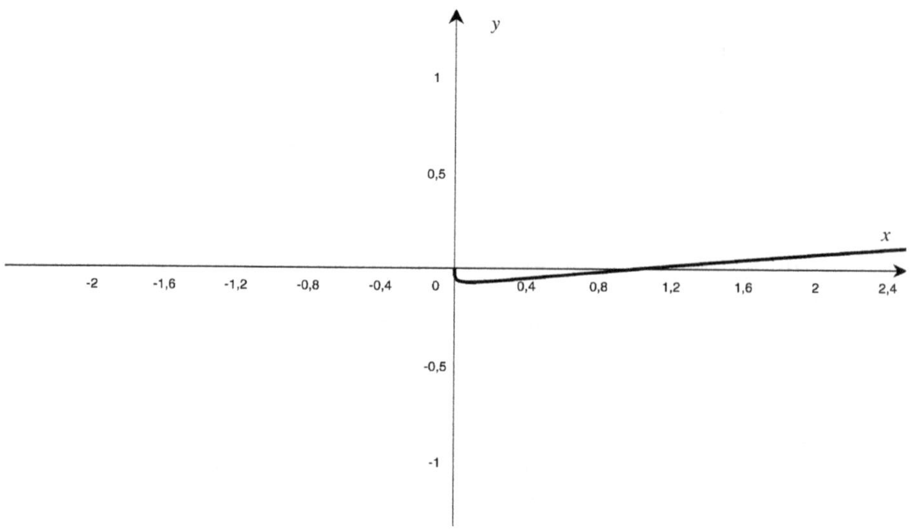

FIGURE 2.4: A special case of the function defined in Example 6(iii)

eigenvalue of problem (2.32) *and the corresponding set of eigenvectors is a cone. Moreover, no* $\lambda \in (0, \lambda_1)$ *is an eigenvalue of problem* (2.32). *However, there exists* $\mu_1 > \lambda_1$ *such that any* $\lambda \in (\mu_1, \infty)$ *is an eigenvalue of problem* (2.32).

Proof. For any $u \in H_0^1(\Omega)$ we denote

$$u_\pm(x) = \max\{\pm u(x), 0\}, \quad \forall\, x \in \Omega\,.$$

Then u_+, $u_- \in H_0^1(\Omega)$ and

$$\nabla u_+ = \begin{cases} 0, & \text{if } [u \le 0] \\ \nabla u, & \text{if } [u > 0]\,, \end{cases} \qquad \nabla u_- = \begin{cases} 0, & \text{if } [u \ge 0] \\ \nabla u, & \text{if } [u < 0]\,. \end{cases}$$

Thus, problem (2.32) with f given by relation (2.33) becomes

$$\begin{cases} -\Delta u = \lambda[h(x,u_+) - u_-], & \text{in } \Omega \\ u = 0, & \text{on } \partial\Omega\,. \end{cases} \tag{2.34}$$

Then $\lambda > 0$ is an eigenvalue of problem (2.34) if there exists $u \in H_0^1(\Omega) \setminus \{0\}$ such that for all $v \in H_0^1(\Omega)$

$$\int_\Omega \nabla u_+ \nabla v\, dx - \int_\Omega \nabla u_- \nabla v\, dx - \lambda \int_\Omega [h(x,u_+) - u_-] v\, dx = 0\,. \tag{2.35}$$

Lemma 11 *No $\lambda \in (0, \lambda_1)$ can be an eigenvalue of problem* (2.34).

Proof. Assume that $\lambda > 0$ is an eigenvalue of problem (2.34) with the corresponding eigenfunction u. Letting $v = u_+$ and $v = u_-$ in the definition of the eigenvalue λ we find that the following two relations hold:

$$\int_\Omega |\nabla u_+|^2\, dx = \lambda \int_\Omega h(x,u_+)u_+\, dx \tag{2.36}$$

and

$$\int_\Omega |\nabla u_-|^2\, dx = \lambda \int_\Omega u_-^2\, dx\,. \tag{2.37}$$

In this context, hypothesis (H1) and relations (2.31), (2.36) and (2.37) imply

$$\lambda_1 \int_\Omega u_+^2\, dx \le \int_\Omega |\nabla u_+|^2\, dx = \lambda \int_\Omega h(x,u_+)u_+\, dx \le \lambda \int_\Omega u_+^2\, dx$$

and

$$\lambda_1 \int_\Omega u_-^2\, dx \le \int_\Omega |\nabla u_-|^2\, dx = \lambda \int_\Omega u_-^2\, dx\,.$$

If λ is an eigenvalue of problem (2.34) then $u \ne 0$ and thus, at least one of the functions u_+ and u_- is not the zero function. Therefore, the last two inequalities show that λ is an eigenvalue of problem (2.34) only if $\lambda \ge \lambda_1$. □

Lemma 12 *λ_1 is an eigenvalue of problem* (2.34). *Moreover, the set of eigenvectors corresponding to λ_1 is a cone.*

Proof. Indeed, as we have already pointed out, λ_1 is the lowest eigenvalue of problem (2.30). It is simple, that is, all the associated eigenfunctions are simply multiples of each other and the corresponding eigenfunctions of λ_1 never change sign in Ω. In other words, there exists $e_1 \in H_0^1(\Omega) \setminus \{0\}$, with $e_1(x) < 0$ for any $x \in \Omega$ such that

$$\int_\Omega \nabla e_1 \nabla v \, dx - \lambda_1 \int_\Omega e_1 v \, dx = 0 \, ,$$

for any $v \in H_0^1(\Omega)$. Thus, we have $(e_1)_+ = 0$ and $(e_1) = -e_1$ and we deduce that relation (2.35) holds with $u = e_1 \in H_0^1(\Omega) \setminus \{0\}$ and $\lambda = \lambda_1$. In other words, λ_1 is an eigenvalue of problem (2.34) and undoubtedly, the set of its corresponding eigenvectors forms a cone of $H_0^1(\Omega)$. The proof of Lemma 12 is complete. □

Lemma 13 *λ_1 is isolated in the set of eigenvalues of problem* (2.34).

Proof. By Lemma 11 we know that there is no eigenvalue of problem (2.34) in the interval $(0, \lambda_1)$. On the other hand, hypothesis (H1) and relations (2.31) and (2.36) show that if λ is an eigenvalue of problem (2.34) for which the positive part of its corresponding eigenfunction, that is u_+, is not identically zero then

$$\lambda_1 \int_\Omega u_+^2 \, dx \leq \int_\Omega |\nabla u_+|^2 \, dx = \lambda \int_\Omega h(x, u_+) u_+ \, dx \leq \lambda C \int_\Omega u_+^2 \, dx \, ,$$

and thus, since $C \in (0, 1)$ we infer $\lambda \geq \frac{\lambda_1}{C} > \lambda_1$. We deduce that for any eigenvalue $\lambda \in (0, \lambda_1/C)$ of problem (2.34) we must have $u_+ = 0$. It follows that if $\lambda \in (0, \lambda_1/C)$ is an eigenvalue of problem (2.34) then it is also an eigenvalue of problem (2.30) with the corresponding eigenfunction being negative in Ω. But, we already noticed that the set of eigenvalues of problem (2.30) is discrete and $\lambda_1 < \lambda_2$. In other words, taking $\delta = \min\{\lambda_1/C, \lambda_2\}$ we find that $\delta > \lambda_1$ and no $\lambda \in (\lambda_1, \delta)$ can be an eigenvalue of problem (2.30). Consequently, no $\lambda \in (\lambda_1, \delta)$ is an eigenvalue of problem (2.34). We conclude that λ_1 is isolated in the set of eigenvalues of problem (2.34). The proof of Lemma 13 is complete. □

Next, we show that there exists $\mu_1 > 0$ such that any $\lambda \in (\mu_1, \infty)$ is an eigenvalue of problem (2.34). For this purpose, we consider the eigenvalue problem

$$\begin{cases} -\Delta u = \lambda h(x, u_+), & \text{in} \quad \Omega \\ u = 0, & \text{on} \quad \partial\Omega \, . \end{cases} \tag{2.38}$$

We say that λ is an eigenvalue of problem (2.38) if there exists $u \in H_0^1(\Omega) \setminus \{0\}$ such that

$$\int_\Omega \nabla u \nabla v \, dx - \lambda \int_\Omega h(x, u_+) v \, dx = 0 \, ,$$

for any $v \in H_0^1(\Omega)$.

We notice that if λ is an eigenvalue for (2.38) with the corresponding eigenfunction u, then taking $v = u_-$ in the above relation we deduce that $u_- = 0$, and thus, we find $u \geq 0$. In other words, the eigenvalues of problem (2.38) possesses nonnegative corresponding eigenfunctions. Moreover, the above discussion shows that an eigenvalue of problem (2.38) is also an eigenvalue of problem (2.34).

For each $\lambda > 0$ we define the energy functional associated to problem (2.38) by $I_\lambda : H_0^1(\Omega) \to \mathbb{R}$,

$$I_\lambda(u) = \frac{1}{2}\int_\Omega |\nabla u|^2\, dx - \lambda \int_\Omega H(x, u_+)\, dx\,,$$

where $H(x,t) = \int_0^t h(x,s)\, ds$. Then $I_\lambda \in C^1(H_0^1(\Omega), \mathbb{R})$ and we have

$$\langle I'_\lambda(u), v\rangle = \int_\Omega \nabla u \nabla v\, dx - \lambda \int_\Omega h(x, u_+) v\, dx\,,$$

for any u, $v \in H_0^1(\Omega)$. Thus, $\lambda > 0$ is an eigenvalue of problem (2.38) if and only if there exists a critical nontrivial point of functional I_λ.

Lemma 14 *The functional I_λ is bounded from below and coercive.*

Proof. Invoking hypothesis (H3) we can deduce that

$$\lim_{t\to\infty} \frac{H(x,t)}{t^2} = 0, \quad \text{uniformly in } \Omega\,.$$

Then for a given $\lambda > 0$ there exists a positive constant $C_\lambda > 0$ such that

$$\lambda H(x,t) \leq \frac{\lambda_1}{4} t^2 + C_\lambda, \quad \forall\, t \geq 0, \text{ a.e. } x \in \Omega\,,$$

where λ_1 is given by relation (2.31).

Thus, we see that for any $u \in H_0^1(\Omega)$ the following holds

$$I_\lambda(u) \geq \frac{1}{2}\int_\Omega |\nabla u|^2\, dx - \frac{\lambda_1}{4}\int_\Omega u^2\, dx - C_\lambda|\Omega| \geq \frac{1}{4}\|u\|^2 - C_\lambda|\Omega|\,,$$

where $\|\cdot\|$ denotes the norm on $H_0^1(\Omega)$, that is $\|u\| = (\int_\Omega |\nabla u|^2\, dx)^{1/2}$. This shows that I_λ is bounded from below and coercive. The proof of Lemma 14 is complete. □

Lemma 15 *There exists $\lambda^\star > 0$ such that assuming that $\lambda \geq \lambda^\star$ we have $\inf_{H_0^1(\Omega)} I_\lambda < 0$.*

Proof. Hypothesis (H2) implies that there exists $t_0 > 0$ such that

$$H(x, t_0) > 0 \quad \text{a.e. } x \in \overline{\Omega}\,.$$

Let $\Omega_1 \subset \Omega$ be a sufficiently large compact subset and pick $u_0 \in C_0^1(\Omega) \subset H_0^1(\Omega)$ such that $u_0(x) = t_0$ for any $x \in \Omega_1$ and $0 \leq u_0(x) \leq t_0$ for any $x \in \Omega \setminus \Omega_1$.

Thus, by hypothesis (H1) we have

$$\begin{aligned}\int_\Omega H(x,u_0)\,dx &\geq \int_{\Omega_1} H(x,t_0)\,dx - \int_{\Omega\setminus\Omega_1} Cu_0^2\,dx \\ &\geq \int_{\Omega_1} H(x,t_0)\,dx - Ct_0^2|\Omega\setminus\Omega_1| > 0\,.\end{aligned}$$

We conclude that $I_\lambda(u_0) < 0$ for sufficiently large $\lambda > 0$, and thus, $\inf_{H_0^1(\Omega)} I_\lambda < 0$. The proof of Lemma 15 is complete. □

Lemmas 14 and 15 show that for any large enough $\lambda > 0$, the functional I_λ possesses a negative global minimum, hence any large enough $\lambda > 0$ is an eigenvalue of problem (2.38) and consequently also of problem (2.34). Combining this fact with the assertions of Lemmas 11, 12 and 13 we conclude that Theorem 4 holds. □

We refer to Pucci and Rădulescu [82] for an extension of this result to the case of poly-harmonic operators.

2.5 Variable exponents versus Morse theory and local linking

In this section we investigate the existence of nontrivial solutions of the following $p(x)$-Laplacian problem:

$$\begin{cases} -\Delta_{p(x)}u + |u|^{p(x)-2}u = f(x,u) & \text{in } \Omega, \\ u = 0 & \text{on } \partial\Omega, \end{cases} \tag{2.39}$$

where Ω is a bounded domain in $\mathbb{R}^n$ with smooth boundary.

Let $p \in C(\overline{\Omega})$ and $1 < p_- := \min_{x\in\bar\Omega} p(x) \leq p(x) \leq p_+ := \max_{x\in\bar\Omega} p(x) < \infty$ and $F(x,t) = \int_0^t f(x,s)ds$, $\mathcal{F}(x,t) = f(x,t)t - p_+F(x,t)$.

We assume that the reaction term $f(x,u)$ satisfies the following hypotheses:
(H1) $f \in C(\overline{\Omega}\times\mathbb{R})$ with $f(x,0) = 0$ and there exists $C_1 > 0$ such that

$$|f(x,t)| \leq C_1(1+|t|^{q(x)-1}), \quad \forall (x,t) \in \Omega\times\mathbb{R},$$

where $q(x) \in C(\overline{\Omega})$, $1 < q(x) < p^*(x)$ for all $x \in \overline{\Omega}$ and $p^* = \frac{Np(x)}{N-p(x)}$ if $p(x) < N$, $p^*(x) = +\infty$ if $p(x) \geq N$;
(H2) $\lim_{|t|\to\infty} \frac{F(x,t)}{|t|^{p_+}} = +\infty$ uniformly with respect to $x \in \bar\Omega$;
(H3) there exists $\theta \geq 1$ such that $\theta\mathcal{F}(x,t) \geq \mathcal{F}(x,st)$ for $(x,t) \in \Omega\times\mathbb{R}$ and

TABLE 2.2: Admissible nonlinearity not satisfying the Ambrosetti-Rabinowitz condition

$f(x,t) \equiv f(t)$	$F(x,t) \equiv F(t)$
$\|t\|^{p_+-2}t\left(p_+\log(1+\|t\|)+\dfrac{\|t\|}{1+\|t\|}\right)$	$\|t\|^{p_+}\log(1+\|t\|)$

$s \in [0,1]$;
(H4) there exists $\nu > 0$ such that

$$\frac{f(x,t)}{|t|^{p_+-2}t} \quad \text{is increasing in } t \geq \nu \text{ and decreasing in } t \leq -\nu;$$

(H5) there are small enough constants r and R with $0 < r < R$ such that for all $t \in \mathbb{R}$ with $r \leq |t| \leq R$

$$C_2|t|^{\alpha(x)} \leq p(x)F(x,t) \leq C_3|t|^{p(x)}, \text{ a.e. } x \in \Omega, \tag{1.2}$$

where C_2 and C_3 are constants with $0 < C_2 < C_3 < 1$, $\alpha(x) \in C(\bar{\Omega})$ and $1 < \alpha(x) < p(x)$. Moreover, there exists $C_4 > 0$ such that

$$F(x,t) \geq -C_4|t|^{p_+} \quad \text{for all } (x,t) \in \Omega \times \mathbb{R}. \tag{1.3}$$

The assumption (H2) implies that problem (2.39) is superlinear at infinity. Several papers concerning superlinear elliptic boundary value problems have used the condition which for the case $p = 2$ is originally due to Ambrosetti and Rabinowitz [4], that is,

(AR): there exist $\mu > p_+$ and $M > 0$ such that

$$0 < \mu F(x,t) \leq f(x,t)t \text{ for all } x \in \Omega \text{ and } |t| \geq M. \tag{2.40}$$

It follows from (2.40) that for some a, $b > 0$,

$$F(x,t) \geq a|t|^{\mu} - b \quad \text{for } (x,t) \in \Omega \times \mathbb{R}. \tag{2.41}$$

Obviously, relation (2.41) implies the weaker hypothesis (H2).

In Table 2.1 we give an example of a function that does not satisfy the Ambrosetti-Rabinowitz condition (for simplicity we drop the x-dependence). We point out that f does not satisfy (AR) condition for any $\mu > p_+$, but it satisfies hypotheses (H2) and (H3). Furthermore, we can show that the function fulfills all hypotheses (H1)–(H5).

We establish the following two existence properties.

Theorem 5 *Assume that hypotheses (H1), (H2), (H3) and (H5) are fulfilled. Then problem (2.39) has at least one nontrivial weak solution in* $W_0^{1,p(x)}(\Omega)$.

Theorem 6 *Assume that hypotheses (H1), (H2), (H4) and (H5) are fulfilled. Then problem* (2.39) *has at least one nontrivial weak solution in* $W_0^{1,p(x)}(\Omega)$.

Before we state Theorem 7, we recall some results about the eigenvalues of $-\Delta_{p(x)}$ on $W_0^{1,p(x)}(\Omega)$. We say that λ is an eigenvalue of $-\Delta_{p(x)}$ with Dirichlet boundary conditions if that the following boundary value problem:

$$\begin{cases} -\Delta_{p(x)}u = \lambda|u|^{p(x)-2}u & x \in \Omega, \\ u = 0 & x \in \partial\Omega, \end{cases} \tag{2.42}$$

has a nonzero solution.

The existence of the principal eigenvalue $\lambda_* > 0$ was obtained in [32] under the following assumption:

(P): there exists a vector $l \in \mathbb{R}^N \setminus \{0\}$ such that for any $x \in \Omega$, $c(t) = p(x+tl)$ is monotone in $t \in I_x = \{t : x + tl \in \Omega\}$.

Theorem 7 *Assume that conditions (P), (H1), (H5) are fulfilled and additionally, that*
(H6): $\limsup_{|t|\to\infty} \frac{p(x)F(x,t)}{|t|^{p(x)}} < \frac{p_-}{p_+}\lambda_* + 1$ *uniformly on* $x \in \bar{\Omega}$;
(H7): $F(x,t) \geq 0$ *for all* $x \in \Omega$ *and* $|t| \leq r$.

Then problem (2.39) *has at least two nontrivial weak solutions in* $W_0^{1,p(x)}(\Omega)$.

The following example illustrates the importance of various hypotheses of Theorem 7. Let

$$\Omega = B(x_0, 1) := \{x \in \mathbb{R}^N : |x - x_0| < 1\},$$

where $x_0 = (x_{01}, x_{02}, \ldots, x_{0N})$, $x_{0k} > 1, k = 1, 2, \ldots, N$, $p(x) = |x|$ for all $x \in \Omega$. Then $p(x)$ satisfies the condition (P). If we take $f(x,t) = \beta|t|^{p_- - 2}t$, $0 < \beta < \frac{p_-}{p_+}\lambda_* + 1$, then $f(x,t)$ satisfies the hypotheses of Theorem 7.

Let $\mathcal{P}(\Omega)$ be the set of all Lebesgue measurable functions $p : \Omega \to (1, +\infty)$ and

$$||u||_{p(x)} = \inf\left\{t > 0 : \int_\Omega \left|\frac{u}{t}\right|^{p(x)} dx \leq 1\right\}.$$

We recall that the variable exponent Lebesgue space $L^{p(x)}(\Omega)$ is the class of all functions u such that $\int_\Omega |u(x)|^{p(x)}dx < \infty$. Then the variable exponent Sobolev space $W^{1,p(x)}(\Omega)$ is the class of all functions $u \in L^{p(x)}(\Omega)$ such that $|\nabla u| \in L^{p(x)}(\Omega)$ with the norm

$$||u||_{1,p(x)} = ||u||_{p(x)} + ||\nabla u||_{p(x)}.$$

For $u \in W^{1,p(x)}(\Omega)$, we define

$$|||u||| = \inf\left\{t > 0 : \int_\Omega \frac{|\nabla u|^{p(x)} + |u|^{p(x)}}{t^{p(x)}} dx \leq 1\right\}.$$

We deduce that

$$\frac{1}{2}||u||_{1,p(x)} \leq |||u||| \leq 2||u||_{1,p(x)}.$$

We recall that $I(u) = \int_\Omega \left(|\nabla u|^{p(x)} + |u|^{p(x)}\right)dx$. If $u \in W^{1,p(x)}(\Omega)$, then the following properties hold:
(i) If $u \neq 0$, then $|||u||| = l \Leftrightarrow I(\frac{u}{l}) = 1$.
(ii) If $|||u||| \geq 1$, then $|||u|||^{p_-} \leq I(u) \leq |||u|||^{p_+}$.
(iii) If $|||u||| \leq 1$, then $|||u|||^{p_-} \leq I(u) \leq |||u|||^{p_+}$.
(iv) If $|||u||| = 1$, then $I(u) = 1$.

Finally, we recall the definition of the *Nemytskii operator*. Let $F : \Omega \times \mathbb{R} \to \mathbb{R}$ be a Carathéodory function and assume that

$$|F(x,t)| \leq a(x) + b|t|^{\frac{p_1(x)}{p_2(x)}}, \ \forall (x,t) \in \Omega \times \mathbb{R},$$

where $a(x) \in L^\infty(\Omega)$, b is a positive constant, and $p_1(x), p_2(x) \in \mathcal{P}(\Omega) \cap L^\infty(\Omega)$.

We define the Nemytskii operator as follows:

$$(N_F(u))(x) = F(x, u(x)) .$$

An elementary computation shows that the mapping $N_F : L^{p_1(x)}(\Omega) \to L^{p_2(x)}(\Omega)$ is continuous and bounded.

2.5.1 Verification of the compactness condition

From the assumption (H1) we deduce that the energy functional $\Phi : W_0^{1,p(x)}(\Omega) \to \mathbb{R}$ given by

$$\Phi(u) = \int_\Omega \frac{1}{p(x)}(|\nabla u|^{p(x)} + |u|^{p(x)})dx - \int_\Omega F(x,u)dx$$

is well defined and belongs to the class C^1. The derivative of Φ at u is given by

$$\langle \Phi'(u), v\rangle = \int_\Omega (|\nabla u|^{p(x)-2}\nabla u \cdot \nabla v + |u|^{p(x)-2}u \cdot v)dx - \int_\Omega f(x,u)v dx\,,$$

for $v \in W_0^{1,p(x)}(\Omega)$.

Set

$$I(u) = \int_\Omega \frac{1}{p(x)}(|\nabla u|^{p(x)} + |u|^{p(x)})dx, \quad J(u) = \int_\Omega F(x,u)dx.$$

Then $\Phi(u) = I(u) - J(u)$. We have $I \in C^1(W_0^{1,p(x)}(\Omega), \mathbb{R})$ and

$$\langle I'(u), \varphi\rangle = \int_\Omega (|\nabla u|^{p(x)-2}\nabla u \nabla \varphi + |u|^{p(x)-2}u\varphi)dx, \quad \forall u, \varphi \in W_0^{1,p(x)}(\Omega).$$

Moreover, the mapping $I' : W_0^{1,p(x)}(\Omega) \to (W_0^{1,p(x)}(\Omega))^*$ is a bounded homeomorphism and is of type (S^+), namely

$$u_n \rightharpoonup u \text{ and } \limsup_{n\to\infty}\langle I'(u_n) - I'(u), u_n - u\rangle \le 0 \text{ imply } u_n \to u.$$

Again by hypothesis (H1), we have $J \in C^1(W_0^{1,p(x)}(\Omega), \mathbb{R})$ and

$$\langle J'(u), \varphi\rangle = \int_\Omega f(x,u)\varphi dx \quad \forall u, \varphi \in W_0^{1,p(x)}(\Omega).$$

Moreover, the mapping $I' : W_0^{1,p(x)}(\Omega) \to (W_0^{1,p(x)}(\Omega))^*$ is weakly-strongly continuous, namely,

$$u_n \rightharpoonup u \text{ implies } u_n \to u,$$

where $\rightharpoonup$ and $\to$ denote the weak and strong convergence in $W_0^{1,p(x)}(\Omega)$, respectively.

We denote $X = W_0^{1,p(x)}(\Omega)$ and $X^* = (W_0^{1,p(x)}(\Omega))^*$.

The function $u \in X$ is called a *weak solution* of problem (2.39) if for any $\phi \in X$

$$\int_\Omega (|\nabla u|^{p(x)-2}\nabla u \cdot \nabla\phi + |u|^{p(x)-2}u \cdot \phi)dx - \int_\Omega f(x,u)\phi dx = 0.$$

In what follows, we are concerned in finding nontrivial weak solutions of problem (2.39), that is, nontrivial critical points of Φ.

We recall that the functional Φ satisfies the (C) condition if for $c \in \mathbb{R}$, any sequence $\{u_n\} \subset X$ such that $\Phi(u) \to c, (1 + ||u_n||)||\Phi'(u_n)||_{X^*} \to 0$ has a convergent subsequence. The functional Φ satisfies the (PS) condition if any sequence $\{u_n\} \subset X$ such that $\Phi(u_n)$ is bounded and $\Phi'(u_n) \to 0$ has a convergent subsequence.

Theorem 8 *Suppose that Φ satisfies conditions (H1), (H2) and (H3). Then Φ satisfies also the (C) condition.*

Proof. We first verify the boundedness of (C) sequences. Assume Φ has an unbounded (C) sequence $\{u_n\}$. Up to a subsequence we may assume that

$$\Phi(u) \to c, \quad |||u_n||| \to \infty, \quad \langle\Phi'(u_n), u_n\rangle \to 0 \text{ as } n \to \infty.$$

Let $v_n = |||u_n|||^{-1}u_n$, then $|||v_n||| = 1$. By reflexivity, up to a subsequence, we have

$$v_n \rightharpoonup v \text{ in } X, \; v_n \to v \text{ in } L^{q(x)}(\Omega), \; v_n \to v \;\text{ a.e. } x \in \Omega. \tag{2.43}$$

If $v = 0$, we choose a sequence $\{t_n\} \subset [0,1]$ such that $\Phi(t_n u_n) = \max_{t\in[0,1]} \Phi(tu_n)$.

For any positive integer m, we can choose $r = (2mp_+)^{\frac{1}{p-}}$ such that $r|||u_n|||^{-1} \in (0,1)$ for large enough n. Since $|||v_n||| = 1$, we have

$$\int_\Omega (|\nabla v_n|^{p(x)} + |v_n|^{p(x)})dx = 1. \tag{2.44}$$

Since $v_n \to 0$ in $L^{q(x)}(\Omega)$ and (H1), we deduce from the continuity of the Nemytskii operator that $F(\cdot, rv_n) \to 0$ in $L^1(\Omega)$. Thus

$$\lim_{n\to\infty} \int_\Omega F(x, rv_n)dx = 0. \tag{2.45}$$

Hence, for large enough n, relations (2.44) and (2.45) yield

$$\Phi(t_n u_n) \geq \Phi(r|||u_n|||^{-1}u_n) = \Phi(rv_n) \geq \frac{r^{p-}}{p^+} - \int_\Omega F(x, rv_n)dx \geq m.$$

We deduce that $\Phi(t_n u_n) \to +\infty$. However, $\Phi(0) = 0, \Phi(u_n) \to c$, so $t_n \in (0,1)$ and

$$\langle \Phi'(t_n u_n), t_n u_n \rangle = t_n \frac{d}{dt}\Big|_{t=t_n} \Phi(tu_n) = 0.$$

Now using (H3), we obtain

$$\begin{aligned}\frac{1}{\theta}\Phi(t_n u_n) &= \frac{1}{\theta}(\Phi(t_n u_n) - \frac{1}{p_+}\langle \Phi'(t_n u_n), t_n u_n\rangle) \\ &\leq \int_\Omega t^{p(x)}(\frac{1}{p(x)} - \frac{1}{p_+})(|\nabla u_n|^{p(x)} + |u_n|^{p(x)})dx + \frac{1}{p_+}\int_\Omega \frac{\mathcal{F}(x, t_n u_n)}{\theta}dx \\ &\leq \int_\Omega (\frac{1}{p(x)} - \frac{1}{p_+})(|\nabla u_n|^{p(x)} + |u_n|^{p(x)})dx + \frac{1}{p_+}\int_\Omega \mathcal{F}(x, u_n)dx \\ &= \Phi(u_n) - \frac{1}{p_+}\langle \Phi'(u_n), u_n\rangle \to c.\end{aligned} \tag{2.46}$$

This contradicts the fact that $\Phi(t_n u_n) \to +\infty$.

If $v \neq 0$, we have $|u_n(x)| \to +\infty, \forall x \in \Omega'$, for the set $\Omega' := \{x \in \Omega : v(x) \neq 0\}$. Since the set Ω' has positive Lebesgue measure and $|||u_n||| > 1$ for large n, using (H2) we obtain

$$\begin{aligned}\frac{1}{p_-} &\geq \frac{1}{p_-}\int_\Omega \frac{1}{|||u_n|||^{p(x)}}(|\nabla u_n|^{p(x)} + |u_n|^{p(x)})dx \\ &\geq \int_\Omega \frac{1}{p(x)|||u_n|||^{p(x)}}(|\nabla u_n|^{p(x)} + |u_n|^{p(x)})dx \\ &\geq \frac{\Phi(u_n)}{|||u_n|||^{p_+}} + \int_{\Omega'} \frac{F(x, u_n)}{|||u_n|||^{p_+}}dx + \int_{\Omega\setminus\Omega'} \frac{F(x, u_n)}{|||u_n|||^{p_+}}dx \\ &\geq \frac{\Phi(u_n)}{|||u_n|||^{p_+}} + \int_{\Omega'} \frac{F(x, u_n)}{|u_n|^{p_+}}|v_n|^{p_+}dx - C_4\int_{\Omega\setminus\Omega'} |v_n|^{p_+}dx \to +\infty.\end{aligned}$$

This contradiction shows that $\{u_n\}$ is bounded in X.

Next, with standard arguments as in [4], we prove that any (C) sequence has a convergent subsequence. Since $\{u_n\}$ is bounded in X, we can assume by reflexivity that, up to a subsequence,

$$u_n \rightharpoonup u \text{ in } X, \quad u_n \to u \text{ in } L^{q(x)}(\Omega). \tag{2.47}$$

By the boundedness of $\{u_n\}$ in $L^{p(x)}(\Omega)$, we have

$$\sup_n \int_\Omega |u_n|^{p(x)} dx < \infty. \tag{2.48}$$

By (H1), (2.47) and (2.48), we have

$$\int_\Omega (f(x,u_n) - f(x,u))(u_n - u)dx \to 0 \text{ as } n \to \infty.$$

Note that $\Phi'(u_n) \to 0$, hence

$$\begin{aligned} \langle I'(u_n) - I'(u), u_n - u\rangle &= \langle \Phi'(u_n) - \Phi'(u), u_n - u\rangle \\ &+ \int_\Omega (f(x,u_n) - f(x,u))(u_n - u)dx \to 0. \end{aligned}$$

We deduce that $u_n \to u$ in X. The proof is complete. □

Theorem 9 *Suppose that Φ satisfies conditions (H1), (H2) and (H4). Then Φ satisfies also the (C) condition.*

Proof. We only need to verify that condition (2.46) is fulfilled, the rest of the proof being very similar to the proof of Theorem 8. Due to (H4), there exists a constant $C > 0$ such that

$$\mathcal{F}(x,s) \leq \mathcal{F}(x,t) \text{ for } x \in \Omega \text{ and } 0 \leq s \leq t \text{ or } t \leq s \leq 0.$$

Therefore

$$\begin{aligned} \Phi(t_n u_n) &= \Phi(t_n u_n) - \frac{1}{p_+}\langle \Phi'(t_n u_n), t_n u_n\rangle \\ &\leq \int_\Omega t^{p(x)}\left(\frac{1}{p(x)} - \frac{1}{p_+}\right)(|\nabla u_n|^{p(x)} + |u_n|^{p(x)})dx + \frac{1}{p_+}\int_\Omega \mathcal{F}(x, t_n u_n)dx \\ &\leq \int_\Omega \left(\frac{1}{p(x)} - \frac{1}{p_+}\right)(|\nabla u_n|^{p(x)} + |u_n|^{p(x)})dx + \frac{1}{p_+}\int_\Omega \mathcal{F}(x,u_n)dx + \frac{C}{p_+}|\Omega| \\ &= \Phi(u_n) - \frac{1}{p_+}\langle \Phi'(u_n), u_n\rangle + \frac{C}{p_+}|\Omega| \to c + \frac{C}{p_+}|\Omega| \end{aligned}$$

as $n \to \infty$, where $|\Omega|$ denotes the measure of the domain Ω. This completes the proof. □

Theorem 10 *Suppose that Φ satisfies conditions (P), (H1) and (H6). Then Φ satisfies also the (PS) condition.*

Proof. We first prove that Φ is coercive. It follows from (H1) and (H6) that for some small enough $\varepsilon > 0$, there exists a constant $C_\varepsilon > 0$ such that

$$|F(x,u)| \leq \frac{1}{p(x)}\big[\frac{p_-}{p_+}(\lambda_* - \varepsilon) + 1\big]|u|^{p(x)} + C_\varepsilon. \tag{2.49}$$

Hence by (2.49), for $u \in W_0^{1,p(x)}(\Omega)$ and $|||u||| \geq 2$ we have

$$\begin{aligned}
\Phi(u) &= \int_\Omega \frac{1}{p(x)}(|\nabla u|^{p(x)} + |u|^{p(x)})dx - \int_\Omega F(x,u)dx \\
&\geq \int_\Omega \frac{1}{p(x)}(|\nabla u|^{p(x)} + |u|^{p(x)})dx - \int_\Omega \frac{1}{p(x)}\big[\frac{p_-}{p_+}(\lambda_* - \varepsilon) + 1\big]|u|^{p(x)}dx - C_\varepsilon|\Omega| \\
&= \int_\Omega \frac{1}{p(x)}|\nabla u|^{p(x)}dx - \frac{p_-}{p_+}(\lambda_* - \varepsilon)\int_\Omega \frac{1}{p(x)}|u|^{p(x)}dx - C_\varepsilon|\Omega| \\
&\geq \int_\Omega \frac{1}{p(x)}|\nabla u|^{p(x)}dx - \frac{\lambda_* - \varepsilon}{\lambda_*}\int_\Omega \frac{1}{p(x)}|\nabla u|^{p(x)}dx - C_\varepsilon|\Omega| \\
&\geq C(\varepsilon)\,|||u|||^{p_-} - C_\varepsilon|\Omega| \to \infty
\end{aligned}$$

as $|||u||| \to \infty$. This completes the proof. □

2.5.2 Computation of critical groups

Let X be a real Banach space and $\Phi \in C^1(X,\mathbb{R})$, $K = \{u \in X : \Phi'(u) = 0\}$. Then the qth *critical group* of Φ at an isolated critical point $u \in K$ with $\Phi(u) = c$ is defined by

$$C_q(\Phi, u) := H_q(\Phi^c \cap U, \Phi^c \cap U \setminus \{u\}), \quad q \in \mathbb{N} := \{0, 1, 2, \cdots\},$$

where $\Phi^c = \{u \in X : \Phi(u) \leq c\}$, U is any neighborhood of u, containing the unique critical point, and H_* is the singular relative homology with coefficients in an Abelian group G.

We say that $u \in K$ is a *homologically nontrivial critical point* of Φ if at least one of its critical groups is nontrivial.

Proposition 4 *Let X be a real Banach space and let $\Phi \in C^1(X,\mathbb{R})$ satisfy the (PS) condition and be bounded from below. If Φ has a critical point that is homologically nontrivial and is not a minimizer of Φ, then Φ has at least three critical points.*

If Φ satisfies the condition (C) and the critical values of Φ are bounded from below by some $a < \inf \Phi(K)$, then the critical groups of Φ at infinity are defined by

$$C_q(\Phi,\infty) := H_q(X,\Phi^a), \quad q \in \mathbb{N}. \tag{2.50}$$

If Φ satisfies the condition (C), then Φ satisfies the deformation condition. By the deformation lemma, the right-hand side of (2.50) does not depend on the choice of a.

Morse theory tells us that if $K = \{0\}$ then $C_q(\Phi,\infty) = C_q(\Phi,0)$ for all $q \in \mathbb{N}$. It follows that if $C_q(\Phi,\infty) \neq C_q(\Phi,0)$ for some $q \in \mathbb{N}$, then Φ must have a nontrivial critical point. So we must compute the critical groups at zero and at infinity.

In what follows we may assume for the proofs of our results in this section, that Φ has only finitely many critical points. Since Φ satisfies the condition (C), the critical groups $C_q(\Phi,\infty)$ at infinity make sense.

Theorem 11 *Suppose that Φ satisfies conditions (H1), (H2) and (H3). Then $C_q(\Phi,\infty) = 0$ for all $q \in \mathbb{N}$.*

Proof. Let $S = \{u \in X : |||u||| = 1\}$. For $u \in S$, by Fatou's lemma and hypothesis (H2) we have

$$\lim_{t\to+\infty}\int_\Omega \frac{F(x,tu)}{|t|^{p_+}}dx \geq \int_\Omega \lim_{t\to+\infty}\frac{F(x,tu)}{|tu|^{p_+}}|u|^{p_+}dx = +\infty.$$

Therefore

$$\Phi(tu) \leq \frac{t^{p_+}}{p_-} - \int_\Omega F(x,tu)dx \leq t^{p_+}\Big(\frac{1}{p_-} - \int_\Omega \frac{F(x,tu)}{|t|^{p_+}}dx\Big) \to -\infty \text{ as } t \to +\infty.$$

Choose $a < \min\{\inf_{\|u\|\leq 1}\Phi(u), 0\}$. Then for any $u \in S$, there exists $t_0 > 1$ such that $\Phi(t_0 u) \leq a$. By (H3), we have

$$\mathcal{F}(x,m) \geq 0 \quad for \ (x,m) \in \Omega \times \mathbb{R}. \tag{2.51}$$

Therefore, if

$$\Phi(tu) = \int_\Omega \frac{t^{p(x)}}{p(x)}(|\nabla u|^{p(x)} + |u|^{p(x)})dx - \int_\Omega F(x,tu)dx \leq a,$$

then

$$\int_\Omega (t^{p(x)}(|\nabla u|^{p(x)} + |u|^{p(x)}))dx \leq p_+ a + \int_\Omega p_+ F(x,tu)dx.$$

Using (2.51), we obtain

$$\begin{aligned}\frac{d}{dt}\Phi(tu) &= \frac{1}{t}\Big[\int_\Omega t^{p(x)}(|\nabla u|^{p(x)} + |u|^{p(x)})dx - \int_\Omega f(x,tu)tudx\Big]\\ &\leq \frac{1}{t}\Big[p_+ a - \int_\Omega \mathcal{F}(x,tu)dx\Big] < 0.\end{aligned}$$

Then by the implicit function theorem, there exists a unique $T \in C(S, \mathbb{R})$ such that $\Phi(T(u)u) = a$. Next, we use the function T to construct a strong deformation retract from $X \setminus \{0\}$ to Φ^a. Therefore, we can deduce

$$C_q(\Phi, \infty) = H_q(X, \Phi^a) = H_q(X, X \setminus \{0\}) = 0, \forall q \in \mathbb{N}.$$

The proof is now complete. □

In a similar manner we can prove the following result.

Theorem 12 *Suppose that Φ satisfies conditions (H1), (H2) and (H4). Then $C_q(\Phi, \infty) = 0$ for all $q \in \mathbb{N}$.*

Since X is a separable and reflexive Banach space, there exist $\{e_n\}_{n=1}^{\infty} \subset X$ and $\{f_n\}_{n=1}^{\infty} \subset X^*$ such that

$$f_n(e_m) = \delta_{n,m} = \begin{cases} 1, & n \neq m \\ 0, & n = m \end{cases}$$

and

$$X = \overline{\operatorname{span}\{e_n : n = 1, 2, \ldots\}},\ X^* = \overline{\operatorname{span}\{f_n : n = 1, 2, \ldots\}}.$$

For $k = 1, 2, \ldots$, we denote $Y_k = \operatorname{span}\{e_1, e_2, \ldots, e_k\}$, hence Y_k has a closed complementary subspace Z_k in X. Thus, $X = Y_k \oplus Z_k$.

Theorem 13 *Supposed that Φ satisfies conditions (H1) and (H5). Then there exists $k_0 \in \mathbb{N}$ such that $C_{k_0}(\Phi, 0) \neq 0$.*

Proof. Since $f(x, 0) = 0$, the zero function 0 is a critical point of Φ. So we only need to prove that Φ has a local linking at 0 with respect to $X = Y_k \oplus Z_k$. We split the proof into the following two parts.

Step 1. Take $u \in Y_k$. Since Y_k is finite dimensional, we have that for given $R > 0$, there exists small enough $0 < \rho < 1$ such that

$$u \in Y_k, |||u||| < \rho \Rightarrow |u(x)| < R, \quad \forall x \in \Omega.$$

For $0 < r < R$, let $\Omega_1 = \{x \in \Omega : |u(x)| < r\}$, $\Omega_2 = \{x \in \Omega : r \leq |u(x)| \leq R\}$, and $\Omega_3 = \{x \in \Omega : |u(x)| > R\}$. Then $\Omega = \cup_{i=1}^{3} \Omega_i$ and Ω_i are pairwise disjoint. For the sake of simplicity, let $G(x, u) = F(x, u) - \frac{C_2}{p(x)}|u|^{\alpha(x)}$. Therefore it follows from (H5) that

$$\begin{aligned}
\Phi(u) &= \int_\Omega \frac{1}{p(x)}(|\nabla u|^{p(x)} + |u|^{p(x)})dx - (\int_{\Omega_1} + \int_{\Omega_2} + \int_{\Omega_3})F(x, u)dx \\
&= \int_\Omega \frac{1}{p(x)}(|\nabla u|^{p(x)} + |u|^{p(x)})dx - \int_\Omega \frac{C_2}{p(x)}|u|^{\alpha(x)}dx \\
&\quad - \int_{\Omega_1} G(x, u)dx - \int_{\Omega_2} G(x, u)dx - \int_{\Omega_3} G(x, u)dx \\
&\leq \int_\Omega (|\nabla u|^{p(x)} + |u|^{p(x)})dx - \frac{C_2}{p_+}\int_\Omega |u|^{\alpha(x)}dx - \int_{\Omega_1} G(x, u)dx.
\end{aligned}$$

Due to the assumptions on $\alpha(x)$ and since Ω is a bounded domain in $\mathbb{R}^N$, the embedding $L^{p(x)}(\Omega) \to L^{\alpha(x)}(\Omega)$ is continuous. This implies that there exists a constant $C > 1$ such that

$$||u||_{\alpha(x)} \leq C||u||_{p(x)} \leq C||u||_{1,p(x)} \leq 2C|||u||| \leq 2C\rho.$$

If $\rho \leq \frac{1}{2C}$, then $||u||_{\alpha(x)} \leq 1$. Note that the norms on Y_k are equivalent, so in particular, $||u||_{\alpha(x)}$ is equivalent to $|||u|||$. Since $\alpha(x), p(x) \in C(\bar{\Omega})$ and $\alpha(x) < p(x)$, for each $x \in \Omega$, there exists an open subset $B_\delta(x)$ of $\bar{\Omega}$ such that

$$\alpha_x := \sup_{x\in B_\delta(x)} \alpha(x) < \inf_{x\in B_\delta(x)} p(x) := p_x.$$

Then $\{B_\delta(x)\}_{x\in\bar{\Omega}}$ is an open covering of $\bar{\Omega}$. Since $\bar{\Omega}$ is compact, there is a finite subcovering $\{B_\delta(x_i)\}_{i=1}^m$. We can use all hyperplanes, for each there exists at least one hypersurface of some $\{B_\delta(x_i)\}_{i=1}^m$ lying on it, in order to divide $\{B_\delta(x_i)\}_{i=1}^m$ into finitely many pairwise disjoint hypercubes $\{\mathcal{Q}\}_{j=1}^n$. It is obvious that $\overline{\Omega} = \cup_{j=1}^n \overline{\mathcal{Q}}_j$ and

$$\alpha_{j+} := \sup_{x\in\mathcal{Q}_j} \alpha(x) < \inf_{x\in\mathcal{Q}_j} p(x) := p_{j-}.$$

Notice that $\int_{\Omega_1} G(x,u)dx \to 0$ as $r \to 0$ and $|||u|||_{\mathcal{Q}_j} \leq |||u|||$. Therefore, there is a constant $C > 0$ such that for small enough $\rho > 0$ and r

$$\begin{aligned}\Phi(u) \leq &\sum_{j=1}^n \left[\int_{\mathcal{Q}_j} (|\nabla u|^{p(x)} + |u|^{p(x)})dx - \frac{C_2}{p_-}\int_{\mathcal{Q}_j} |u|^{\alpha(x)}dx\right] + \int_{\Omega_1} G(x,u)dx \\ \leq &\sum_{j=1}^n (|||u|||_{\mathcal{Q}_j}^{p_{j-}} - C|||u|||_{\mathcal{Q}_j}^{\alpha_{j+}}) - \int_{\Omega_1} G(x,u)dx \leq 0.\end{aligned}$$

Step 2. By (H1) and the Young inequality, there exists $C > 0$ such that

$$|F(x,u)| \leq \frac{C_3}{p_+}|u|^{p(x)} + C|u|^{s(x)}, \text{ for all } x \in \Omega \text{ and } |u| \geq R, \tag{2.52}$$

where $s(x) \in C(\bar{\Omega})$ and $p(x) < s(x) < p^*(x)$. For the sake of simplicity, let $H(x,u) = F(x,u) - \frac{C_3}{p(x)}|u|^{p(x)}$. Therefore, if $u \in Z_k$ and $|||u||| \leq 1$, we can deduce from (H5) and relation (2.52) that

$$\begin{aligned}\Phi(u) &= \int_\Omega \frac{1}{p(x)}(|\nabla u|^{p(x)} + |u|^{p(x)})dx - (\int_{\Omega_1} + \int_{\Omega_2} + \int_{\Omega_3})F(x,u)dx \\ &= \int_\Omega \frac{1}{p(x)}(|\nabla u|^{p(x)} + |u|^{p(x)})dx - \int_\Omega \frac{C_3}{p(x)}|u|^{p(x)}dx \\ &\quad - \int_{\Omega_1} H(x,u)dx - \int_{\Omega_2} H(x,u)dx - \int_{\Omega_3} H(x,u)dx \\ &\geq \frac{1-C_3}{p_+}\int_\Omega (|\nabla u|^{p(x)} + |u|^{p(x)})dx - C\int_\Omega |u|^{s(x)}dx - \int_{\Omega_1} H(x,u)dx.\end{aligned}$$

Consider the functional $\varphi : X \to \mathbb{R}$ given by $\varphi(u) = \int_\Omega |u|^{s(x)}dx$. Since the embedding $X \to L^{s(x)}(\Omega)$ is compact, we deduce that φ is weakly-strongly continuous. Hence by [33, Lemma 3.3], we have

$$\eta_k = \sup_{u\in Z_k, ||u||\le 1} |\varphi(u)| \to 0 \quad \text{as } k \to \infty. \tag{2.53}$$

Note that $\int_{\Omega_1} H(x,u)dx \to 0$ as $r \to 0$. Therefore using the same argument as in Step 1, we obtain

$$\begin{aligned}\Phi(u) &\ge \sum_{i=1}^{l}\Big[\frac{1-C_3}{p_+}\int_{\mathbb{Q}_i}(|\nabla u|^{p(x)} + |u|^{p(x)})dx - C\int_{\mathbb{Q}_i}|u|^{s(x)}dx\Big] + \int_{\Omega_1} H(x,u)dx \\ &\ge \sum_{i=1}^{l}\Big[\frac{1-C_3}{p_+}|||u|||_{\mathbb{Q}_i}^{p_{i+}} - C\eta_k|||u|||_{\mathbb{Q}_i}^{s_{i-}}\Big] - \int_{\Omega_1} H(x,u)dx.\end{aligned}$$

Due to relation (2.53), there exists $k_0 \in \mathbb{N}$ such that $\eta_{k_0} \le \frac{1-C_3}{2Cp_+}$. Therefore

$$\Phi(u) \ge \sum_{i=1}^{l}\frac{1-C_3}{2Cp_+}(|||u|||_{\mathbb{Q}_i}^{p_{i+}} - |||u|||_{\mathbb{Q}_i}^{s_{i-}}) - \int_{\Omega_1} H(x,u)dx > 0$$

if $0 < \rho < 1$ and r are sufficiently small. Thus, there exists $k_0 \in \mathbb{N}$ such that $\Phi(u) > 0$ when $u \in Z_{k_0}$ and $0 < |||u||| \le \rho$.

Combining Step 1 and Step 2, we can complete the proof of Theorem 13 by invoking [47, Proposition 2.1]. □

The above arguments show that the conclusion of Theorem 13 remains valid under the assumptions (H1), (H5) and (H7).

We now deduce the main existence results of this section.

Proof of Theorem 5. By Theorem 11 and Theorem 13, we have $C_{k_0}(\Phi, \infty) \neq C_{k_0}(\Phi, 0)$ for some $k_0 \in \mathbb{N}$. Now Theorem 5 immediately follows from Theorem 8. □

Proof of Theorem 6. By Theorem 12 and Theorem 13, we have $C_{k_0}(\Phi, \infty) \neq C_{k_0}(\Phi, 0)$ for some $k_0 \in \mathbb{N}$. Now Theorem 6 immediately follows from Theorem 8. □

Proof of Theorem 7. By Theorem 10, Φ satisfies the (PS) condition and is bounded from below. By assumption (H7), the trivial solution $u = 0$ is homologically nontrivial and is not a minimizer. The conclusion now follows from [47, Theorem 2.1].

Finally, we remark that the methods developed in this section can also be applied to study the following more general problem:

$$\begin{cases} -\text{div}(|\nabla u|^{p(x)-2}\nabla u) + b(x)|u|^{p(x)-2}u = f(x,u), & x \in \Omega \\ u = 0, & x \in \partial\Omega \end{cases}$$

where $b(x) \in C(\bar{\Omega})$ and $1 < b_- \le b_+ < \infty$.

2.6 The Caffarelli-Kohn-Nirenberg inequality with variable exponents

Caffarelli, Kohn and Nirenberg proved the following important inequality in [14]: given $p \in (1, N)$, there exists a positive constant $C_{a,b}$ such that for all $u \in C_c^1(\Omega)$,

$$\left(\int_\Omega |x|^{-bq}|u|^q\,dx\right)^{p/q} \leq C_{a,b}\int_\Omega |x|^{-ap}|\nabla u|^p\,dx\,, \tag{2.54}$$

where

$$-\infty < a < \frac{N-p}{p}, \quad a \leq b \leq a+1, \quad q = \frac{Np}{N-p(1+a-b)}\,,$$

and $\Omega \subseteq \mathbb{R}^N$ is an arbitrary open domain. Note that the Caffarelli-Kohn-Nirenberg inequality (2.54) reduces to the classical Sobolev inequality (if $a = b = 0$) and to the Hardy inequality (if $a = 0$ and $b = 1$).

Inequality (2.54) is a basic tool in the qualitative analysis of degenerate elliptic problems. It is also related to the understanding of some important phenomena such as best constants, existence or nonexistence of extremal functions, symmetry properties of minimizers, compactness of minimizing sequences, concentration phenomena, etc. An important consequence of the Caffarelli-Kohn-Nirenberg inequality is that it enables the study of some degenerate elliptic equations which involve differential operators of the type

$$\text{div}\,(a(x)|\nabla u|^{p-2}\nabla u)\,,$$

where $a(x)$ is a nonnegative function satisfying $\inf_x a(x) = 0$. Thus, the resulting operator is not uniformly elliptic and consequently some of the techniques that can be applied for solving equations involving uniformly elliptic operators fail in this new context. Degenerate differential operators involving a nonnegative weight that is allowed to have zeros at some points or even to be unbounded are used in the study of many physical phenomena related to equilibrium of anisotropic continuous media.

In this section, we obtain inequalities of type (2.54) in the case when the constant p is replaced by a function $p(x)$ of class C^1 and we use them for studying some degenerate elliptic equations involving variable exponent growth conditions.

Assume that $\Omega \subset \mathbb{R}^N$ $(N \geq 2)$ is an open, bounded set with smooth boundary.

For each $x \in \Omega$, $x = (x_1, ..., x_N)$ and $i \in \{1, ..., N\}$ we denote

$$m_i = \inf_{x\in\Omega} x_i \qquad M_i = \sup_{x\in\Omega} x_i\,.$$

For $i \in \{1, ..., N\}$, let $a_i : [m_i, M_i] \to \mathbb{R}$ be functions of class C^1. In particular, the functions a_i are allowed to vanish.

Let $\overrightarrow{a} : \Omega \to \mathbb{R}^N$ be defined by

$$\overrightarrow{a}(x) = (a_1(x_1), ..., a_N(x_N)) .$$

We assume that there exists a constant $a_0 > 0$ such that

$$\operatorname{div} \overrightarrow{a}(x) \geq a_0 > 0, \quad \forall\, x \in \overline{\Omega} . \tag{2.55}$$

Next, we consider a function of class C^1 $p : \overline{\Omega} \to (1, N)$ satisfying

$$\overrightarrow{a}(x) \cdot \nabla p(x) = 0, \quad \forall\, x \in \Omega . \tag{2.56}$$

The main result of this section is the following inequality.

Theorem 14 *Assume that $\overrightarrow{a}(x)$ and $p(x)$ are defined as above and satisfy conditions* (2.55) *and* (2.56). *Then there exists a positive constant C such that*

$$\int_\Omega |u(x)|^{p(x)}\, dx \leq C \int_\Omega |\overrightarrow{a}(x)|^{p(x)} |\nabla u(x)|^{p(x)}\, dx, \quad \forall\, u \in C_c^1(\Omega) . \tag{2.57}$$

Proof. Using hypothesis (2.56) we obtain that for all $u \in C_c^1(\Omega)$

$$\begin{aligned}
&\operatorname{div}(|u(x)|^{p(x)} \overrightarrow{a}(x)) = \sum_{i=1}^N \frac{\partial}{\partial x_i} \left(|u(x)|^{p(x)} a_i(x_i)\right) = \\
&|u(x)|^{p(x)} \operatorname{div} \overrightarrow{a}(x) + \\
&\sum_{i=1}^N a_i(x_i) \left[p(x)|u(x)|^{p(x)-2} u(x) \frac{\partial u}{\partial x_i} + |u(x)|^{p(x)} \log(|u(x)|) \frac{\partial p}{\partial x_i} \right] = \\
&|u(x)|^{p(x)} \operatorname{div} \overrightarrow{a}(x) + p(x)|u(x)|^{p(x)-2} u(x) \nabla u(x) \cdot \overrightarrow{a}(x) + \\
&|u(x)|^{p(x)} \log(|u(x)|) \nabla p(x) \cdot \overrightarrow{a}(x) = \\
&|u(x)|^{p(x)} \operatorname{div} \overrightarrow{a}(x) + p(x)|u(x)|^{p(x)-2} u(x) \nabla u(x) \cdot \overrightarrow{a}(x).
\end{aligned}$$

On the other hand, the flux-divergence theorem implies that for each $u \in C_c^1(\Omega)$ we have

$$\int_\Omega \operatorname{div}(|u(x)|^{p(x)} \overrightarrow{a}(x))\, dx = \int_{\partial\Omega} |u(x)|^{p(x)} \overrightarrow{a}(x) \cdot \overrightarrow{n}\, d\sigma(x) = 0 .$$

Thus, we obtain for all $u \in C_c^1(\Omega)$

$$\int_\Omega |u(x)|^{p(x)} \operatorname{div} \overrightarrow{a}(x)\, dx \leq p^+ \int_\Omega |u(x)|^{p(x)-1} |\nabla u(x)| |\overrightarrow{a}(x)|\, dx .$$

Fix $\epsilon > 0$, $x \in \Omega$, and A, $B \geq 0$. Then by the Young inequality,

$$AB \le \epsilon A^{\frac{p(x)}{p(x)-1}} + \frac{1}{\epsilon^{p(x)-1}} B^{p(x)} .$$

We fix $\epsilon > 0$ such that

$$p^+ \epsilon < a_0 ,$$

where a_0 is given by relation (2.55).

The above facts and relation (2.55) yield

$$a_0 \int_\Omega |u(x)|^{p(x)} \, dx \le$$
$$p^+ \left[\epsilon \int_\Omega |u(x)|^{p(x)} \, dx + \int_\Omega \left(\frac{1}{\epsilon}\right)^{p(x)-1} |\overrightarrow{a}(x)|^{p(x)} |\nabla u(x)|^{p(x)} \, dx \right] ,$$

for any $u \in C_c^1(\Omega)$. Therefore

$$(a_0 - \epsilon p^+) \int_\Omega |u(x)|^{p(x)} \, dx \le$$
$$\left[\left(\frac{1}{\epsilon}\right)^{p^- - 1} + \left(\frac{1}{\epsilon}\right)^{p^+ - 1} \right] p^+ \int_\Omega |\overrightarrow{a}(x)|^{p(x)} |\nabla u(x)|^{p(x)} \, dx ,$$

for any $u \in C_c^1(\Omega)$. This completes the proof of Theorem 14. □

The result of Theorem 14 implies that under the hypotheses (2.55) and (2.56), there exists a positive constant D such that

$$\int_\Omega |u(x)|^{p(x)} \, dx \le D \int_\Omega |\nabla u(x)|^{p(x)} \, dx, \quad \forall\, u \in C_c^1(\Omega) .$$

Thus, we deduce that in the hypotheses of Theorem 14 we have

$$\inf_{u \in W_0^{1,p(x)}(\Omega) \setminus \{0\}} \frac{\displaystyle\int_\Omega |\nabla u|^{p(x)} \, dx}{\displaystyle\int_\Omega |u|^{p(x)} \, dx} > 0 . \tag{2.58}$$

This relation asserts that the first eigenvalue of the $p(x)$-Laplace operator (that is, $\Delta_{p(x)} u := \operatorname{div}(|\nabla u|^{p(x)-2} \nabla u)$) is positive.

Example 7 *We give an example of functions $\overrightarrow{a}(x)$ and $p(x)$ satisfying conditions* (2.55) *and* (2.56) *for the case when $\overrightarrow{a}(x)$ can vanish at some points of Ω. Let $N \ge 3$ and let $\Omega = B_{\frac{1}{\sqrt{N}}}(0)$ denote the ball centered in the origin of radius $\frac{1}{\sqrt{N}}$. We define $\overrightarrow{a}(x) : \Omega \to \mathbb{R}^N$ by*

$$\overrightarrow{a}(x) = (-x_1, x_2, x_3, ..., x_{N-1}, x_N) ,$$

(more exactly, function $\vec{a}(x)$ *is associated to a vector* $x \in \Omega$ *obtained from* x *by changing in the first coordinate* x_1 *by* $-x_1$ *and keeping unchanged* x_i *for* $i \in \{2, ..., N\}$*). Clearly,* $\vec{a}(x)$ *is of class* C^1, $\vec{a}(0) = 0$ *and we have*

$$\operatorname{div}(\vec{a}(x)) = N - 2 \geq 1, \quad \forall\, x \in \Omega\,.$$

Thus, condition (2.55) *is satisfied.*

Next, we define $p : \overline{\Omega} \to (1, N)$ *by*

$$p(x) = x_1(x_2 + x_3 + ... + x_{N-1} + x_N) + 2, \quad \forall\, x \in \overline{\Omega}\,.$$

Then p *is of class* C^1 *and we have*

$$\nabla p(x) \cdot \vec{a}(x) = (x_2 + ... + x_N)(-x_1) + x_1x_2 + ... + x_1x_N = 0, \quad \forall\, x \in \Omega\,.$$

This means that condition (2.56) *is satisfied, too.*

Example 8 *We now give a related example for* $N = 2$*. Taking* $\Omega = B_{\frac{1}{3^{1/3}}}(0)$, $\vec{a}(x) = (-x_1, 2x_2)$ *and* $p(x) = x_1^2x_2 + \frac{3}{2}$*, we see that relations* (2.55) *and* (2.56) *are fulfilled.*

If N, a and p are as in the above examples, then the assertion of Theorem 14 reads as follows: there exists a positive constant $C > 0$ such that

$$\int_\Omega |u(x)|^{p(x)}\, dx \leq C \int_\Omega |x|^{p(x)}|\nabla u(x)|^{p(x)}\, dx, \quad \forall\, u \in C_c^1(\Omega)\,. \tag{2.59}$$

2.6.1 A compact embedding

We assume that N, Ω, $\vec{a}(x)$ and $p(x)$ are as in Examples 7 and 8.

Let $\mathcal{D}_0^{1,p(x)}(\Omega)$ denote the closure of $C_c^1(\Omega)$ under the norm

$$\|u\| = |\ |x \cdot \nabla u(x)|\ |_{p(x)}\ .$$

Then $(\mathcal{D}_0^{1,p(x)}(\Omega), \|\cdot\|)$ is a reflexive Banach space.

In what follows, we establish the following embedding property.

Theorem 15 *Assume that* N, Ω, $\vec{a}(x)$ *and* $p(x)$ *are as in Examples 7 and 8 and* $p^- > \frac{2N}{2N-1}$*. Then* $\mathcal{D}_0^{1,p(x)}(\Omega)$ *is compactly embedded in* $L^q(\Omega)$ *for each* $q \in \left(1, \frac{2Np^-}{2N+p^-}\right)$.

Proof. Let $\{u_n\}$ be a bounded sequence in $\mathcal{D}_0^{1,p(x)}(\Omega)$. There exists $\epsilon_0 \in (0,1)$ such that $\overline{B}_{\epsilon_0}(0) \subset \Omega$. Let $\epsilon \in (0, \epsilon_0)$ be arbitrary but fixed. By Theorem 14 it follows that $\{u_n\}$ is also a bounded sequence in $L^{p(x)}(\Omega)$. Consequently, $\{u_n\} \subset W^{1,p(x)}(\Omega \setminus \overline{B}_\epsilon(0))$ is a bounded sequence. Since

$$W^{1,p(x)}(\Omega \setminus \overline{B}_\epsilon(0)) \subset W^{1,p^-}(\Omega \setminus \overline{B}_\epsilon(0)),$$

we deduce that $\{u_n\}$ is a bounded sequence in $W^{1,p^-}(\Omega \setminus \overline{B}_\epsilon(0))$. The classical compact embedding theorem implies that there exists a convergent subsequence of $\{u_n\}$, still denoted by $\{u_n\}$, in $L^q(\Omega \setminus \overline{B}_\epsilon(0))$. Thus, for any large enough n and m we have

$$\int_{\Omega\setminus\overline{B}_\epsilon(0)} |u_n - u_m|^q \, dx < \epsilon \,. \tag{2.60}$$

On the other hand, the Hölder inequality for variable exponent spaces implies

$$\begin{aligned}\int_{B_\epsilon(0)} |u_n - u_m|^q \, dx &= \int_{B_\epsilon(0)} |x|^{-\frac{q}{2}} |x|^{\frac{q}{2}} |u_n - u_m|^q \, dx \\ &\leq D_1 \left| |x|^{-\frac{q}{2}} \chi_{B_\epsilon(0)} \right|_{\left(\frac{p(x)}{q}\right)'} \left| |x|^{\frac{q}{2}} |u_n - u_m|^q \right|_{\frac{p(x)}{q}} ,\end{aligned}$$

where D_1 is a positive constant.

Furthermore, inequality (2.59) and relations (1.4) and (1.5) imply

$$\begin{aligned}&\left| |x|^{\frac{q}{2}} |u_n - u_m|^q \right|_{\frac{p(x)}{q}} \leq \\ &\left(\int_\Omega |x|^{\frac{p(x)}{2}} |u_n - u_m|^{p(x)} \, dx \right)^{\frac{q}{p^-}} + \left(\int_\Omega |x|^{\frac{p(x)}{2}} |u_n - u_m|^{p(x)} \, dx \right)^{\frac{q}{p^+}} \leq \\ &\left[(\sup_{x\in\Omega} |x| + 1)^{\frac{qp^+}{2p^-}} + (\sup_{x\in\Omega} |x| + 1)^{\frac{q}{2}} \right] \left[\rho_{p(x)}(u_n - u_m)^{\frac{q}{p^-}} + \rho_{p(x)}(u_n - u_m)^{\frac{q}{p^+}} \right] \leq \\ &D_2 \left[\left(\int_\Omega |x|^{p(x)} |\nabla(u_n - u_m)|^{p(x)} \, dx \right)^{\frac{q}{p^-}} + \left(\int_\Omega |x|^{p(x)} |\nabla(u_n - u_m)|^{p(x)} \, dx \right)^{\frac{q}{p^+}} \right]\end{aligned}$$

where D_2 is a positive constant.

Combining these relations, we conclude that there exists a positive constant M such that

$$\int_{B_\epsilon(0)} |u_n - u_m|^q \, dx \leq M \left| |x|^{-\frac{q}{2}} \chi_{B_\epsilon(0)} \right|_{\left(\frac{p(x)}{q}\right)'} .$$

Using again relations (1.4) and (1.5) we deduce that

$$\begin{aligned}\left| |x|^{-\frac{q}{2}} \chi_{B_\epsilon(0)} \right|_{\left(\frac{p(x)}{q}\right)'} \leq \; & \rho_{\left(\frac{p(x)}{q}\right)'} \left(|x|^{-\frac{q}{2}} \chi_{B_\epsilon(0)} \right)^{\left(\left(\frac{p(x)}{q}\right)'\right)^+} + \\ & \rho_{\left(\frac{p(x)}{q}\right)'} \left(|x|^{-\frac{q}{2}} \chi_{B_\epsilon(0)} \right)^{\left(\left(\frac{p(x)}{q}\right)'\right)^-} ,\end{aligned}$$

where $\left(\frac{p(x)}{q}\right)' = \frac{p(x)}{p(x)-q}$, and assuming $\epsilon \in (0,1)$

$$\begin{aligned}\int_{B_\epsilon(0)} |x|^{\frac{-qp(x)}{2(p(x)-q)}}\, dx &\leq \int_{B_\epsilon(0)} |x|^{\frac{-qp^-}{2(p^--q)}}\, dx \\ &= \int_0^\epsilon \omega_N r^{N-1} r^{\frac{-qp^-}{2(p^--q)}}\, dr \\ &= \omega_N \frac{1}{\alpha}\epsilon^\alpha ,\end{aligned}$$

where $\alpha = N - \frac{qp^-}{2(p^--q)} > 0$ and ω_N is the area of the unit ball in $\mathbb{R}^N$.

Consequently,

$$\int_{B_\epsilon(0)} |u_n - u_m|^q\, dx \leq M_1(\epsilon^{\alpha_1} + \epsilon^{\alpha_2}),$$

with α_1, $\alpha_2 > 0$ and a constant $M_1 > 0$.

The above inequality and relation (2.60) show that for any large enough n and m we have

$$\int_\Omega |u_n - u_m|^q\, dx \leq M_2(\epsilon + \epsilon^{\alpha_1} + \epsilon^{\alpha_2}),$$

where M_2 is a positive constant. We infer that $\{u_n\}$ is a Cauchy sequence in $L^q(\Omega)$ and consequently $\mathcal{D}_0^{1,p(x)}(\Omega)$ is compactly embedded in $L^q(\Omega)$. The proof of Theorem 15 is complete. □

Remark 2 *The proof of Theorem 15 still works if we replace the space $L^q(\Omega)$ by $L^{q(x)}(\Omega)$, where $q : \overline{\Omega} \to (1,\infty)$ is a continuous function satisfying $1 < q^- \leq q^+ < \frac{2Np^-}{2N+p^-}$.*

2.6.2 Application to a nonlinear problem with three variable exponents

Assume that $q(x)$ is a function satisfying the hypotheses given in Remark 2. We investigate the existence of solutions of the following problem

$$\begin{cases} -\mathrm{div}(|x|^{p(x)}|\nabla u(x)|^{p(x)-2}\nabla u(x)) = \lambda |u(x)|^{q(x)-2}u(x), & x \in \Omega \\ u(x) = 0, & x \in \partial\Omega , \end{cases} \tag{2.61}$$

where λ is a positive constant.

We say that $u \in \mathcal{D}_0^{1,p(x)}(\Omega)$ is a weak solution of problem (2.61) if

$$\int_\Omega |x|^{p(x)}|\nabla u|^{p(x)-2}\nabla u \nabla v\, dx - \lambda \int_\Omega |u|^{q(x)-2}uv\, dx = 0, \quad \forall\, v \in \mathcal{D}_0^{1,p(x)}(\Omega).$$

We now prove the following existence result concerning problem (2.61).

Theorem 16 *For each $\lambda > 0$ problem* (2.61) *has a nontrivial weak solution.*

Proof. For all $\lambda > 0$, we define $J_\lambda : \mathcal{D}_0^{1,p(x)}(\Omega) \to \mathbb{R}$ by

$$J_\lambda(u) = \int_\Omega \frac{|x|^{p(x)}}{p(x)} |\nabla u|^{p(x)}\, dx - \lambda \int_\Omega \frac{1}{q(x)} |u|^{q(x)}\, dx\,.$$

Then $J_\lambda \in C^1(\mathcal{D}_0^{1,p(x)}(\Omega), \mathbb{R})$ and

$$\langle J_\lambda'(u), v\rangle = \int_\Omega |x|^{p(x)} |\nabla u|^{p(x)-2} \nabla u \nabla v\, dx - \lambda \int_\Omega |u|^{q(x)-2} uv\, dx\,,$$

for all $u,\ v \in \mathcal{D}_0^{1,p(x)}(\Omega)$. We infer that u is a solution of problem (2.61) if and only if u is a critical point of J_λ. Consequently, we focus on finding critical points of J_λ. In this context, we prove the following assertions:

(a) the functional J_λ is weakly lower semi-continuous;

(b) the functional J_λ is bounded from below and coercive;

(c) there exists $\psi \in \mathcal{D}_0^{1,p(x)}(\Omega) \setminus \{0\}$ such that $J_\lambda(\psi) < 0$.

(a): This follows by standard arguments, already used earlier in this chapter.

(b): For all $u \in \mathcal{D}_0^{1,p(x)}(\Omega)$ we have

$$\begin{aligned} J_\lambda(u) &\geq \frac{1}{p^+} \int_\Omega |x|^{p(x)} |\nabla u|^{p(x)}\, dx - \frac{\lambda}{q^-} \int_\Omega |u|^{q(x)}\, dx \\ &\leq \frac{1}{p^+} \int_\Omega |x|^{p(x)} |\nabla u|^{p(x)}\, dx - \frac{\lambda}{q^-} (|u|_{q(x)}^{q^-} + |u|_{q(x)}^{q^+})\,. \end{aligned}$$

If $\|u\| > 1$, the above inequality and Theorem 15 imply the existence of some $K > 0$ such that

$$J_\lambda(u) \geq \frac{1}{p^+} \|u\|^{p^-} - \frac{K\lambda}{q^-} (\|u\|^{q^-} + \|u\|^{q^+})\,.$$

Taking into account that $1 < q^- \leq q^+ < \frac{2Np^-}{2N+p^-} < p^-$ the above inequality shows that $\lim_{\|u\| \to \infty} J_\lambda(u) = \infty$, that is, J_λ is coercive.

On the other hand, for all $u \in \mathcal{D}_0^{1,p(x)}(\Omega)$ we have

$$J_\lambda(u) \geq \frac{1}{p^+} \min\{\|u\|^{p^-}, \|u\|^{p^+}\} - \frac{K\lambda}{q^-} (\|u\|^{q^-} + \|u\|^{q^+})\,,$$

and thus, we deduce that J_λ is bounded from below.

(c): We fix $\phi \in C_c^1(\Omega)$, $\phi \neq 0$. Then for each $t \in (0,1)$ we have

$$\begin{aligned} J_\lambda(t\phi) &= \int_\Omega \frac{|x|^{p(x)} t^{p(x)}}{p(x)} |\nabla \phi|^{p(x)}\, dx - \lambda \int_\Omega \frac{t^{q(x)}}{q(x)} |\phi|^{q(x)}\, dx \\ &\leq t^{p^-} \int_\Omega \frac{|x|^{p(x)}}{p(x)} |\nabla \phi|^{p(x)}\, dx - \lambda t^{q^+} \int_\Omega \frac{1}{q(x)} |\phi|^{q(x)}\, dx\,. \end{aligned}$$

Thus, there exist positive constants L_1 and L_2 such that for each $t \in (0,1)$ we have

$$J_\lambda(t\phi) \leq L_1 t^{p^-} - L_2 t^{q^+}.$$

Taking into account that $q^+ < p^-$, we infer by the above inequality that for any $t \in (0, \min\{1, (\frac{L_2}{L_1})^{1/(p^- - q^+)}\})$ we have

$$J_\lambda(t\phi) < 0.$$

Next, we deduce from (a) and (b) that J_λ is weakly lower semi-continuous, bounded from below and coercive. Thus, there exists a global minimum point $u_\lambda \in \mathcal{D}_0^{1,p(x)}(\Omega)$ of J_λ. Moreover, since (c) holds, it follows that $u_\lambda \neq 0$. Standard arguments based on Theorem 15 show that u_λ is actually a critical point of J_λ and thus, a nontrivial weak solution of problem (2.61). The proof of Theorem 16 is complete. □

2.7 Glossary

Direct Method of the Calculus of Variations: It is a general method for constructing a proof of existence of a minimizer for a given functional. This fundamental tool was introduced by Stanislaw Zaremba and David Hilbert around 1900.

Dirichlet Boundary Condition: It is a type of boundary condition which specifies the values that a solution of a partial differential equation can assume along the boundary of the domain. It was named after the German mathematician Peter Gustav Lejeune Dirichlet (1805–1859).

Ekeland Variational Principle: This is the nonlinear version of the Bishop-Phelps theorem and it is a theorem that asserts the existence of nearly optimal solutions of some optimization problems. This result was named after the French mathematician Ivar Ekeland (1974).

Electrorheological Fluid: A fluid whose viscosity is inversely proportional to the strength of the applied electric field.

Energy Functional: It is associated to an appropriate function space and measures the total energy of a certain system. The system tends to develop from the state with higher energy (higher cost) to the state with lower energy, thus local minima of this functional are usually related to the stable stationary states.

Flux-Divergence Theorem: It is also known as Gauss's theorem or Ostrogradsky's theorem and it is a result that relates the flow (that is, flux) of a vector field through a surface to the behavior of the vector field inside the surface.

Lithium Polymetachrylate: Example of electrorheological fluid.

Mean Value Theorem: A basic result in analysis which roughly states that given a planar differentiable arc between two endpoints, there is at least one point at which the tangent to the arc is parallel to the secant through its endpoints.

Morse Theory: A method in differential topology which enables the study of the critical points of the energy functional on paths. The theory was named after the American mathematician Marston Morse (1892–1977).

Mountain Pass Theorem: A central existence theorem in nonlinear analysis, which establishes natural conditions for the existence of a saddle point of a functional. This result is due to Antonio Ambrosetti and Paul Rabinowitz (1973).

Nemytskii Operator: Nemytskii operators are a class of nonlinear operators on L^p spaces with good continuity and boundedness properties. They got their name from the Russian mathematician Viktor V. Nemytskii (1900–1967).

Plateau Problem: It consists of finding a surface with least area spanning a given simple closed curve in $\mathbb{R}^3$.

$p(x)$-Laplace Operator: A differential operator with variable exponent.

Rayleigh Quotient: It is used in the min-max theory to get exact values of all eigenvalues in relationship with the minimization of a certain functional. The result was named after the British scientist John William Strutt, third Baron Rayleigh (1842–1919), who won the Nobel Prize in Physics 1904.

Chapter 3

Spectral Theory for Differential Operators with Variable Exponents

3.1 Continuous spectrum for differential operators with two variable exponents 76
3.2 A nonlinear eigenvalue problem with three variable exponents and lack of compactness 82
3.3 Concentration phenomena: the case of several variable exponents and indefinite potential 92
3.3.1 Concentration of the spectrum 95
3.3.2 Optimization of the eigenvalues 102
3.3.3 Further properties on closed balls 103
3.4 Anisotropic problems with lack of compactness and nonlinear boundary condition 107
3.5 Glossary 115

> Talent hits a target no one else can hit; genius hits a target no one else can see.
>
> Arthur Schopenhauer (1788–1860)

In this chapter, we are concerned with the study of the spectrum of some differential nonhomogeneous operators with one or several variable exponents. We establish results that are no longer valid in the case of homogeneous operators. Essentially, these new results are possible due to the *gap* between the infimum and the maximum of the nonconstant exponents. The analysis carried out in this chapter takes into account the two possible Rayleigh quotients associated to these problems. The content of this chapter is related to the papers [52, 54, 56, 27, 70, 65, 66, 69, 71].

3.1 Continuous spectrum for differential operators with two variable exponents

In this section, we are concerned with the study of the eigenvalue problem

$$\begin{cases} -\mathrm{div}((|\nabla u|^{p_1(x)-2} + |\nabla u|^{p_2(x)-2})\nabla u) = \lambda |u|^{q(x)-2}u, & x \in \Omega \\ u = 0, & x \in \partial\Omega\,, \end{cases} \tag{3.1}$$

where $\Omega \subset \mathbb{R}^N$ $(N \geq 3)$ is a bounded domain with smooth boundary, λ is a positive real number, and p_1, p_2, q are continuous functions on $\overline{\Omega}$.

We study problem (3.1) under the following assumptions:

$$1 < p_2(x) < \min_{y\in\overline{\Omega}} q(y) \leq \max_{y\in\overline{\Omega}} q(y) < p_1(x), \quad \forall\, x \in \overline{\Omega} \tag{3.2}$$

and

$$\max_{y\in\overline{\Omega}} q(y) < p_2^\star(x), \quad \forall\, x \in \overline{\Omega}\,, \tag{3.3}$$

where $p_2^\star(x) := \frac{Np_2(x)}{N-p_2(x)}$ if $p_2(x) < N$ and $p_2^\star(x) := +\infty$ if $p_2(x) \geq N$.

Under these assumptions, we establish the existence of positive constants λ_0 and λ_1 with $\lambda_0 \leq \lambda_1$ such that any $\lambda \in [\lambda_1, \infty)$ is an eigenvalue of problem (3.1) while no $\lambda \in (0, \lambda_0)$ is an eigenvalue of problem (3.1). An important consequence of our study is that, under hypotheses (3.2) and (3.3), we have

$$\inf_{u\in W_0^{1,p_1(x)}(\Omega)\setminus\{0\}} \frac{\displaystyle\int_\Omega \frac{1}{p_1(x)}|\nabla u|^{p_1(x)}\,dx + \int_\Omega \frac{1}{p_2(x)}|\nabla u|^{p_2(x)}\,dx}{\displaystyle\int_\Omega \frac{1}{q(x)}|u|^{q(x)}\,dx} > 0\,.$$

This fact is proved by using the Lagrange multipliers theorem. The absence of homogeneity will be balanced by the fact that assumptions (3.2) and (3.3) yield

$$\lim_{\|u\|_{p_1(x)}\to 0} \frac{\displaystyle\int_\Omega \frac{1}{p_1(x)}|\nabla u|^{p_1(x)}\,dx + \int_\Omega \frac{1}{p_2(x)}|\nabla u|^{p_2(x)}\,dx}{\displaystyle\int_\Omega \frac{1}{q(x)}|u|^{q(x)}\,dx} = \infty$$

and

$$\lim_{\|u\|_{p_1(x)}\to \infty} \frac{\displaystyle\int_\Omega \frac{1}{p_1(x)}|\nabla u|^{p_1(x)}\,dx + \int_\Omega \frac{1}{p_2(x)}|\nabla u|^{p_2(x)}\,dx}{\displaystyle\int_\Omega \frac{1}{q(x)}|u|^{q(x)}\,dx} = \infty\,,$$

where $\|\cdot\|_{p_1(x)}$ stands for the norm in the variable exponent Sobolev space $W_0^{1,p_1(x)}(\Omega)$.

Since $p_2(x) < p_1(x)$ for any $x \in \overline{\Omega}$, it follows that $W_0^{1,p_1(x)}(\Omega)$ is continuously embedded in $W_0^{1,p_2(x)}(\Omega)$. Thus, a solution of problem (3.1) can be searched in the variable exponent space $W_0^{1,p_1(x)}(\Omega)$.

We say that $\lambda \in \mathbb{R}$ is an *eigenvalue* of problem (3.1) if there exists $u \in W_0^{1,p_1(x)}(\Omega) \setminus \{0\}$ such that

$$\int_\Omega (|\nabla u|^{p_1(x)-2} + |\nabla u|^{p_2(x)-2})\nabla u \nabla v \, dx - \lambda \int_\Omega |u|^{q(x)-2} uv \, dx = 0 ,$$

for all $v \in W_0^{1,p_1(x)}(\Omega)$. We point out that if λ is an eigenvalue of problem (3.1) then the corresponding eigenfunction $u \in W_0^{1,p_1(x)}(\Omega) \setminus \{0\}$ is a *weak solution* of problem (3.1).

Define the *first Rayleigh quotient* by

$$\lambda_1 := \inf_{u \in W_0^{1,p_1(x)}(\Omega)\setminus\{0\}} \frac{\displaystyle\int_\Omega \frac{1}{p_1(x)}|\nabla u|^{p_1(x)} \, dx + \int_\Omega \frac{1}{p_2(x)}|\nabla u|^{p_2(x)} \, dx}{\displaystyle\int_\Omega \frac{1}{q(x)}|u|^{q(x)} \, dx} .$$

The main result of this section establishes the concentration of the spectrum at infinity, while no eigenvalue exists in a neighborhood of the origin.

Theorem 17 *Assume that conditions* (3.2) *and* (3.3) *are fulfilled. Then* $\lambda_1 > 0$. *Moreover, any* $\lambda \in [\lambda_1, \infty)$ *is an eigenvalue of problem* (3.1). *Furthermore, there exists a positive constant* λ_0 *such that* $\lambda_0 \leq \lambda_1$ *and no* $\lambda \in (0, \lambda_0)$ *is an eigenvalue of problem* (3.1).

Proof. Let E denote the generalized Sobolev space $W_0^{1,p_1(x)}(\Omega)$. We denote the norm on $W_0^{1,p_1(x)}(\Omega)$ by $\|\cdot\|$ and the norm on $W_0^{1,p_2(x)}(\Omega)$ by $\|\cdot\|_1$.

Define the functionals J, I, J_1, $I_1 : E \to \mathbb{R}$ by

$$J(u) = \int_\Omega \frac{1}{p_1(x)}|\nabla u|^{p_1(x)} \, dx + \int_\Omega \frac{1}{p_2(x)}|\nabla u|^{p_2(x)} \, dx,$$

$$I(u) = \int_\Omega \frac{1}{q(x)}|u|^{q(x)} \, dx,$$

$$J_1(u) = \int_\Omega |\nabla u|^{p_1(x)} \, dx + \int_\Omega |\nabla u|^{p_2(x)} \, dx,$$

$$I_1(u) = \int_\Omega |u|^{q(x)} \, dx.$$

Then J, $I \in C^1(E, \mathbb{R})$ and for all u, $v \in E$,

$$\langle J'(u), v\rangle = \int_\Omega (|\nabla u|^{p_1(x)-2} + |\nabla u|^{p_2(x)-2})\nabla u \nabla v \, dx,$$

$$\langle I'(u), v\rangle = \int_\Omega |u|^{q(x)-2}uv\, dx.$$

We split the proof of Theorem 17 into four steps.

Step 1. We show that $\lambda_1 > 0$.

Since for any $x \in \overline{\Omega}$ we have $p_1(x) > q^+ \geq q(x) \geq q^- > p_2(x)$, we can deduce that for any $u \in E$,

$$2(|\nabla u(x)|^{p_1(x)} + |\nabla u(x)|^{p_2(x)}) \geq |\nabla u(x)|^{q^+} + |\nabla u(x)|^{q^-}$$

and

$$|u(x)|^{q^+} + |u(x)|^{q^-} \geq |u(x)|^{q(x)}.$$

Integrating the above inequalities we obtain

$$2\int_\Omega (|\nabla u|^{p_1(x)} + |\nabla u|^{p_2(x)})\, dx \geq \int_\Omega (|\nabla u|^{q^+} + |\nabla u|^{q^-})\, dx, \qquad \forall\, u \in E \quad (3.4)$$

and

$$\int_\Omega (|u|^{q^+} + |u|^{q^-})\, dx \geq \int_\Omega |u|^{q(x)}\, dx, \qquad \forall\, u \in E. \quad (3.5)$$

By Sobolev embeddings properties, there exist positive constants λ_{q^+} and λ_{q^-} such that

$$\int_\Omega |\nabla u|^{q^+}\, dx \geq \lambda_{q^+} \int_\Omega |u|^{q^+}\, dx, \qquad \forall\, u \in W_0^{1,q^+}(\Omega) \quad (3.6)$$

and

$$\int_\Omega |\nabla u|^{q^-}\, dx \geq \lambda_{q^-} \int_\Omega |u|^{q^-}\, dx, \qquad \forall\, u \in W_0^{1,q^-}(\Omega). \quad (3.7)$$

Using again the fact that $q^- \leq q^+ < p_1(x)$ for any $x \in \overline{\Omega}$ we deduce that E is continuously embedded in $W_0^{1,q^+}(\Omega)$ and $W_0^{1,q^-}(\Omega)$. Thus, inequalities (3.6) and (3.7) hold for any $u \in E$.

Using inequalities (3.5), (3.6) and (3.7), we deduce that there is a positive constant μ such that

$$\int_\Omega (|\nabla u|^{q^+} + |\nabla u|^{q^-})\, dx \geq \mu \int_\Omega |u|^{q(x)}\, dx, \qquad \forall\, u \in E. \quad (3.8)$$

Next, inequalities (3.4) and (3.8) yield

$$\int_\Omega (|\nabla u|^{p_1(x)} + |\nabla u|^{p_2(x)})\, dx \geq \frac{\mu}{2} \int_\Omega |u|^{q(x)}\, dx, \qquad \forall\, u \in E. \quad (3.9)$$

We deduce from relation (3.9) that the *second Rayleigh quotient* is positive

$$\lambda_0 := \inf_{v \in E\setminus\{0\}} \frac{J_1(v)}{I_1(v)} > 0. \quad (3.10)$$

Therefore

$$J_1(u) \geq \lambda_0 I_1(u), \quad \forall\, u \in E. \quad (3.11)$$

The above inequality yields

$$p_1^+ \cdot J(u) \geq J_1(u) \geq \lambda_0 I_1(u) \geq \lambda_0 I(u) \quad \forall\, u \in E. \tag{3.12}$$

The last inequality assures that $\lambda_1 > 0$ and thus Step 1 is verified.

Step 2. We show that λ_1 is an eigenvalue of problem (3.1).

Lemma 16 *The following relations hold:*

$$\lim_{\|u\| \to \infty} \frac{J(u)}{I(u)} = \infty \tag{3.13}$$

and

$$\lim_{\|u\| \to 0} \frac{J(u)}{I(u)} = \infty. \tag{3.14}$$

Proof. Since E is continuously embedded in $L^{q^\pm}(\Omega)$ it follows that there exist positive constants c_1 and c_2 such that

$$\|u\| \geq c_1 \cdot |u|_{q^+}, \quad \forall\, u \in E \tag{3.15}$$

and

$$\|u\| \geq c_2 \cdot |u|_{q^-}, \quad \forall\, u \in E. \tag{3.16}$$

From relations (1.4), (3.5), (3.15) and (3.16), we deduce that for any $u \in E$ with $\|u\| > 1$

$$\frac{J(u)}{I(u)} \geq \frac{\dfrac{\|u\|^{p_1^-}}{p_1^+}}{\dfrac{|u|_{q^+}^{q^+} + |u|_{q^-}^{q^-}}{q^-}} \geq \frac{\dfrac{\|u\|^{p_1^-}}{p_1^+}}{\dfrac{c_1^{-q^+}\|u\|^{q^+} + c_2^{-q^-}\|u\|^{q^-}}{q^-}}.$$

Since $p_1^- > q^+ \geq q^-$, passing to the limit as $\|u\| \to \infty$ in the above inequality we deduce that relation (3.13) holds.

Next, let us remark that since $p_1(x) > p_2(x)$ for any $x \in \overline{\Omega}$, the space $W_0^{1,p_1(x)}(\Omega)$ is continuously embedded in $W_0^{1,p_2(x)}(\Omega)$. Thus, if $\|u\| \to 0$ then $\|u\|_1 \to 0$.

The above remarks enable us to affirm that for any $u \in E$ with small enough $\|u\| < 1$ we have $\|u\|_1 < 1$.

On the other hand, by hypothesis (3.3) we deduce that $W_0^{1,p_2(x)}(\Omega)$ is continuously embedded in $L^{q^\pm}(\Omega)$. It follows that there exist positive constants d_1 and d_2 such that

$$\|u\|_1 \geq d_1 \cdot |u|_{q^+}, \quad \forall\, u \in W_0^{1,p_2(x)}(\Omega) \tag{3.17}$$

and

$$\|u\|_1 \geq d_2 \cdot |u|_{q^-}, \quad \forall\, u \in W_0^{1,p_2(x)}(\Omega). \tag{3.18}$$

Thus, for any $u \in E$ with small enough $\|u\| < 1$, relations (1.5), (3.5), (3.17) and (3.18) imply

$$\frac{J(u)}{I(u)} \geq \frac{\dfrac{\int_\Omega |\nabla u|^{p_2(x)}\,dx}{p_2^+}}{\dfrac{|u|_{q^+}^{q^+} + |u|_{q^-}^{q^-}}{q^-}} \geq \frac{\dfrac{\|u\|_1^{p_2^+}}{p_2^+}}{\dfrac{d_1^{-q^+}\|u\|_1^{q^+} + d_2^{-q^-}\|u\|_1^{q^-}}{q^-}}.$$

Since $p_2^+ < q^- \leq q^+$, passing to the limit as $\|u\| \to 0$ (and thus, $\|u\|_1 \to 0$) in the above inequality we deduce that relation (3.14) holds. The proof of Lemma 16 is complete. $\square$

Lemma 17 *There exists $u \in E \setminus \{0\}$ such that $\frac{J(u)}{I(u)} = \lambda_1$.*

Proof. Let $\{u_n\} \subset E \setminus \{0\}$ be a minimizing sequence for λ_1, that is,

$$\lim_{n\to\infty} \frac{J(u_n)}{I(u_n)} = \lambda_1 > 0. \tag{3.19}$$

By relation (3.13) it is clear that $\{u_n\}$ is bounded in E. Since E is reflexive it follows that there exists $u \in E$ such that, up to a subsequence, $\{u_n\}$ weakly converges to u in E. On the other hand, similar arguments as those used in the proof of Lemma 3 show that the functional J is weakly lower semi-continuous. Thus, we can conclude that

$$\liminf_{n\to\infty} J(u_n) \geq J(u). \tag{3.20}$$

By the compact embedding theorem for spaces with variable exponent and the assumption $1 \leq \max_{y\in\overline{\Omega}} q(y) < p_1(x)$ for all $x \in \overline{\Omega}$ (see hypothesis (3.2)), it follows that E is compactly embedded in $L^{q(x)}(\Omega)$. Thus, $\{u_n\}$ strongly converges in $L^{q(x)}(\Omega)$. It then follows by relation (1.6) that

$$\lim_{n\to\infty} I(u_n) = I(u). \tag{3.21}$$

Relations (3.20) and (3.21) imply that if $u \not\equiv 0$ then

$$\frac{J(u)}{I(u)} = \lambda_1.$$

Thus, in order to conclude that the lemma holds, it is enough to show that u is not trivial. Assume to the contrary. Then u_n weakly converges to 0 in E and strongly in $L^{q(x)}(\Omega)$. In other words, we have

$$\lim_{n\to\infty} I(u_n) = 0. \tag{3.22}$$

Letting $\epsilon \in (0, \lambda_1)$ be fixed by relation (3.19) we deduce that for large enough n we have

$$|J(u_n) - \lambda_1 I(u_n)| < \epsilon I(u_n),$$

or

$$(\lambda_1 - \epsilon)I(u_n) < J(u_n) < (\lambda_1 + \epsilon)I(u_n).$$

Passing to the limit in the above inequalities and taking into account that relation (3.22) holds we find

$$\lim_{n\to\infty} J(u_n) = 0.$$

This fact combined with relation (1.6) implies that u_n actually strongly converges to 0 in E, that is, $\lim_{n\to\infty} \|u_n\| = 0$. From this information and relation (3.14) we obtain

$$\lim_{n\to\infty} \frac{J(u_n)}{I(u_n)} = \infty,$$

which is a contradiction. Thus, $u \not\equiv 0$. The proof of Lemma 17 is complete. □

By Lemma 17 we conclude that there exists $u \in E \setminus \{0\}$ such that

$$\frac{J(u)}{I(u)} = \lambda_1 = \inf_{w\in E\setminus\{0\}} \frac{J(w)}{I(w)}. \tag{3.23}$$

Then for any $v \in E$ we have

$$\frac{d}{d\epsilon}\frac{J(u+\epsilon v)}{I(u+\epsilon v)}\Big|_{\epsilon=0} = 0\ .$$

A simple computation yields

$$\begin{aligned}&\int_\Omega (|\nabla u|^{p_1(x)-2} + |\nabla u|^{p_2(x)-2})\nabla u \nabla v\, dx \cdot I(u) - \\ &J(u)\cdot \int_\Omega |u|^{q(x)-2}uv\, dx = 0, \quad \forall\, v \in E.\end{aligned} \tag{3.24}$$

Relation (3.24) combined with the fact that $J(u) = \lambda_1 I(u)$ and $I(u) \neq 0$ implies the fact that λ_1 is an eigenvalue of problem (3.1). Thus, Step 2 is verified.

Step 3. We show that any $\lambda \in (\lambda_1, \infty)$ is an eigenvalue of problem (3.1). Let $\lambda \in (\lambda_1, \infty)$ be arbitrary but fixed. Define $T_\lambda : E \to \mathbb{R}$ by

$$T_\lambda(u) = J(u) - \lambda I(u).$$

Clearly, $T_\lambda \in C^1(E, \mathbb{R})$ with

$$\langle T'_\lambda(u), v\rangle = \langle J'(u), v\rangle - \lambda\langle I'(u), v\rangle, \quad \forall\, u \in E.$$

Thus, λ is an eigenvalue of problem (3.1) if and only if there exists a critical point $u_\lambda \in E \setminus \{0\}$ of T_λ.

With similar arguments as in the proof of relation (3.13) we deduce that T_λ is coercive, that is, $\lim_{\|u\|\to\infty} T_\lambda(u) = \infty$. On the other hand, similar arguments as those used in the proof of Lemma 3 show that the functional T_λ is weakly lower semi-continuous. Thus, there exists a global minimum point $u_\lambda \in E$ of T_λ, hence a critical point of T_λ. In order to complete Step 4, it is enough to show that u_λ is not trivial. Indeed, since $\lambda_1 = \inf_{u\in E\setminus\{0\}} \frac{J(u)}{I(u)}$ and $\lambda > \lambda_1$ it follows that there exists $v_\lambda \in E$ such that

$$J(v_\lambda) < \lambda I(v_\lambda),$$

or

$$T_\lambda(v_\lambda) < 0.$$

Thus

$$\inf_E T_\lambda < 0$$

and we conclude that u_λ is a nontrivial critical point of T_λ, or λ is an eigenvalue of problem (3.1). Thus Step 3 is verified.

Step 4. No $\lambda \in (0, \lambda_0)$, where λ_0 is given by (3.10), is an eigenvalue of problem (3.1).

Indeed, assuming by contradiction that there exists an eigenvalue $\lambda \in (0, \lambda_0)$ of problem (3.1), it follows that there exists $u_\lambda \in E \setminus \{0\}$ such that

$$\langle J'(u_\lambda), v\rangle = \lambda\langle I'(u_\lambda), v\rangle, \quad \forall\, v \in E.$$

Thus for $v = u_\lambda$ we can find

$$\langle J'(u_\lambda), u_\lambda\rangle = \lambda\langle I'(u_\lambda), u_\lambda\rangle,$$

that is,

$$J_1(u_\lambda) = \lambda I_1(u_\lambda).$$

The fact that $u_\lambda \in E \setminus \{0\}$ assures that $I_1(u_\lambda) > 0$. Since $\lambda < \lambda_0$, the above information yields

$$J_1(u_\lambda) \geq \lambda_0 I_1(u_\lambda) > \lambda I_1(u_\lambda) = J_1(u_\lambda).$$

Clearly, the above inequalities lead to a contradiction. Thus Step 4 is verified.

We deduce from Steps 2, 3 and 4 that $\lambda_0 \leq \lambda_1$. The proof of Theorem 17 is now complete. □

3.2 A nonlinear eigenvalue problem with three variable exponents and lack of compactness

Let Ω be a smooth *exterior domain* in $\mathbb{R}^N$ ($N \geq 3$), that is, Ω is the complement of a bounded domain with Lipschitz boundary. In this section, we are concerned with the study of the eigenvalue problem

$$\begin{cases} -\text{div}(|\nabla u|^{p(x)-2}\nabla u) + |u|^{p(x)-2}u + |u|^{q(x)-2}u = \lambda g(x)|u|^{r(x)-2}u, \ x \in \Omega \\ u = 0, \ x \in \partial\Omega\,. \end{cases} \tag{3.25}$$

The mappings p, q, $r : \overline{\Omega} \to [2, \infty)$ are Lipschitz continuous functions while $g : \overline{\Omega} \to [0, \infty)$ is a measurable function for which there exists a nonempty set $\Omega_0 \subset \Omega$ such that $g(x) > 0$ for any $x \in \Omega_0$, and $\lambda \geq 0$ is a real number. We point out that the *smoothness assumption* on the boundary of Ω is needed only to ensure the existence of Sobolev embeddings.

We study problem (3.25) assuming that the functions p, q and r satisfy the hypothesis

$$2 \leq p^- \leq p^+ < N\,, \tag{3.26}$$

$$p^+ < r^- \leq r^+ < q^- \leq q^+ < \frac{Np^-}{N - p^-}\,. \tag{3.27}$$

Furthermore, we assume that the function $g(x)$ satisfies the hypotheses

$$g \in L^\infty(\Omega) \cap L^{p_0(x)}(\Omega)\,, \tag{3.28}$$

where $p_0(x) = p^\star(x)/(p^\star(x) - r^-)$ for any $x \in \overline{\Omega}$.

We say that $\lambda \in \mathbb{R}$ is an eigenvalue of problem (3.25) if there exists $u \in W_0^{1,p(x)}(\Omega) \setminus \{0\}$ such that

$$\int_\Omega (|\nabla u|^{p(x)-2}\nabla u \nabla v + |u|^{p(x)-2}uv + |u|^{q(x)-2}uv)\, dx - \lambda \int_\Omega g(x)|u|^{r(x)-2}uv\, dx = 0,$$

for all $v \in W_0^{1,p(x)}(\Omega)$.

We point out that if λ is an eigenvalue of the problem (3.25) then the corresponding $u \in W_0^{1,p(x)}(\Omega) \setminus \{0\}$ is a weak solution of (3.25).

Define the Rayleigh quotients as follows:

$$\lambda_1 := \inf_{u \in W_0^{1,p(x)}(\Omega)\setminus\{0\}} \frac{\displaystyle\int_\Omega \frac{1}{p(x)}(|\nabla u|^{p(x)} + |u|^{p(x)})\, dx + \int_\Omega \frac{1}{q(x)}|u|^{q(x)}\, dx}{\displaystyle\int_\Omega \frac{g(x)}{r(x)}|u|^{r(x)}\, dx}$$

and

$$\lambda_0 := \inf_{u \in W_0^{1,p(x)}(\Omega)\setminus\{0\}} \frac{\displaystyle\int_\Omega (|\nabla u|^{p(x)} + |u|^{p(x)})\, dx + \int_\Omega |u|^{q(x)}\, dx}{\displaystyle\int_\Omega g(x)|u|^{r(x)}\, dx}\,.$$

The main result in this section establishes the following qualitative property.

Theorem 18 *Let Ω be an exterior domain with Lipschitz boundary in $\mathbb{R}^N$, where $N \geq 3$. Suppose that p, q, $r : \overline{\Omega} \to [2, \infty)$ are Lipschitz continuous functions and $g : \overline{\Omega} \to [0, \infty)$ is a measurable function for which there exists a nonempty set $\Omega_0 \subset \Omega$ such that $g > 0$ in Ω_0. Assume conditions (3.26), (3.27) and (3.28) are fulfilled.*

Then

$$0 < \lambda_0 \leq \lambda_1 .$$

Furthermore, each $\lambda \in [\lambda_1, \infty)$ is an eigenvalue of problem (3.25) while no $\lambda \in (0, \lambda_0)$ is an eigenvalue of problem (3.25).

We point out that a result similar to Theorem 18 can be proved for the following problem

$$\begin{cases} -\Delta u + u + |u|^{q(x)-2}u = \lambda g(x)|u|^{r(x)-2}u & \text{for} \quad x \in \Omega \\ u = 0 \qquad \text{for} \quad x \in \partial\Omega \, , \end{cases}$$

where Ω is a smooth exterior domain in $\mathbb{R}^2$. The mappings q and $r : \overline{\Omega} \to [2, \infty)$ are still Lipschitz continuous functions while $g : \overline{\Omega} \to [0, \infty)$ is a function for which there exists a nonempty set $\Omega_0 \subset \Omega$ such that $g(x) > 0$ for any $x \in \Omega_0$, and $\lambda \geq 0$ is a real number. This time conditions (3.26), (3.27) and (3.28) should be replaced by the following conditions:

$$2 < r^- \leq r^+ < q^- \leq q^+ < \infty$$

and

$$g \in L^\infty(\Omega) \cap L^1(\Omega) .$$

Returning to the proof of Theorem 18, we denote by E the generalized Sobolev space $W_0^{1,p(x)}(\Omega)$. For all $u \in E$ we set

$$J_1(u) = \int_\Omega \frac{1}{p(x)} (|\nabla u|^{p(x)} + |u|^{p(x)}) \, dx + \int_\Omega \frac{1}{q(x)} |u|^{q(x)} \, dx \, ,$$

$$I_1(u) = \int_\Omega \frac{g(x)}{r(x)} |u|^{r(x)} \, dx \, ,$$

$$J_0(u) = \int_\Omega (|\nabla u|^{p(x)} + |u|^{p(x)}) \, dx + \int_\Omega |u|^{q(x)} \, dx \, ,$$

$$I_0(u) = \int_\Omega g(x)|u|^{r(x)} \, dx \, .$$

Then $J_1,\ I_1 \in C^1(E, \mathbb{R})$ and for all $u,\ v \in E$,

$$\langle J_1'(u), v\rangle = \int_\Omega (|\nabla u|^{p(x)-2}\nabla u \nabla v + |u|^{p(x)-2}uv + |u|^{q(x)-2}uv)\, dx\,,$$

$$\langle I_1'(u), v\rangle = \int_\Omega g(x)|u|^{r(x)-2}uv\, dx\,.$$

For any $\lambda > 0$ we define the energy functional $T_\lambda : E \to \mathbb{R}$ by

$$T_\lambda(u) = J_1(u) - \lambda \cdot I_1(u), \quad \forall\, u \in E\,.$$

Then λ is an eigenvalue of problem (3.25) if and only if there exists a critical point $u_\lambda \in E \setminus \{0\}$ of the functional T_λ.

We split the proof of Theorem 18 into four steps.

Step 1. We show that $\lambda_0,\ \lambda_1 > 0$.

Indeed, since by relation (3.27) we have $p(x) < r(x) < q(x)$ for any $x \in \overline{\Omega}$ we deduce that

$$|u(x)|^{p(x)} + |u(x)|^{q(x)} \geq |u(x)|^{r(x)}, \quad \forall\, u \in E \text{ and } \forall\, x \in \overline{\Omega}\,.$$

The above relation shows that

$$\int_\Omega (|u|^{p(x)} + |u|^{q(x)})\, dx \geq \frac{1}{|g|_\infty} \cdot \int_\Omega g(x)|u|^{r(x)}\, dx, \quad \forall\, u \in E\,, \tag{3.29}$$

or

$$J_0(u) \geq \frac{1}{|g|_\infty} \cdot I_0(u), \quad \forall\, u \in E\,.$$

We deduce that $\lambda_0 > 0$.

On the other hand, by relation (3.29) we have

$$\int_\Omega \frac{1}{p(x)}|u|^{p(x)}\, dx + \int_\Omega \frac{1}{q(x)}|u|^{q(x)}\, dx \geq \frac{r^-}{q^+ \cdot |g|_\infty} \cdot \int_\Omega \frac{g(x)}{r(x)}|u|^{r(x)}\, dx, \quad \forall\, u \in E\,,$$

and thus,

$$J_1(u) \geq \frac{r^-}{q^+ \cdot |g|_\infty} \cdot I_1(u), \quad \forall\, u \in E\,.$$

Consequently, $\lambda_1 > 0$ and Step 1 is completed.

Step 2. We show that no $\lambda \in (0, \lambda_0)$ is an eigenvalue of problem (3.25).

Indeed, assuming by contradiction that there exists an eigenvalue $\lambda \in (0, \lambda_0)$ of problem (3.25), we deduce that there exists $u_\lambda \in E \setminus \{0\}$ such that

$$\langle J_1'(u_\lambda), v\rangle = \lambda \cdot \langle I_1'(u_\lambda), v\rangle, \quad \forall\, v \in E\,.$$

Taking $v = u_\lambda$ in the above equality we find

$$\langle J_1'(u_\lambda), u_\lambda\rangle = \lambda \cdot \langle I_1'(u_\lambda), u_\lambda\rangle\,,$$

or

$$J_0(u_\lambda) = \lambda \cdot I_0(u_\lambda)\,.$$

Since $u_\lambda \in E\setminus\{0\}$ it follows that $J_0(u_\lambda) > 0$ and $I_0(u_\lambda) > 0$. This information combined with the above inequality, the definition of λ_0 and the fact that $\lambda \in (0, \lambda_0)$ yield

$$J_0(u_\lambda) \geq \lambda_0 \cdot I_0(u_\lambda) > \lambda \cdot I_0(u_\lambda) = J_0(u_\lambda)\,,$$

which is a contradiction. Thus Step 2 is completed.

Step 3. We show that any $\lambda \in (\lambda_1, \infty)$ is an eigenvalue of problem (3.25).

In order to verify that the conclusion of Step 3 holds we first prove two auxiliary results.

Lemma 18 *Assume that the hypotheses of Theorem 18 are satisfied and s is a real number such that*

$$r^+ < s < (p^-)^\star\,,$$

where $(p^-)^\star = Np^-/(N-p^-)$. *Then* $g \in L^{\frac{s}{s-r^-}}(\Omega) \cap L^{\frac{s}{s-r^+}}(\Omega)$ *and*

$$\int_\Omega g(x)|u|^{r(x)} \leq |g|_{\frac{s}{s-r^-}} |u|_s^{r^-} + |g|_{\frac{s}{s-r^+}} |u|_s^{r^+}, \quad \forall\, u \in E\,. \tag{3.30}$$

Proof. First, we remark that the following inequalities hold

$$\frac{s}{s-r^+} \geq \frac{s}{s-r^-} > \frac{(p^-)^\star}{(p^-)^\star - r^-} \geq \frac{p^\star(x)}{p^\star(x) - r^-} = p_0(x), \quad \forall\, x \in \overline{\Omega}\,,$$

and

$$p_0^+ = \frac{(p^-)^\star}{(p^-)^\star - r^-}\,.$$

On the other hand, taking into account the above information and the fact that relation (3.28) holds we have

$$\begin{aligned}
\int_\Omega [g(x)]^{\frac{s}{s-r^-}}\, dx &= \int_\Omega [g(x)]^{p_0(x)} \cdot [g(x)]^{\frac{s}{s-r^-} - p_0(x)}\, dx \\
&\leq \int_\Omega [g(x)]^{p_0(x)} \cdot |g|_\infty^{\frac{s}{s-r^-} - p_0(x)}\, dx \\
&\leq [|g|_\infty^{\frac{s}{s-r^-} - p_0^+} + |g|_\infty^{\frac{s}{s-r^-} - p_0^-}] \cdot \int_\Omega [g(x)]^{p_0(x)}\, dx < \infty\,.
\end{aligned}$$

Thus we found that $g \in L^{\frac{s}{s-r^-}}(\Omega)$. Similar arguments show that $g \in L^{\frac{s}{s-r^+}}(\Omega)$.

Inequality (3.30) follows from the remark that

$$|u(x)|^{r(x)} \leq |u(x)|^{r^-} + |u(x)|^{r^+}, \quad \forall\, u \in E \text{ and } \forall\, x \in \overline{\Omega}\,,$$

or

$$\begin{aligned}\int_\Omega g(x)|u|^{r(x)}\,dx &\leq \int_\Omega g(x)|u|^{r^-}\,dx+\int_\Omega g(x)|u|^{r^+}\,dx\\ &\leq |g|_{\frac{s}{s-r^-}}|u|_s^{r^-}+|g|_{\frac{s}{s-r^+}}|u|_s^{r^+},\quad \forall\, u\in E\,.\end{aligned}$$

The proof of Lemma 18 is complete. □

Lemma 19 *For any $\lambda>0$ we have*

$$\lim_{\|u\|\to\infty} T_\lambda(u)=\infty\,.$$

Proof. We fix $\lambda>0$ and s such that

$$r^+<s<q^-<(p^-)^\star\,.$$

Then, using relation (3.30), we deduce that the following inequalities hold for any $u\in E$ with $\|u\|>1$

$$\begin{aligned}T_\lambda(u) &= \int_\Omega \frac{1}{p(x)}(|\nabla u|^{p(x)}+|u|^{p(x)})\,dx\\ &+ \int_\Omega\frac{1}{q(x)}|u|^{q(x)}\,dx-\lambda\int_\Omega\frac{g(x)}{r(x)}|u|^{r(x)}\,dx\\ &\geq \frac{1}{2p^+}\int_\Omega(|\nabla u|^{p(x)}+|u|^{p(x)})\,dx\\ &+ \frac{1}{\max\{2p^+,q^+\}}\int_\Omega(|u|^{p(x)}+|u|^{q(x)})\,dx\\ &- \frac{\lambda}{r^-}\int_\Omega g(x)(|u|^{r^+}+|u|^{r^-})\,dx\\ &\geq \frac{1}{2p^+}\int_\Omega(|\nabla u|^{p(x)}+|u|^{p(x)})\,dx\\ &+ \frac{1}{\max\{2p^+,q^+\}}\int_\Omega|u|^s\,dx-C_1\cdot|u|_s^{r^-}-C_2\cdot|u|_s^{r^+}\\ &\geq C_0\cdot\|u\|^{p^-}+\left(\frac{1}{2\max\{2p^+,q^+\}}\int_\Omega|u|^s\,dx-C_1\cdot|u|_s^{r^-}\right)+\\ &\quad\left(\frac{1}{2\max\{2p^+,q^+\}}\int_\Omega|u|^s\,dx-C_2\cdot|u|_s^{r^+}\right),\end{aligned}$$

where C_0, C_1 and C_2 are positive constants.

Actually, by the above inequality we found that there exist four positive constants C_0, C_1, C_2 and C_3 (with $C_3=\frac{1}{2\max\{2p^+,q^+\}}$) such that

$$T_\lambda(u)\geq C_0\cdot\|u\|^{p^-}+\left(C_3\cdot|u|_s^s-C_1\cdot|u|_s^{r^-}\right)+\left(C_3\cdot|u|_s^s-C_2\cdot|u|_s^{r^+}\right),\quad(3.31)$$

for any $u\in E$ with $\|u\|>1$.

Next, we show that for any $u \in E$ there exist positive constants $M_1 = M_1(r^-, s, C_1, C_3)$ and $M_2 = M_2(r^+, s, C_2, C_3)$ such that

$$C_3 \cdot |u|_s^s - C_1 \cdot |u|_s^{r^-} \geq -M_1 \tag{3.32}$$

and

$$C_3 \cdot |u|_s^s - C_2 \cdot |u|_s^{r^+} \geq -M_2 \,. \tag{3.33}$$

In order to prove this, we point out that the functional $\Phi : (0, \infty) \to \infty$ defined by

$$\Phi(t) = \alpha \cdot t^a - \beta \cdot t^b \,,$$

where α, β, a, b are positive constants with $a < b$, achieves its positive global maximum

$$\Phi(t_0) = \frac{b-a}{a} \cdot \left(\frac{a}{b}\right)^{a/(b-a)} \cdot \alpha^{b/(b-a)} \cdot \beta^{a/(a-b)} > 0 \,,$$

where $t_0 = ((\alpha \cdot a)/(\beta \cdot b))^{1/(b-a)} > 0$. Thus, we deduce that

$$\alpha \cdot t^a - \beta \cdot t^b \leq C(a, b) \cdot \alpha^{b/(b-a)} \cdot \beta^{a/(a-b)}, \quad \forall\, t > 0 \,, \tag{3.34}$$

where $C(a, b) = (b - a) \cdot \left(\frac{a^a}{b^b}\right)^{1/(b-a)}$.

Finally, we notice that taking $a = r^-$, $b = s$, $\alpha = C_1$ and $\beta = C_3$ in (3.34), we deduce that relation (3.32) holds with $M_1 = C(r^-, s) \cdot C_1^{s/(s-r^-)} \cdot C_3^{r^-/(r^- - s)}$. Similarly, taking $a = r^+$, $b = s$, $\alpha = C_2$ and $\beta = C_3$ in (3.34), we deduce that (3.33) is valid with $M_2 = C(r^+, s) \cdot C_2^{s/(s-r^+)} \cdot C_3^{r^+/(r^+ - s)}$.

Combining relations (3.31), (3.32) and (3.33) we deduce that Lemma 19 holds. □

Now, we return to the proof of Step 3. First, we fix $\lambda \in (\lambda_1, \infty)$. By Lemma 19 we deduce that $\lim_{\|u\| \to \infty} T_\lambda(u) = \infty$, that is, T_λ is coercive. On the other hand, similar arguments as those used in the proof of Lemma 3 show that the functional T_λ is weakly lower semi-continuous. Thus, there exists a global minimum point $u_\lambda \in E$ of T_λ and thus, a critical point of T_λ. In order to conclude that Step 3 holds, it is enough to show that u_λ is not trivial. Indeed, since $\lambda_1 = \inf_{u \in E \setminus \{0\}} \frac{J_1(u)}{I_1(u)}$ and $\lambda > \lambda_1$ it follows that there exists $v_\lambda \in E$ such that

$$J_1(v_\lambda) < \lambda I_1(v_\lambda),$$

or

$$T_\lambda(v_\lambda) < 0.$$

Thus

$$\inf_E T_\lambda < 0$$

and we conclude that u_λ is a nontrivial critical point of T_λ, or λ is an eigenvalue of problem (3.25). Thus Step 3 is verified.

Step 4. We show that λ_1 is an eigenvalue of problem (3.25).

We begin by proving two auxiliary results.

Lemma 20 *The following relation holds:*

$$\lim_{\|u\|\to 0} \frac{J_0(u)}{I_0(u)} = +\infty\,.$$

Proof. Let s be any real number satisfying the following inequality

$$r^+ < s < q^- < (p^-)^\star\,.$$

Then E is continuously embedded in $L^s(\Omega)$. It follows that there exists a positive constant C such that

$$|u|_s \leq C \cdot \|u\|, \quad \forall\, u \in E\,.$$

Using the above inequality and relation (3.30) from Lemma 18 we find that for any $u \in E$ with $\|u\| < 1$ we have

$$\begin{aligned}
\frac{J_0(u)}{I_0(u)} &= \frac{\displaystyle\int_\Omega (|\nabla u|^{p(x)} + |u|^{p(x)})\,dx + \int_\Omega |u|^{q(x)}\,dx}{\displaystyle\int_\Omega g(x)|u|^{r(x)}\,dx} \\
&\geq \frac{\displaystyle\int_\Omega (|\nabla u|^{p(x)} + |u|^{p(x)})\,dx}{\left(\displaystyle\int_\Omega g(x)|u|^{r^-}\,dx + \int_\Omega g(x)|u|^{r^+}\,dx\right)} \\
&\geq \frac{\|u\|^{p^+}}{|g|_{\frac{s}{s-r^-}}|u|_s^{r^-} + |g|_{\frac{s}{s-r^+}}|u|_s^{r^+}} \\
&\geq \frac{\|u\|^{p^+}}{|g|_{\frac{s}{s-r^-}} \cdot C^{r^-} \cdot \|u\|^{r^-} + |g|_{\frac{s}{s-r^+}} \cdot C^{r^+} \cdot \|u\|^{r^+}}\,.
\end{aligned}$$

Since $r^+ \geq r^- > p^+$, by passing to the limit in the above inequality we deduce that $\lim_{\|u\|\to 0} \frac{J_0(u)}{I_0(u)} = +\infty$ and thus Lemma 20 holds. □

Lemma 21 *Assume that $\{u_n\}$ weakly converges to u in E. Then the following relations hold:*

$$\lim_{n\to\infty} I_0(u_n) = I_0(u)\,, \tag{3.35}$$

$$\lim_{n\to\infty} \langle I_1'(u_n), u_n - u\rangle = 0\,. \tag{3.36}$$

Proof. We prove only relation (3.35). Relation (3.36) can be obtained by similar arguments.

Since $\{u_n\}$ weakly converges to u in E and since E is continuously embedded in $L^{p^\star(x)}(\Omega)$ it follows that the sequence $\{|u_n|_{p^\star(x)}\}$ is bounded. Using this fact we deduce that there exists a positive constant M such that

$$\max_n \left\{ \left||u_n|^{r^+}\right|_{\frac{p^\star(x)}{r^+}}, \left||u|^{r^+}\right|_{\frac{p^\star(x)}{r^+}}, \left||u_n|^{r^-}\right|_{\frac{p^\star(x)}{r^-}}, \left||u|^{r^-}\right|_{\frac{p^\star(x)}{r^-}} \right\} \leq M\,. \tag{3.37}$$

On the other hand, let k be a positive integer such that

$$\mathbb{R}^N \setminus \Omega \subset B_k(0)\,,$$

where $B_k(0) = \{x \in \mathbb{R}^N;\ |x| < k\}$ is the ball centered at the origin of radius k.

Let $\epsilon > 0$ be fixed. Since by relation (3.28) we have $g \in L^{p_0(x)}(\Omega)$, we can consider k defined above to be sufficiently large so that

$$|g|_{L^{p_0(x)}(\Omega\setminus B_k(0))} < \frac{\epsilon}{8M}\,. \tag{3.38}$$

Relations (3.37), (3.38) and the Hölder inequality imply that

$$\begin{aligned}
&\int_{\Omega\setminus B_k(0)} g(x)\big||u_n|^{r(x)} - |u|^{r(x)}\big|\,dx \leq \\
&\int_{\Omega\setminus B_k(0)} g(x)(|u_n|^{r^+} + |u_n|^{r^-} + |u|^{r^+} + |u|^{r^-})\,dx \leq \\
&|g|_{L^{p_0(x)}(\Omega\setminus B_k(0))}\cdot \\
&\max\left\{\left||u_n|^{r^+}\right|_{\frac{p^\star(x)}{r^+}}, \left||u|^{r^+}\right|_{\frac{p^\star(x)}{r^+}}, \left||u_n|^{r^-}\right|_{\frac{p^\star(x)}{r^-}}, \left||u|^{r^-}\right|_{\frac{p^\star(x)}{r^-}}\right\} \leq \frac{\epsilon}{2}\,.
\end{aligned}$$

On the other hand, since $W_0^{1,p(x)}(B_k(0))$ is compactly embedded in $L^{r(x)}(B_k(0))$ and $g \in L^\infty(\Omega)$ we obtain

$$\lim_{n\to\infty}\int_{B_k(0)} g(x)|u_n|^{r(x)}\,dx = \int_{B_k(0)} g(x)|u|^{r(x)}\,dx\,,$$

or

$$\left|\int_{B_k(0)} g(x)|u_n|^{r(x)}\,dx - \int_{B_k(0)} g(x)|u|^{r(x)}\,dx\right| < \frac{\epsilon}{2}\,,$$

for large enough n.

The above piece of information assures that relation (3.35) holds. The proof of Lemma 21 is complete. □

We now return to the proof of Step 4. Let $\lambda_n \searrow \lambda_1$. By Step 3 we deduce that for each n there exists $u_n \in E \setminus \{0\}$ such that

$$\langle J_1'(u_n), v\rangle = \lambda_n \cdot \langle I_1'(u_n), v\rangle, \quad \forall\, v \in E\,. \tag{3.39}$$

Taking $v = u_n$ we find

$$J_0(u_n) = \lambda_n \cdot I_0(u_n)\,. \tag{3.40}$$

Passing to the limit as $n \to \infty$ in relation (3.40) and taking into account that relation (3.35) holds, we deduce that

$$\lim_{n\to\infty} J_0(u_n) = \lambda_1 \cdot I_0(u)\,.$$

The above equality and a similar argument as those used in the proof of Lemma 19 show that the sequence $\{u_n\}$ is bounded in E. Since E is a reflexive Banach space it follows that there exists $u \in E$ such that, up to a subsequence, $\{u_n\}$ weakly converges to u in E. Then it follows by relations (3.35) and (3.36) that

$$\lim_{n\to\infty} I_0(u_n) = I_0(u)$$

and

$$\lim_{n\to\infty} \langle I_1'(u_n), u_n - u\rangle = 0\,.$$

Next, we use the following elementary inequality (see Lemma 4.2 in Lindqvist [46]): for all $\theta \geq 2$ and any $\xi,\ \eta \in \mathbb{R}^N$ we have

$$\frac{2}{2^{\theta-1}-1}|\xi-\eta|^\theta \leq \theta(|\xi|^{\theta-2}\xi - |\eta|^{\theta-2}\eta)\cdot(\xi-\eta)\,. \qquad (3.41)$$

Using inequality (3.41) and the above relations we deduce that there exist two positive constants L_1 and L_2 such that

$$L_1\int_\Omega (|\nabla(u_n-u)|^{p(x)} + |u_n-u|^{p(x)})\,dx \leq$$
$$\int_\Omega (|\nabla u_n|^{p(x)-2}\nabla u_n - |\nabla u|^{p(x)-2}\nabla u)\cdot(\nabla u_n - \nabla u)\,dx +$$
$$\int_\Omega (|u_n|^{p(x)-2}u_n - |u|^{p(x)-2}u)\cdot(u_n-u)\,dx$$

and

$$L_2\int_\Omega |u_n-u|^{q(x)}\,dx \leq \int_\Omega (|u_n|^{q(x)-2}u_n - |u|^{q(x)-2}u)\cdot(u_n-u)\,dx\,.$$

Adding the two relations above, using relations (3.39) and (3.36) and the fact that $\{u_n\}$ weakly converges to u in E, we deduce that

$$L_1\int_\Omega (|\nabla(u_n-u)|^{p(x)} + |u_n-u|^{p(x)})\,dx \leq$$
$$\langle J_1'(u_n) - J_1'(u), u_n - u\rangle =$$
$$|\langle J_1'(u_n), u_n-u\rangle| + |\langle J_1'(u), u_n-u\rangle| =$$
$$|\lambda_n \cdot \langle I_1'(u_n), u_n-u\rangle| + |\langle J_1'(u), u_n-u\rangle| \to 0\,,$$

as $n\to\infty$.

The above inequalities and relations (1.6) and (1.9) show that u_n strongly converges to u in E. Then passing to the limit as $n\to\infty$ in (3.39) it follows that

$$\langle J_1'(u), v\rangle = \lambda_1\cdot\langle I_1'(u), v\rangle, \quad \forall\, v\in E\,.$$

Thus, u is a critical point for T_{λ_1}.

In order to prove that λ_1 is an eigenvalue of problem (3.25), it remains to show that $u \neq 0$. Indeed, passing to the limit as $n \to \infty$ in relation (3.40) we find that

$$\lim_{n\to\infty} \frac{J_0(u_n)}{I_0(u_n)} = \lambda_1 .$$

On the other hand, if we assume by contradiction that $u = 0$ then we have $u_n \to 0$ in E, or $\|u_n\| \to 0$. However, by Lemma 20 we deduce that

$$\lim_{n\to\infty} \frac{J_0(u_n)}{I_0(u_n)} = \infty ,$$

which yields a contradiction. Consequently, $u \neq 0$ and thus, λ_1 is an eigenvalue of problem (3.25).

From Steps 2, 3 and 4 we deduce that $\lambda_0 \leq \lambda_1$. The proof of Theorem 18 is now complete. □

3.3 Concentration phenomena: the case of several variable exponents and indefinite potential

Let $\Omega \subset \mathbb{R}^N$ ($N \geq 3$) be a bounded domain with smooth boundary. In this section we study the following eigenvalue problem

$$\begin{cases} -\mathrm{div}((|\nabla u|^{p_1(x)-2} + |\nabla u|^{p_2(x)-2})\nabla u) + V(x)|u|^{m(x)-2}u = \\ \lambda(|u|^{q_1(x)-2} + |u|^{q_2(x)-2})u, \quad \text{if } x \in \Omega \\ u = 0, \quad \text{if } x \in \partial\Omega , \end{cases} \tag{3.42}$$

where λ is a real number, V is an indefinite sign-changing weight, and p_1, p_2, q_1, q_2, m are continuous functions on $\overline{\Omega}$.

In what follows, we assume that $p_1, p_2, q_1, q_2, m : \overline{\Omega} \to (1, \infty)$ are continuous functions satisfying the following hypotheses:

$$\begin{aligned} \max_{\overline{\Omega}} p_2 &< \min_{\overline{\Omega}} q_2 \leq \max_{\overline{\Omega}} q_2 \leq \min_{\overline{\Omega}} m \leq \max_{\overline{\Omega}} m \\ &\leq \min_{\overline{\Omega}} q_1 \leq \max_{\overline{\Omega}} q_1 < \min_{\overline{\Omega}} p_1 , \end{aligned} \tag{3.43}$$

$$\max_{\overline{\Omega}} q_1 < p_2^\star(x) := \begin{cases} \dfrac{Np_2(x)}{N - p_2(x)} & \text{if } p_2(x) < N \\ +\infty & \text{if } p_2(x) \geq N. \end{cases} \tag{3.44}$$

We suppose that the potential $V : \Omega \to \mathbb{R}$ satisfies

$$V \in L^{r(x)}(\Omega), \quad \text{with } r \in C(\overline{\Omega}) \text{ and } r(x) > \frac{N}{\min_{\overline{\Omega}} m} \quad \forall\, x \in \overline{\Omega}. \tag{3.45}$$

Condition (3.43), which describes the competition between the growth rates involved in equation (3.42), represents the *key* for the present study since it establishes a balance between all variable exponents involved in the problem. Such a balance is essential since our setting considers a nonhomogeneous eigenvalue problem for which a minimization technique based on the Lagrange multipliers theorem cannot be applied in order to find (principal) eigenvalues (unlike the case offered by the homogeneous operators). Thus, in the case of nonlinear nonhomogeneous eigenvalue problems the classical theory used in the homogeneous case does not work entirely, but some of its ideas can still be useful and some particular results can still be obtained in some cases, while in other cases entirely new phenomena can occur. To focus on our case, we observe that condition (3.43) together with conditions (3.44) and (3.45) imply

$$\lim_{\|u\|_{p_1(x)}\to 0}\frac{\displaystyle\int_\Omega \frac{1}{p_1(x)}|\nabla u|^{p_1(x)}\,dx+\int_\Omega \frac{1}{p_2(x)}|\nabla u|^{p_2(x)}\,dx+\int_\Omega \frac{V(x)}{m(x)}|u|^{m(x)}\,dx}{\displaystyle\int_\Omega \frac{1}{q_1(x)}|u|^{q_1(x)}\,dx+\int_\Omega \frac{1}{q_2(x)}|u|^{q_2(x)}\,dx}=\infty$$

and

$$\lim_{\|u\|_{p_1(x)}\to \infty}\frac{\displaystyle\int_\Omega \frac{1}{p_1(x)}|\nabla u|^{p_1(x)}\,dx+\int_\Omega \frac{1}{p_2(x)}|\nabla u|^{p_2(x)}\,dx+\int_\Omega \frac{V(x)}{m(x)}|u|^{m(x)}\,dx}{\displaystyle\int_\Omega \frac{1}{q_1(x)}|u|^{q_1(x)}\,dx+\int_\Omega \frac{1}{q_2(x)}|u|^{q_2(x)}\,dx}=\infty\,.$$

In other words, the absence of homogeneity is balanced by the behavior (actually, the blow-up) of the Rayleigh quotient associated to problem (3.42) in the origin and at infinity. The consequences of the above remarks is that the infimum of the Rayleigh quotient associated to problem (3.42) is a real number, that is,

$$\inf_{u\in W_0^{1,p_1(x)}(\Omega)\setminus\{0\}}\frac{\displaystyle\int_\Omega \frac{1}{p_1(x)}|\nabla u|^{p_1(x)}\,dx+\int_\Omega \frac{1}{p_2(x)}|\nabla u|^{p_2(x)}\,dx+\int_\Omega \frac{V(x)}{m(x)}|u|^{m(x)}\,dx}{\displaystyle\int_\Omega \frac{1}{q_1(x)}|u|^{q_1(x)}\,dx+\int_\Omega \frac{1}{q_2(x)}|u|^{q_2(x)}\,dx}\in\mathbb{R}\,, \tag{3.46}$$

and it is attained for a function $u_0 \in W_0^{1,p_1(x)}(\Omega) \setminus \{0\}$. Moreover, the value in (3.46) represents an eigenvalue of problem (3.42) with the corresponding eigenfunction u_0.

Related to the above ideas, we also discuss the *optimization* of the eigenvalues described by relation (3.46) with respect to the potential V, providing that V belongs to a bounded, closed and convex subset of $L^{r(x)}(\Omega)$ (where $r(x)$ is given by relation (3.45)). By optimization, we understand the existence of some potentials $V_\star$ and $V^\star$ such that the eigenvalue described in relation (3.46) is minimal or maximal with respect to the set in which V lies. The results that we will obtain in the context of optimization of eigenvalues are motivated by the above advances in this field in the case of homogeneous (linear or nonlinear) eigenvalue problems. We refer mainly to the studies developed in Asbaugh-Harrell [5] and Egnell [25], where different optimization problems of the principal eigenvalue of some homogeneous operators were investigated.

Since $p_2(x) < p_1(x)$ for any $x \in \overline{\Omega}$ it follows that $W_0^{1,p_1(x)}(\Omega)$ is continuously embedded in $W_0^{1,p_2(x)}(\Omega)$. Thus, a solution of a problem of type (3.42) will be found in the variable exponent space $W_0^{1,p_1(x)}(\Omega)$.

We say that $\lambda \in \mathbb{R}$ is an *eigenvalue* of problem (3.42) if there exists $u \in W_0^{1,p_1(x)}(\Omega) \setminus \{0\}$ such that for all $v \in W_0^{1,p_1(x)}(\Omega)$

$$\int_\Omega (|\nabla u|^{p_1(x)-2} + |\nabla u|^{p_2(x)-2})\nabla u \nabla v \, dx + \int_\Omega V(x)|u|^{m(x)-2}uv \, dx- \\ -\lambda \int_\Omega (|u|^{q_1(x)-2} + |u|^{q_2(x)-2})uv \, dx = 0\,.$$

We note that if λ is an eigenvalue of problem (3.42), then the corresponding *eigenfunction* $u \in W_0^{1,p_1(x)}(\Omega) \setminus \{0\}$ is a *weak solution* of problem (3.42).

For each potential $V \in L^{r(x)}(\Omega)$ we define the following Rayleigh quotients:

$$E(V) \\ := \inf_{u \in W_0^{1,p_1(x)}(\Omega)\setminus\{0\}} \frac{\displaystyle\int_\Omega \frac{1}{p_1(x)}|\nabla u|^{p_1(x)}\, dx + \int_\Omega \frac{1}{p_2(x)}|\nabla u|^{p_2(x)}\, dx + \int_\Omega \frac{V(x)}{m(x)}|u|^{m(x)}\, dx}{\displaystyle\int_\Omega \frac{1}{q_1(x)}|u|^{q_1(x)}\, dx + \int_\Omega \frac{1}{q_2(x)}|u|^{q_2(x)}\, dx}$$

and

$$F(V) := \inf_{u \in W_0^{1,p_1(x)}(\Omega)\setminus\{0\}} \frac{\displaystyle\int_\Omega |\nabla u|^{p_1(x)}\, dx + \int_\Omega |\nabla u|^{p_2(x)}\, dx + \int_\Omega V(x)|u|^{m(x)}\, dx}{\displaystyle\int_\Omega |u|^{q_1(x)}\, dx + \int_\Omega |u|^{q_2(x)}\, dx}\,.$$

Thus, we can define a function $E : L^{r(x)}(\Omega) \to \mathbb{R}$.

The next result establishes the following properties: (i) that the Rayleigh quotient $E(V)$ is an eigenvalue of problem (3.42); (ii) a relationship between the Rayleigh quotients; (iii) the existence of eigenvalues for all λ bigger than $E(V)$; (iv) no eigenvalues for parameters less than $F(V)$.

Theorem 19 *Assume that conditions* (3.43), (3.44) *and* (3.45) *are fulfilled. Then $E(V)$ is an eigenvalue of problem* (3.42). *Moreover, there exists an eigenfunction $u \in W_0^{1,p_1(x)}(\Omega)\setminus\{0\}$ corresponding to the eigenvalue $E(V)$ such that*

$$E(V) = \frac{\displaystyle\int_\Omega \frac{1}{p_1(x)}|\nabla u|^{p_1(x)}\, dx + \int_\Omega \frac{1}{p_2(x)}|\nabla u|^{p_2(x)}\, dx + \int_\Omega \frac{V(x)}{m(x)}|u|^{m(x)}\, dx}{\displaystyle\int_\Omega \frac{1}{q_1(x)}|u|^{q_1(x)}\, dx + \int_\Omega \frac{1}{q_2(x)}|u|^{q_2(x)}\, dx}\,.$$

Furthermore, $F(V) \le E(V)$, each $\lambda \in (E(V), \infty)$ is an eigenvalue of problem (3.42), *while no $\lambda \in (-\infty, F(V))$ is an eigenvalue of problem* (3.42).

Next, we show that on each convex, bounded and closed subset of $L^{r(x)}(\Omega)$, the function E is bounded from below and attains its minimum.

Theorem 20 *Assume that conditions* (3.43), (3.44) *and* (3.45) *are fulfilled. Assume that S is a convex, bounded and closed subset of $L^{r(x)}(\Omega)$. Then there exists $V_\star \in S$ which minimizes $E(V)$ on S, that is,*

$$E(V_\star) = \inf_{V \in S} E(V)\,.$$

Finally, we focus on the special case when the set S from Theorem 20 is a ball in $L^{r(x)}(\Omega)$. Thus, we denote each closed ball centered at the origin, of radius R, in $L^{r(x)}(\Omega)$ by $\overline{B}_R(0)$, that is,

$$\overline{B}_R(0) := \{u \in L^{r(x)}(\Omega);\ \ |u|_{r(x)} \le R\}\,.$$

By Theorem 20, we can define the function $E_\star : [0,\infty) \to \mathbb{R}$ by

$$E_\star(R) = \min_{V \in \overline{B}_R(0)} E(V)\,.$$

Our next result establishes some qualitative properties of the function $E_\star$.

Theorem 21 *a) The function $E_\star$ is not constant and decreases monotonically.*
b) The function $E_\star$ is continuous.

We point out that results similar to Theorems 20 and 21 can be obtained if we recall that on each convex, bounded and closed subset of $L^{r(x)}(\Omega)$ the function E defined in Theorem 19 is also bounded from above and attains its maximum. We also observe that we can define a function $E^\star : [0,\infty) \to \mathbb{R}$ by

$$E^\star(R) = \max_{V \in \overline{B}_R(0)} E(V)\,,$$

which has similar properties as $E_\star$.

3.3.1 Concentration of the spectrum

In this section we give the proof of Theorem 19.

Let X denote the generalized Sobolev space $W_0^{1,p_1(x)}(\Omega)$. We denote by $\|\cdot\|$ the norm on $W_0^{1,p_1(x)}(\Omega)$ and by $\|\cdot\|_1$ the norm on $W_0^{1,p_2(x)}(\Omega)$.

Define the functionals J_V, $I : X \to \mathbb{R}$ by

$$J_V(u) = \int_\Omega \frac{1}{p_1(x)} |\nabla u|^{p_1(x)}\,dx + \int_\Omega \frac{1}{p_2(x)} |\nabla u|^{p_2(x)}\,dx + \int_\Omega \frac{V(x)}{m(x)} |u|^{m(x)}\,dx\,,$$

$$I(u) = \int_\Omega \frac{1}{q_1(x)} |u|^{q_1(x)}\,dx + \int_\Omega \frac{1}{q_2(x)} |u|^{q_2(x)}\,dx\,.$$

We notice that for any V satisfying condition (3.45) we have

$$J_V(u) = J_0(u) + \int_\Omega \frac{V(x)}{m(x)}|u|^{m(x)}\,dx, \quad \forall\, u \in X\,,$$

where J_0 is obtained in the case when $V = 0$ in Ω.

We have $J_V,\ I \in C^1(X, \mathbb{R})$ and for all $u,\ v \in X$,

$$\langle J_V'(u), v\rangle = \int_\Omega (|\nabla u|^{p_1(x)-2} + |\nabla u|^{p_2(x)-2})\nabla u \nabla v\,dx + \int_\Omega V(x)|u|^{m(x)-2}uv\,dx\,,$$

$$\langle I'(u), v\rangle = \int_\Omega |u|^{q_1(x)-2}uv\,dx + \int_\Omega |u|^{q_2(x)-2}uv\,dx\,.$$

Lemma 22 *Assume that conditions* (3.43), (3.44) *and* (3.45) *are fulfilled. Then for each $\epsilon > 0$ there exists $C_\epsilon > 0$ such that*

$$\left|\int_\Omega \frac{V(x)}{m(x)}|u|^{m(x)}\,dx\right| \le \epsilon \int_\Omega \left(\frac{1}{p_1(x)}|\nabla u|^{p_1(x)}\,dx + \frac{1}{p_2(x)}|\nabla u|^{p_2(x)}\right)\,dx$$
$$+C_\epsilon |V|_{r(x)} \int_\Omega (|u|^{m^-} + |u|^{m^+})\,dx\,,$$

for all $u \in X$.

Proof. First, we point out that since $r(x) > r^-$ on $\overline{\Omega}$ it follows that $L^{r(x)}(\Omega) \subset L^{r^-}(\Omega)$. On the other hand, since $r(x) > \frac{N}{m^-}$ for each $x \in \overline{\Omega}$ it follows that $r^- > \frac{N}{m^-}$. Thus, we infer that $V \in L^{r^-}(\Omega)$ and $r^- > \frac{N}{m^-}$.

Now, let $\epsilon > 0$ be fixed. We claim that there exists $D_\epsilon > 0$ such that for all $u \in W_0^{1,m^-}(\Omega)$

$$\int_\Omega |V(x)| \cdot |u|^{m^-}\,dx \le \epsilon \int_\Omega |\nabla u|^{m^-}\,dx + D_\epsilon |V|_{r^-} \int_\Omega |u|^{m^-}\,dx\,. \tag{3.47}$$

In order to establish relation (3.47), we first show that for each $s \in (1, \frac{Nm^-}{N-m^-})$ there exists $D'_\epsilon > 0$ such that

$$|v|_s \le \epsilon |\ |\nabla v|\ |_{m^-} + D'_\epsilon |v|_{m^-}, \quad \forall\, u \in W_0^{1,m^-}(\Omega)\,. \tag{3.48}$$

Indeed, assume to the contrary that relation (3.48) does not hold for each $\epsilon > 0$. Then there exist $\epsilon_0 > 0$ and a sequence $(v_n) \subset W_0^{1,m^-}(\Omega)$ such that $|v_n|_s = 1$ and

$$\epsilon_0 |\ |\nabla v_n|\ |_{m^-} + n|v_n|_{m^-} < 1, \quad \forall\, n\,.$$

Then it is clear that (v_n) is bounded in $W_0^{1,m^-}(\Omega)$ and $|v_n|_{m^-} \to 0$. Thus, we deduce that passing eventually to a subsequence we can assume that v_n weakly converges to a function v in $W_0^{1,m^-}(\Omega)$ and actually $v = 0$. Since $s \in (1, \frac{Nm^-}{N-m^-})$ it follows by the Rellich-Kondrachov theorem that $W_0^{1,m^-}(\Omega)$

is compactly embedded in $L^s(\Omega)$ and thus v_n converges to 0 in $L^s(\Omega)$. On the other hand, since $|v_n|_s = 1$ for each n we deduce that $|v|_s = 1$ which is a contradiction. Hence relation (3.48) holds.

Next, we note that since $r^- > \frac{N}{m^-}$, it follows that $m^- \cdot r^{-\prime} < \frac{Nm^-}{N-m^-}$, where $r^{-\prime} = \frac{r^-}{r^- - 1}$. Thus, by the Hölder inequality we have

$$\int_\Omega |V(x)| \cdot |u|^{m^-}\, dx \le |V|_{r^-} \cdot |u|^{m^-}_{m^- \cdot r^{-\prime}}, \quad \forall\, u \in W_0^{1,m^-}(\Omega).$$

Combining the last inequality with relation (3.48), we infer that relation (3.47) holds.

Similar arguments as those used in the proof of relation (3.47), combined with the fact that since $r^- > \frac{N}{m^-}$ we also have $r^- > \frac{N}{m^+}$, imply that there exists $D_\epsilon^{''}$ satisfying for all $u \in W_0^{1,m^+}(\Omega)$ the following inequality

$$\int_\Omega |V(x)| \cdot |u|^{m^+}\, dx \le \epsilon \int_\Omega |\nabla u|^{m^+}\, dx + D_\epsilon^{''} |V|_{r^-} \int_\Omega |u|^{m^+}\, dx. \tag{3.49}$$

Using relation (3.43), we deduce that $m^- \le m^+ < p_1(x)$ for any $x \in \overline{\Omega}$ and thus, $W_0^{1,p_1(x)}(\Omega) \subset W_0^{1,m^\pm}(\Omega)$. In other words, relations (3.47) and (3.49) hold for any $u \in X$. Moreover, on the right-hand sides of inequalities (3.47) and (3.49) we can take $|V|_{r(x)}$ instead of $|V|_{r^-}$ since $L^{r(x)}(\Omega)$ is continuously embedded in $L^{r^-}(\Omega)$, via the Hölder inequality.

Finally, we point out that since by (3.43) we have $p_2(x) < m^- \le m(x) \le m^+ < p_1(x)$ for each $x \in \overline{\Omega}$, we deduce that for all $u \in X$

$$\left| \int_\Omega \frac{V(x)}{m(x)} |u|^{m(x)}\, dx \right| \le \frac{1}{m^-} \int_\Omega |V(x)| \cdot (|u|^{m^-} + |u|^{m^+})\, dx \tag{3.50}$$

and

$$\int_\Omega (|\nabla u|^{m^-} + |\nabla u|^{m^+})\, dx \le 2p_1^+ \cdot \int_\Omega \left(\frac{1}{p_1(x)} |\nabla u|^{p_1(x)}\, dx + \frac{1}{p_2(x)} |\nabla u|^{p_2(x)} \right) dx. \tag{3.51}$$

Relations (3.47), (3.49), (3.50) and (3.51) complete the proof of Lemma 22. □

Lemma 23 *The following relations hold:*

$$\lim_{\|u\| \to \infty} \frac{J_V(u)}{I(u)} = \infty \tag{3.52}$$

and

$$\lim_{\|u\| \to 0} \frac{J_V(u)}{I(u)} = \infty. \tag{3.53}$$

Proof.

First, we note that by (3.43), $q_1(x) < m^{\pm} < q_2(x)$ for any $x \in \overline{\Omega}$. Thus, it is clear that

$$|u(x)|^{m^-} + |u(x)|^{m^+} \leq 2(|u(x)|^{q_1(x)} + |u(x)|^{q_2(x)}), \quad \forall\, x \in \overline{\Omega} \text{ and } \forall\, u \in X\,.$$

Integrating the above inequality over Ω, we infer that for all $u \in X$

$$\frac{\displaystyle\int_\Omega (|u|^{m^-} + |u|^{m^+})\, dx}{\displaystyle\int_\Omega (|u|^{q_1(x)} + |u|^{q_2(x)})\, dx} \leq 2\,. \tag{3.54}$$

Using Lemma 22, we find that for any $\epsilon \in (0,1)$ there exists $C_\epsilon > 0$ such that for all $u \in X$

$$\frac{J_V(u)}{I(u)} \geq \frac{\displaystyle\frac{1-\epsilon}{p_1^+}\int_\Omega (|\nabla u|^{p_1(x)} + |\nabla u|^{p_2(x)})\, dx - C_\epsilon |V|_{r(x)} \int_\Omega (|u|^{m^-} + |u|^{m^+})\, dx}{\displaystyle\frac{1}{q_2^-}\int_\Omega (|u|^{q_1(x)} + |u|^{q_2(x)})\, dx}\,.$$

By the above inequality and relation (3.54), we deduce that there exist some positive constants $\beta > 0$ and $\gamma > 0$ such that for all $u \in X$

$$\frac{J_V(u)}{I(u)} \geq \frac{\beta\displaystyle\int_\Omega (|\nabla u|^{p_1(x)} + |\nabla u|^{p_2(x)})\, dx}{\displaystyle\int_\Omega (|u|^{q_1(x)} + |u|^{q_2(x)})\, dx} - \gamma |V|_{r(x)}\,. \tag{3.55}$$

For any $u \in X$ with $\|u\| > 1$, relation (3.55) implies

$$\frac{J_V(u)}{I(u)} \geq \frac{\beta\displaystyle\int_\Omega |\nabla u|^{p_1(x)}\, dx}{|u|_{q_1^-}^{q_1^-} + |u|_{q_1^+}^{q_1^+} + |u|_{q_2^-}^{q_2^-} + |u|_{q_2^+}^{q_2^+}} - \gamma |V|_{r(x)}, \quad \forall\, u \in X \text{ with } \|u\| > 1\,.$$

Now, invoking the continuous embedding of X in $L^{q_i^\pm}(\Omega)$ for $i = 1,2$ we deduce the existence of a positive constant δ such that

$$\frac{J_V(u)}{I(u)} \geq \frac{\delta \|u\|^{p_1^-}}{\|u\|^{q_1^-} + \|u\|^{q_1^+} + \|u\|^{q_2^-} + \|u\|^{q_2^+}} - \gamma |V|_{r(x)}, \quad \forall\, u \in X \text{ with } \|u\| > 1\,.$$

Since $p_1^- > q_1^+ \geq q_1^- \geq q_2^+ \geq q_2^-$, passing to the limit as $\|u\| \to \infty$ in the above inequality we deduce that relation (3.52) holds.

Next, we remark that since $p_1(x) > p_2(x)$ for any $x \in \overline{\Omega}$, the space $W_0^{1,p_1(x)}(\Omega)$ is continuously embedded in $W_0^{1,p_2(x)}(\Omega)$. Thus, if $\|u\| \to 0$ then $\|u\|_1 \to 0$.

The above remarks enable us to affirm that for any $u \in X$ with small enough $\|u\| < 1$ we have $\|u\|_1 < 1$.

On the other hand, since (3.44) holds, we deduce that $W_0^{1,p_2(x)}(\Omega)$ is continuously embedded in $L^{q_i^{\pm}}(\Omega)$ with $i = 1, 2$. It follows that there exist positive constants d_{i1} and d_{i2} with $i = 1, 2$ such that

$$\|u\|_1 \geq d_{i1} \cdot |u|_{q_i^+}, \quad \forall\, u \in W_0^{1,p_2(x)}(\Omega) \text{ and } i = 1, 2 \tag{3.56}$$

and

$$\|u\|_1 \geq d_{i2} \cdot |u|_{q_i^-}, \quad \forall\, u \in W_0^{1,p_2(x)}(\Omega) \text{ and } i = 1, 2\,. \tag{3.57}$$

Thus, for any $u \in X$ with small enough $\|u\| < 1$, relation (3.55) implies

$$\frac{J_V(u)}{I(u)} \geq \frac{\beta \displaystyle\int_\Omega |\nabla u|^{p_2(x)}\,dx}{|u|_{q_1^-}^{q_1^-} + |u|_{q_1^+}^{q_1^+} + |u|_{q_2^-}^{q_2^-} + |u|_{q_2^+}^{q_2^+}} - \gamma |V|_{r(x)}\,.$$

Next, relations (1.5), (3.56) and (3.57) imply that there exists a constant $\xi > 0$ such that

$$\frac{J_V(u)}{I(u)} \geq \frac{\xi \|u\|_1^{p_2^+}}{\|u\|_1^{q_1^-} + \|u\|_1^{q_1^+} + \|u\|_1^{q_2^-} + \|u\|_1^{q_2^+}} - \gamma |V|_{r(x)}\,,$$

for any $u \in X$ with small enough $\|u\| < 1$. Since $p_2^+ < q_2^- \leq q_2^+ \leq q_1^- \leq q_1^+$, passing to the limit as $\|u\| \to 0$ (and thus, $\|u\|_1 \to 0$) in the above inequality we deduce that relation (3.53) holds. The proof of Lemma 23 is complete. □

Lemma 24 *There exists $u \in X \setminus \{0\}$ such that $\frac{J_V(u)}{I(u)} = E(V)$.*

Proof. Let $(u_n) \subset X \setminus \{0\}$ be a minimizing sequence for $E(V)$, that is,

$$\lim_{n\to\infty} \frac{J_V(u_n)}{I(u_n)} = E(V)\,. \tag{3.58}$$

By relation (3.52), we deduce that $\{u_n\}$ is bounded in X. Since X is reflexive, there exists $u \in X$ such that, up to a subsequence, (u_n) weakly converges to u in X. On the other hand, the functional J_0 (obtained for $V = 0$ on Ω) is weakly lower semi-continuous. Thus,

$$\liminf_{n\to\infty} J_0(u_n) \geq J_0(u)\,. \tag{3.59}$$

By the compact embedding theorem for spaces with variable exponents and assumption (3.43), it follows that X is compactly embedded in $L^{\sigma(x)}(\Omega)$ (where $\sigma(x) = m(x)\cdot r(x)/(r(x)-1)$) and $L^{q_i(x)}(\Omega)$ with $i = 1, 2$. Thus, (u_n) strongly converges in $L^{\sigma(x)}(\Omega)$ and $L^{q_i(x)}(\Omega)$ with $i = 1, 2$. Then it follows by the Hölder inequality and relation (1.6) that

$$\lim_{n\to\infty} I(u_n) = I(u) \tag{3.60}$$

and

$$\lim_{n\to\infty} \int_\Omega V(x)|u_n|^{m(x)}\,dx = \int_\Omega V(x)|u|^{m(x)}\,dx\,. \tag{3.61}$$

Relations (3.59), (3.60) and (3.61) imply that if $u \not\equiv 0$, then

$$\frac{J_V(u)}{I(u)} = E(V)\,.$$

Thus, in order to conclude that the lemma is true, it is enough to show that u is not trivial. Assume to the contrary. Then u_n weakly converges to 0 in X and strongly in $L^{s(x)}(\Omega)$ for any $s(x) \in C(\overline{\Omega})$ with $1 < s(x) < \frac{Np_1(x)}{N-p_1(x)}$ on $\overline{\Omega}$. In other words, we have

$$\lim_{n\to\infty} (I(u_n)) = 0\,, \tag{3.62}$$

and

$$\lim_{n\to\infty} \int_\Omega V(x)|u_n|^{m(x)}\,dx = 0\,. \tag{3.63}$$

If $\epsilon \in (0, |E(V)|)$ is fixed by relation (3.58), we deduce that for large enough n we have

$$|J_V(u_n) - E(V)I(u_n)| < \epsilon I(u_n)\,,$$

or

$$(|E(V)| - \epsilon)I(u_n) < J_V(u_n) < (|E(V)| + \epsilon)I(u_n)\,.$$

Passing to the limit in the above inequalities and taking into account that relation (3.62) holds, we find

$$\lim_{n\to\infty} J_V(u_n) = 0\,.$$

Next, by relation (3.63), we find

$$\lim_{n\to\infty} J_0(u_n) = 0\,.$$

This fact combined with relation (1.6) implies that u_n actually strongly converges to 0 in X, that is, $\lim_{n\to\infty} \|u_n\| = 0$. By this information and relation (3.53) we obtain

$$\lim_{n\to\infty} \frac{J_V(u_n)}{I(u_n)} = \infty,$$

which is a contradiction. Thus, $u \not\equiv 0$. The proof of Lemma 24 is complete. □

By Lemma 24, we conclude that there exists $u \in X \setminus \{0\}$ such that

$$\frac{J_V(u)}{I(u)} = E(V) = \inf_{w\in X\setminus\{0\}} \frac{J_V(w)}{I(w)}.. \tag{3.64}$$

Then for any $w \in X$ we have

$$\frac{d}{d\epsilon}\frac{J_V(u+\epsilon w)}{I(u+\epsilon w)}\Big|_{\epsilon=0} = 0\ .$$

A simple computation yields

$$\langle J_V'(u), w\rangle I(u) - J_V(u)\langle I'(u), w\rangle = 0\,, \tag{3.65}$$

for all $w \in X$. Relation (3.65) combined with the fact that $J_V(u) = E(V)\cdot I(u)$ and $I(u) \neq 0$ implies the fact that $E(V)$ is an eigenvalue of problem (3.42).

Next, we show that any $\lambda \in (E(V), \infty)$ is an eigenvalue of problem (3.42). Let $\lambda \in (E(V), \infty)$ be arbitrary but fixed. Define $T_{V,\lambda} : X \to \mathbb{R}$ by

$$T_{V,\lambda}(u) = J_V(u) - \lambda I(u)\,.$$

Clearly, $T_{V,\lambda} \in C^1(X, \mathbb{R})$ with

$$\langle T_{V,\lambda}'(u), v\rangle = \langle J_V'(u), v\rangle - \lambda\langle I'(u), v\rangle, \quad \forall\, u \in X.$$

Thus, λ is an eigenvalue of problem (3.42) if and only if there exists a critical point $u_\lambda \in X \setminus \{0\}$ of $T_{V,\lambda}$.

With similar arguments as in the proof of relation (3.52), we can show that $T_{V,\lambda}$ is coercive, that is, $\lim_{\|u\|\to\infty} T_{V,\lambda}(u) = \infty$. On the other hand, the functional $T_{V,\lambda}$ is weakly lower semi-continuous. Thus, there exists a global minimum point $u_\lambda \in X$ of $T_{V,\lambda}$, hence a critical point of $T_{V,\lambda}$. It is enough to show that u_λ is not trivial. Indeed, since $E(V) = \inf_{u\in X\setminus\{0\}} \frac{J_V(u)}{I(u)}$ and $\lambda > E(V)$, it follows that there exists $v_\lambda \in X$ such that

$$J_V(v_\lambda) < \lambda I(v_\lambda)\,,$$

or

$$T_{V,\lambda}(v_\lambda) < 0\,.$$

Thus,

$$\inf_X T_{V,\lambda} < 0$$

and we conclude that u_λ is a nontrivial critical point of $T_{V,\lambda}$, or λ is an eigenvalue of problem (3.42).

Finally, we prove that no $\lambda < F(V)$ is an eigenvalue of problem (3.42). With this goal in mind, we assume by contradiction that there exists an eigenvalue $\lambda < F(V)$ of problem (3.42). It follows that there exists $u_\lambda \in X \setminus \{0\}$ such that

$$\langle J_V'(u_\lambda), u_\lambda\rangle = \lambda\langle I'(u_\lambda), u_\lambda\rangle\,.$$

Since $u_\lambda \neq 0$ we have $\langle I'(u_\lambda), u_\lambda\rangle > 0$. Using this fact and the definition of $F(V)$ it follows that the following relations hold

$$\langle J_V'(u_\lambda), u_\lambda\rangle = \lambda\langle I'(u_\lambda), u_\lambda\rangle < F(V)\langle I'(u_\lambda), u_\lambda\rangle \leq \langle J_V'(u_\lambda), u_\lambda\rangle\,.$$

Obviously, this is a contradiction. We deduce that no $\lambda \in (-\infty, F(V))$ is an eigenvalue of problem (3.42). Furthermore, it is clear that $E(V) \geq F(V)$.

The proof of Theorem 19 is complete. □

Remark 3 *We point out that in the case when $V = 0$ in Ω, the same arguments as in the proof of Step 1 in Theorem 18 show that $E(0) > 0$.*

3.3.2 Optimization of the eigenvalues

We now present the proof of Theorem 20.

Let S be a convex, bounded and closed subset of $L^{r(x)}(\Omega)$ and

$$E_\star := \inf_{V\in S} E(V)\,.$$

Clearly, relation (3.55) shows that $E_\star$ is finite.

On the other hand, let $(V_n)\subset S$ be a minimizing sequence for $E_\star$, that is,

$$E(V_n)\to E_\star, \quad \text{as } n\to\infty\,.$$

Obviously, (V_n) is a bounded sequence and thus, there exists $V_\star\in L^{r(x)}(\Omega)$ such that V_n weakly converges to $V_\star$ in $L^{r(x)}(\Omega)$. Moreover, since S is convex and closed, it is also weakly closed; consequently $V_\star\in S$.

Next, we show that $E(V_\star)=E_\star$. Indeed, by Theorem 19, we deduce that for each positive integer n, there exists $u_n\in X\setminus\{0\}$ such that

$$\frac{J_{V_n}(u_n)}{I(u_n)} = E(V_n)\,. \tag{3.66}$$

Since $(E(V_n))$ is a bounded sequence and due to relation (3.55) we have

$$\frac{J_{V_n}(u_n)}{I(u_n)} \geq \beta\frac{J_0(u_n)}{I(u_n)} - C, \quad \text{for any } n\,,$$

where C is a positive constant. We infer that (u_n) is bounded in X and cannot contain a subsequence converging to 0 (otherwise we can obtain a contradiction by applying Lemma 23). Thus, there exists $u_0\in X\setminus\{0\}$ such that (u_n) weakly converges to u_0 in X. Using the Rellich-Kondrachov theorem, we deduce that (u_n) strongly converges to u_0 in $L^{s(x)}(\Omega)$ for any $s(x)\in C(\overline{\Omega})$ satisfying $1<s(x)<\frac{Np_1(x)}{N-p_1(x)}$ for any $x\in\overline{\Omega}$. In particular, using conditions (3.43), (3.44) and (3.45), we deduce that (u_n) converges to u_0 in $L^{m(x)}(\Omega)$ and in $L^{m(x)\cdot r'(x)}(\Omega)$, where $r'(x)=\frac{r(x)}{r(x)-1}$. Using this information in combination with the Hölder inequality and the fact that $V_\star\in L^{r(x)}(\Omega)$ and (V_n) is bounded in $L^{r(x)}(\Omega)$, we find that

$$\lim_{n\to\infty}\int_\Omega \frac{V_\star(x)}{m(x)}|u_n|^{m(x)}\,dx = \int_\Omega \frac{V_\star(x)}{m(x)}|u_0|^{m(x)}\,dx \tag{3.67}$$

and

$$\lim_{n\to\infty}\int_\Omega\left(\frac{V_n(x)}{m(x)}|u_n|^{m(x)} - \frac{V_n(x)}{m(x)}|u_0|^{m(x)}\right)dx = 0\,. \tag{3.68}$$

On the other hand, since (V_n) weakly converges to $V_\star$ in $L^{r(x)}(\Omega)$ and $u_0\in L^{m(x)\cdot r'(x)}(\Omega)$, where $r'(x)=\frac{r(x)}{r(x)-1}$, we deduce that

$$\lim_{n\to\infty}\int_\Omega \frac{V_n(x)}{m(x)}|u_0|^{m(x)}\,dx = \int_\Omega \frac{V_\star(x)}{m(x)}|u_0|^{m(x)}\,dx\,. \tag{3.69}$$

Combining the equality

$$\begin{aligned}
&\int_\Omega \frac{V_\star(x)}{m(x)}|u_n|^{m(x)}\,dx - \int_\Omega \frac{V_n(x)}{m(x)}|u_n|^{m(x)}\,dx = \\
&\int_\Omega \frac{V_\star(x)}{m(x)}|u_n|^{m(x)}\,dx - \int_\Omega \frac{V_\star(x)}{m(x)}|u_0|^{m(x)}\,dx + \\
&\int_\Omega \frac{V_\star(x)}{m(x)}|u_0|^{m(x)}\,dx - \int_\Omega \frac{V_n(x)}{m(x)}|u_0|^{m(x)}\,dx + \\
&\int_\Omega \frac{V_n(x)}{m(x)}|u_0|^{m(x)}\,dx - \int_\Omega \frac{V_n(x)}{m(x)}|u_n|^{m(x)}\,dx\,,
\end{aligned}$$

with relations (3.67), (3.68) and (3.69), we obtain

$$\lim_{n\to\infty}\int_\Omega\left(\frac{V_\star(x)}{m(x)}|u_n|^{m(x)} - \frac{V_n(x)}{m(x)}|u_n|^{m(x)}\right)\,dx = 0\,. \tag{3.70}$$

Since

$$E(V_\star) = \inf_{u\in X\setminus\{0\}}\frac{J_{V_\star}(u)}{I(u)}\,,$$

it follows that

$$E(V_\star) \le \frac{J_{V_\star}(u_n)}{I(u_n)}\,.$$

Combining the above inequality and equality (3.66) we obtain

$$E(V_\star) \le \frac{J_{V_\star}(u_n) - J_{V_n}(u_n)}{I(u_n)} + E(V_n)\,.$$

Taking into account relation (3.70), the fact that $I(u_n)$ is bounded and does not converge to 0 and that $(E(V_n))$ converges to $E_\star$, then passing to the limit as $n \to \infty$ in the last inequality, we infer that

$$E(V_\star) \le E_\star\,.$$

But using the definition of $E_\star$ and the fact that $V_\star \in S$ we conclude that actually

$$E(V_\star) = E_\star\,.$$

The proof of Theorem 20 is complete. □

3.3.3 Further properties on closed balls

In this section we give the proof of Theorem 21.

(a) First, we show that function $E_\star$ is not constant. Indeed, by Remark 3, we have $E_\star(0) = E(0) > 0$. On the other hand, by Theorem 18, it follows that

$$\lambda_m := \inf_{u\in X\setminus\{0\}}\frac{\displaystyle\int_\Omega \frac{1}{p_1(x)}|\nabla u|^{p_1(x)}\,dx + \int_\Omega \frac{1}{p_2(x)}|\nabla u|^{p_2(x)}\,dx}{\displaystyle\int_\Omega \frac{1}{m(x)}|u|^{m(x)}\,dx} > 0\,.$$

Moreover, Lemma 19 implies that there exists $u_m \in X \setminus \{0\}$ such that

$$\lambda_m = \frac{\displaystyle\int_\Omega \frac{1}{p_1(x)} |\nabla u_m|^{p_1(x)}\, dx + \int_\Omega \frac{1}{p_2(x)} |\nabla u_m|^{p_2(x)}\, dx}{\displaystyle\int_\Omega \frac{1}{m(x)} |u_m|^{m(x)}\, dx}.$$

Thus, taking $V_m(x) = -\lambda_m$ for all $x \in \Omega$, we see that $V_m \in L^\infty(\Omega) \subset L^{r(x)}(\Omega)$ and

$$\frac{J_{V_m}(u_m)}{I(u_m)} = 0\,.$$

It follows that

$$E(V_m) \leq 0\,,$$

and we find

$$E_\star(\lambda_m) \leq 0\,.$$

We conclude that $E_\star$ is not constant. Furthermore, we point out that a similar proof as the one presented above shows that function $E_\star$ also assumes negative values. To see this, simply note that by Theorem 18, for each $\lambda > \lambda_m$ there exists $u_\lambda \in X \setminus \{0\}$ such that taking $V_\lambda = -\lambda$ for all $x \in \Omega$ we have

$$\frac{J_{V_\lambda}(u_\lambda)}{I(u_\lambda)} < 0\,.$$

Next, we note that $E_\star$ decreases monotonically. Indeed, if we consider $0 \leq R_1 < R_2$ then it is clear that $\overline{B}_{R_1}(0) \subset \overline{B}_{R_2}(0)$. Then the definition of function $E_\star$ implies $E_\star(R_1) \geq E_\star(R_2)$.

(b) Finally, we show that the function $E_\star$ is continuous. Let $R > 0$ and $t \in (0, R)$ be fixed. We will verify that $\lim_{t\searrow 0} E_\star(R+t) = \lim_{t\searrow 0} E_\star(R-t) = E_\star(R)$.

First, we prove that $\lim_{t\searrow 0} E_\star(R+t) = E_\star(R)$. By Theorem 21 (a), we have

$$E_\star(R) \geq E_\star(R+t)\,.$$

Moreover, by Theorem 20, it follows that there exists $V_{R+t} \in \overline{B}_{R+t}(0)$ (that is, $|V_{R+t}|_{r(x)} \leq R+t$) such that

$$E(V_{R+t}) = E_\star(R+t)\,.$$

Taking now $V_{R,t} := \frac{R}{R+t} V_{R+t}$ we have

$$|V_{R,t}|_{r(x)} = \frac{R}{R+t} |V_{R+t}|_{r(x)} \leq R\,,$$

or $V_{R,t} \in \overline{B}_R(0)$. Therefore clearly, $E(V_{R,t}) \geq E_\star(R)$.

On the other hand by Theorem 19, there exists $u_t \in X \setminus \{0\}$ such that

$$E(V_{R+t}) = \frac{J_{V_{R+t}}(u_0)}{I(u_0)}\,.$$

Combining the above pieces of information, we find

$$\begin{aligned}
E_\star(R+t) = E(V_{R+t}) &= \frac{J_{V_{R+t}}(u_t)}{I(u_t)} \\
&= \frac{J_{\frac{R+t}{R}\cdot V_{R,t}}(u_t)}{I(u_t)} \\
&= \frac{R+t}{R} \cdot \frac{J_{V_{R,t}}(u_t)}{I(u_t)} - \frac{t}{R} \cdot \frac{J_0(u_t)}{I(u_t)} \\
&\geq \frac{R+t}{R} \cdot E_\star(R) - \frac{t}{R} \cdot \frac{J_0(u_t)}{I(u_t)}\,.
\end{aligned}$$

On the other hand, by relation (3.55), we have that for each $t \in (0, R)$,

$$\begin{aligned}
E_\star(R) \geq E_\star(R+t) = E(V_{R+t}) &= \frac{J_{V_{R+t}}(u_t)}{I(u_t)} \\
&\geq \beta_1 \cdot \frac{J_0(u_t)}{I(u_t)} - \gamma \cdot |V_{R+t}|_{r(x)} \\
&= \beta_1 \cdot \frac{J_0(u_t)}{I(u_t)} - \gamma \cdot 2R\,,
\end{aligned}$$

where $\beta_1 > 0$ and $\gamma > 0$ are real constants.

Combining the last two inequalities, we deduce

$$E_\star(R) \geq E_\star(R+t) \geq \frac{R+t}{R} \cdot E_\star(R) - \frac{t}{R} \cdot \frac{E_\star(R) + \gamma \cdot 2R}{\beta_1}\,,$$

for each $t \in (0, R)$.

We conclude that

$$\lim_{t \searrow 0} E_\star(R+t) = E_\star(R)\,.$$

In the sequel, we argue that $\lim_{t\searrow 0} E_\star(R-t) = E_\star(R)$. Obviously,

$$E_\star(R) \leq E_\star(R-t), \quad \forall\, t \in (0, R)\,.$$

By Theorem 20, there exists $V_R \in \overline{B}_R(0)$ such that

$$E_\star(R) = E(V_R)\,.$$

Moreover, by Theorem 19, there exists $u_0 \in X \setminus \{0\}$ such that

$$E(V_R) = \frac{J_{V_R}(u_0)}{I(u_0)}\,.$$

Define now

$$V_t := \frac{R-t}{R} V_R, \quad \forall\, t \in (0, R)\,.$$

Clearly, $V_t \in \overline{B}_{R-t}(0)$. Thus, it is clear that

$$\frac{J_{V_t}(u_0)}{I(u_0)} \geq E_\star(R-t), \quad \forall\, t \in (0,R).$$

Taking into account the above information, we find for all $t \in (0,R)$

$$\begin{aligned} E_\star(R) = E(V_R) = \frac{J_{V_R}(u_0)}{I(u_0)} &= \frac{J_{\frac{R}{R-t}V_t}(u_0)}{I(u_0)} \\ &= \frac{J_{V_t}(u_0)}{I(u_0)} + \frac{t}{R-t}\cdot\frac{\displaystyle\int_\Omega \frac{V_t(x)}{m(x)}|u_0|^{m(x)}\,dx}{I(u_0)} \\ &\geq E_\star(R-t) + \frac{t}{R}\cdot\frac{\displaystyle\int_\Omega \frac{V_R(x)}{m(x)}|u_0|^{m(x)}\,dx}{I(u_0)}. \end{aligned}$$

We infer

$$\lim_{t\searrow 0} E_\star(R-t) = E_\star(R).$$

It follows that the function $E_\star$ is continuous. The proof of Theorem 21 is complete. □

By Theorem 21 (a), we obtain that $E_\star$ decreases monotonically. We note that in the special case when $q_1(x) = m(x) = q_2(x) = q$ for each $x \in \overline{\Omega}$, where $q > 1$ is a real number for which conditions (3.43), (3.44) and (3.45) are fulfilled, the above quoted result can be improved, in the sense that we can show that actually, the function $E_\star$ is strictly decreasing on $[0,\infty)$. Indeed, letting $0 \leq R_1 < R_2$ be given, by Theorem 20, we deduce that there exists $V_1 \in \overline{B}_{R_1}(0)$ such that

$$E(V_1) = E_\star(R_1).$$

Then for each real number $t \in (0, R_2 - R_1)$ we have $V_1 - t \in \overline{B}_{R_2}(0)$, since $|V_1 - t|_{r(x)} \leq |V_1|_{r(x)} + t \leq R_2$. Next, by Theorem 19, there exists $u_1 \in X\setminus\{0\}$ such that

$$E(V_1) = \frac{J_{V_1}(u_1)}{I(u_1)}.$$

Taking into account all the above remarks we infer

$$E_\star(R_1) - \frac{t}{2} = E(V_1) - \frac{t}{2} = \frac{J_{V_1}(u_1)}{I(u_1)} - \frac{t}{2} = \frac{J_{V_1-t}(u_1)}{I(u_1)} \geq E(V_1 - t) \geq E_\star(R_2),$$

or

$$E_\star(R_1) > E_\star(R_2).$$

We observe that the above arguments show that in the case when we manipulate *homogeneous* quantities, we obtain better results than in the case when we deal with *nonhomogeneous* quantities.

We also remark that by Theorem 21 (b), we deduce that

$$E_\star(R) = \inf_{s \le R} E_\star(s) \quad \text{and} \quad E_\star(R) = \sup_{s \ge R} E_\star(s)\,.$$

We further note that the function $E_\star$ can be used to define a *continuous set function* on a subset of $L^{r(x)}(\Omega)$. We still denote each closed ball centered at the origin of radius R in $L^{r(x)}(\Omega)$ by $\overline{B}_R(0)$, that is,

$$\overline{B}_R(0) := \{u \in L^{r(x)}(\Omega);\ \ |u|_{r(x)} \le R\}\,.$$

By Theorem 21 (b), we deduce that $E_\star$ is a continuous function. By the proof of Theorem 21 (a), we have $E_\star(0) > 0$ and there exists $R_1 > 0$ such that $E_\star(R_1) < 0$. Thus, we infer that there exists $R_0 > 0$ such that $E_\star(R_0) = 0$.

We define

$$\Gamma = \{\overline{B}_R(0) \setminus \overline{B}_{R_0}(0);\ R \ge R_0\} \subset L^{r(x)}(\Omega)$$

and $\mu : \Gamma \to [0, \infty)$ by

$$\mu(\overline{B}_R(0) \setminus \overline{B}_{R_0}(0)) = -E_\star(R), \quad \forall\, R \ge R_0\,.$$

By Theorem 21 (a), we find that the function μ has the following properties:

(1) $\mu(\emptyset) = 0$.

(2) For each S_1, $S_2 \in \Gamma$ such that $S_1 \subset S_2$ we have $\mu(S_1) \le \mu(S_2)$.

Thus, μ is a *set function* on Γ. By Theorem 21 (b), we have that for each $S \subset \Gamma$ the following holds

$$\mu(S) = \sup_{T \subseteq S} \mu(T) \quad \text{and} \quad \mu(S) = \inf_{T \supseteq S} \mu(T)\,.$$

We conclude that μ is a *continuous set function* on Γ.

3.4 Anisotropic problems with lack of compactness and nonlinear boundary condition

Let Ω be a smooth exterior domain in $\mathbb{R}^N$, that is, Ω is the complement of a bounded domain with $C^{1,\delta}$ boundary ($0 < \delta < 1$). Assume that p is a real number satisfying $1 < p < N$, $a \in L^\infty(\Omega) \cap C^{0,\delta}(\overline{\Omega})$ is a positive function and $b \in L^\infty(\Omega) \cap C(\Omega)$ is nonnegative. Let $p^* := Np/(N-p)$ denote the critical Sobolev exponent and assume that r is a real number satisfying $p < r < p^*$. Suppose that $g \in L^\infty(\Omega) \cap L^{p_0}(\Omega)$, with $p_0 := p^*/(p^* - r)$, is a nontrivial potential which is positive on some nonempty open subset of Ω. Let E be the completion of $C_0^\infty(\Omega)$ under the norm

$$\|u\|_{a,b} = \left(\int_\Omega [a(x)|Du|^p + b(x)|u|^p]dx \right)^{1/p}\,.$$

In this section, we consider the anisotropic problem

$$\begin{cases} -\mathrm{div}\,(a(x)|Du|^{p-2}Du) + |u|^{q-2}u = \lambda g(x)|u|^{r-2}u & \text{in } \Omega, \\ a(x)|Du|^{p-2}\partial_\nu u + b(x)|u|^{p-2}u = 0 & \text{on } \partial\Omega, \end{cases} \tag{3.71}$$

where λ is a real parameter and ν is the unit vector of the outward normal on $\partial\Omega$.

Our hypotheses are the following:

(H1) $g \in L^\infty(\Omega) \cap L^{p_0}(\Omega)$, *with* $p_0 := p^*/(p^* - r)$, $p < r < q < p^*$, *is a nonnegative function which is positive on a nonempty open subset of* Ω;

(H2) b *is a continuous positive function on* $\Gamma = \partial\Omega$.

Without altering the proof below, the coefficient 1 of the dominating term $|u|^{p-2}u$ can be replaced by any function $f \in L^\infty(\Omega)$, with $\operatorname{infess}_\Omega f > 0$. Hence equation (3.71) is the renormalized form.

Problem (3.71) may be viewed as a prototype of pattern formation in biology and is related to the steady–state problem for a chemotactic aggregation model introduced by Keller and Segel [41]. Problem (3.71) also plays an important role in the study of activator–inhibitor systems modeling of biological pattern formation, as proposed by Gierer and Meinhardt [34].

By a weak (nontrivial) solution of problem (3.71) we mean a nontrivial function $u \in X = E \cap L^q(\Omega)$ satisfying for all $\varphi \in X$ the identity

$$\begin{aligned} \int_\Omega a(x)|Du|^{p-2}Du \cdot D\varphi dx &+ \int_\Gamma b(x)|u|^{p-2}u\varphi d\sigma + \int_\Omega |u|^{q-2}u\varphi dx \\ &= \lambda \int_\Omega g(x)|u|^{r-2}u\varphi dx, \end{aligned} \tag{3.72}$$

where E is now the completion of the restriction of functions of $C_0^\infty(\mathbb{R}^N)$ on Ω with respect to the norm

$$\|u\|_{a,b} = \left(\int_\Omega a(x)|Du|^p dx + \int_\Gamma b(x)|u|^p d\sigma\right)^{1/p},$$

and X is a reflexive Banach space endowed with the norm

$$\|u\| = \left\{\|u\|_{a,b}^p + \|u\|_{L^q(\Omega)}^p\right\}^{1/p}.$$

Hence, by (H1)–(H2), all integrals in (3.72) are well defined and they converge.

The loss of compactness of the Sobolev embeddings on unbounded domains renders variational techniques more delicate. Some of the papers treating problems on unbounded domains use special function spaces where the compactness is preserved, such as spaces of radially symmetric functions. We point out that even if Ω is unbounded, standard compact embeddings still exist, for instance, if Ω is *thin at infinity*, in the sense that

$$\lim_{R\to\infty} \sup\left\{\mu(\Omega \cap B(x,1)) \,:\, x \in \mathbb{R}^N,\ |x| = R\right\} = 0,$$

TABLE 3.1: Existence of solutions depending on the values of the parameter

Value of λ	Existence or nonexistence
$\lambda < \lambda^*$	No solution
$\lambda \geq \lambda^*$	One solution or more

where μ denotes the Lebesgue measure and $B(x, 1)$ is the unit ball centered at x. Such arguments cannot be applied to our general unbounded domain Ω. In this case, since Ω is not *thin* and it looks like $\mathbb{R}^N$ at infinity (because Ω is an exterior domain), the analysis of the compactness failure shows that a Palais–Smale sequence of the associated energy functional differs from its weak limit by "*waves*" that go to infinity. However, the definition of X, combined with the main assumption $p < r < p^*$, ensures that

$$X \text{ is compactly embedded into the weighted Lebesgue space } L^r(\Omega; g). \tag{3.73}$$

Taking $\varphi = u$ in relation (3.72), we deduce that any weak solution u of problem (3.71) satisfies the equality

$$\|u\|_{a,b}^p + \|u\|_{L^q(\Omega)}^q = \lambda \|u\|_{L^r(\Omega;g)}^r, \tag{3.74}$$

so that problem (3.71) does not have any nontrivial solution whenever $\lambda \leq 0$. We first prove that the result still remains true for sufficiently small values of $\lambda > 0$ when $p < r < q < p^*$, that is, the term $|u|^{q-2}u$ *dominates* the right-hand side and makes impossible the existence of a solution of our problem (3.71). On the other hand, if $\lambda > 0$ is sufficiently large, then (3.71) admits weak solutions. The behavior described in Theorem 22 is summarized in Table 3.1.

Theorem 22 *Under the assumptions* (H1) *and* (H2) *there exists* $\lambda^* > 0$ *such that*

(i) if $\lambda < \lambda^*$, *then problem* (3.71) *does not have any weak solution;*

(ii) if $\lambda \geq \lambda^*$, *then problem* (3.71) *has at least one weak solution* u, *with the properties*

(a) $u \in L^\infty_{\text{loc}}(\Omega)$;

(b) $u \in C^{1,\alpha}(\Omega \cap B_R)$, $\alpha = \alpha(R) \in (0, 1)$;

(c) $u > 0$ *in* Ω;

(d) $u \in L^m(\Omega)$ *for all* $p^* \leq m < \infty$ *and* $\lim_{|x|\to\infty} u(x) = 0$.

We start by outlining the main ideas of the proof of Theorem 22.

(a) There is some $\lambda^* > 0$ such that problem (3.71) does not have any solution for any $\lambda < \lambda^*$. This means that if a solution exists then λ must be sufficiently large. One of the key arguments in this proof is based on the assumption that $q > r$. In particular, this proof yields an energy lower bound of solutions in terms of λ which will be useful to conclude that problem (3.71) has a nontrivial solution if $\lambda = \lambda^*$.

(b) There exists $\lambda^{**} > 0$ such that problem (3.71) has at least one solution for any $\lambda > \lambda^{**}$. Next, by the properties of λ^* and λ^{**}, we deduce that $\lambda^{**} = \lambda^*$. The proof uses variational arguments and is based on the coercivity of the corresponding energy functional defined on X by

$$J_\lambda(u) = \frac{1}{p}\|u\|_{a,b}^p + \frac{1}{q}\|u\|_{L^q(\Omega)}^q - \frac{\lambda}{r}\|u\|_{L^r(\Omega;g)}^r.$$

We show that the minimum of J is achieved by a weak solution of (3.71). In order to obtain that this global minimizer is not trivial, we prove that the corresponding energy level is negative, provided that λ is sufficiently large.

Proof of Theorem 22. We split the proof into several steps as follows.

Step 1. Nonexistence for small $\lambda > 0$. It is enough to show that if there is a weak solution of problem (3.71), then λ must be sufficiently large. Assume that u is a weak solution of (3.71). Then by (3.72) we can deduce (3.74). Since by (H1), $r < q$ and $g^{q/(q-r)}$ is in $L^1(\Omega)$, applying the Young inequality we deduce that

$$\lambda\|u\|_{L^r(\Omega;g)}^r \le \frac{(q-r)\lambda^{q/(q-r)}}{q}\int_\Omega g(x)^{q/(q-r)}dx + \frac{r}{q}\|u\|_{L^q(\Omega)}^q. \tag{3.75}$$

Next, by relations (3.74) and (3.75) and the fact that u is nontrivial, we get

$$\begin{aligned} 0 < \|u\|_{a,b}^p &\le \frac{q-r}{q}\lambda^{q/(q-r)}\int_\Omega g(x)^{q/(q-r)}dx + \frac{r-q}{q}\|u\|_{L^q(\Omega)}^q \\ &\le \frac{q-r}{q}\lambda^{q/(q-r)}\int_\Omega g(x)^{q/(q-r)}dx := \lambda^{q/(q-r)}A < \infty. \end{aligned} \tag{3.76}$$

The continuity of the embedding $X \hookrightarrow L^r(\Omega; g)$ implies that there exists $C = C(\Omega, g, p, q, r) > 0$ such that

$$C\|v\|_{L^r(\Omega;g)}^p \le \|v\|_{a,b}^p \tag{3.77}$$

for any $v \in X$. Thus, by relations (3.74) and (3.77), we have $C\|u\|_{L^r(\Omega;g)}^p \le \lambda\|u\|_{L^r(\Omega;g)}^r$. Since by (3.74), $p < r < q$, $\lambda > 0$ and $\|u\|_{L^r(\Omega;g)} > 0$, we deduce that

$$\lambda \ge C\|u\|_{L^r(\Omega;g)}^{p-r} \ge CC^{(-1+r/p)}\|u\|_{a,b}^{p-r} \ge C^{r/p}\lambda^{q(p-r)/p(q-r)}A^{(p-r)/p}.$$

It follows that $\lambda \ge (A^{p-r}C^r)^{(q-r)/r(q-p)}$, which also implies that $\lambda^* \le (A^{p-r}C^r)^{(q-r)/r(q-p)}$. This completes the proof of (i).

In particular, Step 1 shows that if for some $\lambda > 0$ problem (3.71) has a weak solution u, then

$$(C^r/\lambda^p)^{1/(r-p)} \leq \|u\|_{a,b}^p \leq \lambda^{q/(q-r)} A, \tag{3.78}$$

where $C = C(\Omega, g, p, q, r) > 0$ is the constant given in (3.77).

Step 2. Coercivity of J. This follows by (H1). Indeed, for any $u \in X$ and all $\lambda > 0$

$$J_\lambda(u) = \frac{1}{p}\|u\|_{a,b}^p + \frac{1}{2q}\|u\|_{L^q(\Omega)}^q + \frac{1}{2q}\|u\|_{L^q(\Omega)}^q - \frac{\lambda}{r}\|u\|_{L^r(\Omega;g)}^r.$$

By the Hölder inequality and (H1) we have

$$J_\lambda(u) \geq \frac{1}{p}\|u\|_{a,b}^p + \frac{1}{2q}\|u\|_{L^q(\Omega)}^q + \frac{1}{2q}\|u\|_{L^q(\Omega)}^q - \frac{\lambda}{r}\|g\|_{L^{q/(q-r)}(\Omega)}\|u\|_{L^q(\Omega)}^r. \tag{3.79}$$

Now, since for any positive numbers α, β, q and r, with $r < q$, the function $\Phi : \mathbb{R}_0^+ \to \mathbb{R}$ defined by $\Phi(t) = \alpha t^r - \beta t^q$, achieves its positive global maximum

$$\Phi(t_0) = \frac{q-r}{q}\left(\frac{r}{q}\right)^{r/(q-r)} \alpha^{q/(q-r)}\beta^{r/(r-q)} > 0$$

at the point $t_0 = (\alpha r/\beta q)^{1/(q-r)} > 0$, we immediately have $\alpha t^r - \beta t^q \leq C(q,r)\alpha^{q/(q-r)}\beta^{r/(r-q)}$, where $C(q,r) = (q-r)\left(r^r/q^q\right)^{1/(q-r)}$. Returning to (3.79) and using the above inequality, with $t = \|u\|_{L^q(\Omega)}$, $\alpha = \lambda\|g\|_{L^{q/(q-r)}(\Omega)}/r$ and $\beta = 1/2q$, we deduce that

$$J_\lambda(u) \geq \frac{1}{p}\|u\|_{a,b}^p + \frac{1}{2q}\|u\|_{L^q(\Omega)}^q - C(\lambda, q, r, g),$$

where $C(\lambda, q, r, g) = 2^{r/(q-r)}(q-r)\left(\lambda\|g\|_{L^{q/(q-r)}(\Omega)}\right)^{q/(q-r)}/qr$. This implies the claim.

Let $n \mapsto u_n$ be a minimizing sequence of J_λ in X, which is bounded in X by Step 2. Without loss of generality, we may assume that $(u_n)_n$ is nonnegative, weakly converges to some u in X and also converges pointwisely.

Step 3. The nonnegative weak limit $u \in X$ is a weak solution of problem (3.71). To prove this, we show that

$$J_\lambda(u) \leq \liminf_{n\to\infty} J_\lambda(u_n).$$

By the weak lower semi-continuity of the norm $\|\cdot\|$ we have

$$\frac{1}{p}\|u\|_{a,b}^p + \frac{1}{q}\|u\|_{L^q(\Omega)}^q \leq \liminf_{n\to\infty}\left(\frac{1}{p}\|u_n\|_{a,b}^p + \frac{1}{q}\|u_n\|_{L^q(\Omega)}^q\right).$$

Next, the boundedness of $(u_n)_n$ in X implies with the same argument that

$$\|u\|_{L^r(\Omega;g)} = \lim_{n\to\infty} \|u_n\|_{L^r(\Omega;g)}$$

by (3.73). Hence u is a global minimizer of J_λ in X.

Step 4. The weak limit u is a nonnegative weak solution of (3.71) *if* $\lambda > 0$ *is sufficiently large.* Clearly $J_\lambda(0) = 0$. Thus, by Step 3 it is enough to show that there exists $\Lambda > 0$ such that

$$\inf_{u\in X} J_\lambda(u) < 0 \quad \text{for all } \lambda > \Lambda.$$

Consider the constrained minimization problem

$$\Lambda := \inf\left\{\frac{1}{p}\|w\|_{a,b}^p + \frac{1}{q}\|w\|_{L^q(\Omega)}^q : w \in X \text{ and } \|w\|_{L^r(\Omega;g)}^r = r\right\}. \tag{3.80}$$

Let $n \mapsto v_n \in X$ be a minimizing sequence of (3.80), which is clearly bounded in X, so that we can assume, without loss of generality, that it weakly converges to some $v \in X$, with $\|v\|_{L^r(\Omega;g)}^r = r$ and

$$\Lambda = \frac{1}{p}\|v\|_{a,b}^p + \frac{1}{q}\|v\|_{L^q(\Omega)}^q$$

by the weak lower semi-continuity of $\|\cdot\|$. Thus, $J_\lambda(v) = \Lambda - \lambda < 0$ for any $\lambda > \Lambda$.

Now put

$$\begin{aligned}
\lambda^* &:= \sup\{\lambda > 0 : \text{problem (3.71) does not admit any weak solutions}\},\\
\lambda^{**} &:= \inf\{\lambda > 0 : \text{problem (3.71) admits a weak solution}\}.
\end{aligned}$$

Of course, $\Lambda \geq \lambda^{**} \geq \lambda^* > 0$. To complete the proof of Theorem 22, it is enough to prove the following essential facts:
(a) problem (3.71) has a weak solution for any $\lambda > \lambda^{**}$;
(b) $\lambda^{**} = \lambda^*$ and problem (3.71) admits a weak solution when $\lambda = \lambda^*$.

Step 5. Problem (3.71) *has a weak solution for any $\lambda > \lambda^{**}$ and $\lambda^{**} = \lambda^*$.* Fix $\lambda > \lambda^{**}$. By the definition of λ^{**}, there exists $\mu \in (\lambda^{**}, \lambda)$ such that J_μ has a nontrivial critical point $u_\mu \in X$. Of course, u_μ is a sub–solution of problem (3.71). In order to find a super–solution of (3.71) which dominates u_μ, we consider the constrained minimization problem

$$\inf\left\{\frac{1}{p}\|w\|_{a,b}^p + \frac{1}{q}\|w\|_{L^q(\Omega)}^q - \frac{\lambda}{r}\|w\|_{L^r(\Omega;g)}^r : w \in X \text{ and } w \geq u_\mu\right\}.$$

Arguments similar to those used in Step 4 show that the above minimization problem has a solution $u_\lambda \geq u_\mu$ which is also a weak solution of problem (3.71), provided that $\lambda > \lambda^{**}$.

We already know that $\lambda^{**} \geq \lambda^*$. But, by the definition of λ^{**} and the above remark, problem (3.71) has no solutions for any $\lambda < \lambda^{**}$. Passing to the supremum, this forces $\lambda^{**} = \lambda^*$ and completes the proof.

Step 6. Problem (3.71) *admits a nonnegative weak solution when* $\lambda = \lambda^*$. Let $n \mapsto \lambda_n$ be a decreasing sequence converging to λ^* and let $n \mapsto u_n$ be a corresponding sequence of nonnegative weak solutions of (3.71). As noted in Step 2, the sequence $(u_n)_n$ is bounded in X, so that without loss of generality, we may assume that it weakly converges in X, strongly in $L^r(\Omega; g)$, and pointwisely to some $u^* \in X$, with $u^* \geq 0$. By (3.72), for all $\varphi \in X$,

$$\int_\Omega a(x)|Du_n|^{p-2}Du_n \cdot D\varphi dx + \int_\Gamma b(x)|u_n|^{p-2}u_n\varphi d\sigma + \int_\Omega |u_n|^{q-2}u_n\varphi dx = \lambda_n \int_\Omega g(x)|u_n|^{r-2}u_n\varphi dx,$$

and passing to the limit as $n \to \infty$ we deduce that u^* satisfies (3.72) for $\lambda = \lambda^*$, as claimed.

It remains to argue that $u^* \neq 0$. A key ingredient of this argument is the lower bound energy given in (3.78). Hence, since u_n is a nontrivial weak solution of problem (3.71) corresponding to λ_n, we have $\|u_n\|_{a,b}^p \geq (C^r/\lambda^p)^{1/(r-p)}$ by (3.78), where $C > 0$ is the constant given in (3.77) and not depending on λ_n. Next, since $\lambda_n \searrow \lambda^*$ as $n \to \infty$ and $\lambda^* > 0$, it is enough to show that

$$\|u_n - u^*\|_{a,b} \to 0 \quad \text{as } n \to \infty. \tag{3.81}$$

Since u_n and u^* are weak solutions of (3.71) corresponding to λ_n and λ^*, we have by (3.72), with $\varphi = u_n - u^*$,

$$\begin{aligned} &\int_\Omega a(x)\left(|Du_n|^{p-2}Du_n - |Du^*|^{p-2}Du^*\right) \cdot D(u_n - u^*)dx \\ &\qquad + \int_\Gamma b(x)\left(|u_n|^{p-2}u_n - |u^*|^{p-2}u^*\right)(u_n - u^*)d\sigma \\ &\qquad + \int_\Omega (|u_n|^{q-2}u_n - |u^*|^{q-2}u^*)(u_n - u^*)dx \\ &= \int_\Omega g(x)\left(\lambda_n |u_n|^{r-2}u_n - \lambda^* |u^*|^{r-2}u^*\right)(u_n - u^*)dx. \end{aligned} \tag{3.82}$$

Elementary monotonicity properties imply that

$$\int_\Omega (|u_n|^{q-2}u_n - |u^*|^{q-2}u^*)(u_n - u^*)dx \geq 0 \quad \text{and} \quad \langle I'(u_n^*) - I'(u^*), u_n - u^*\rangle \geq 0,$$

where

$$I(u) := \|u\|_{a,b}^p/p.$$

Since $\lambda_n \searrow \lambda^*$ as $n \to \infty$ and X is compactly embedded in $L^r(\Omega; g)$ for all

$p > 1$, relation (3.82) implies

$$\begin{aligned} 0 &\leq \langle I'(u_n^*) - I'(u^*), u_n - u^* \rangle \\ &\leq \int_\Omega g(x) \left[\lambda_n u_n^{r-1} - \lambda^* (u^*)^{r-1}\right] (u_n - u^*) dx \to 0 \end{aligned} \tag{3.83}$$

as $n \to \infty$.

Now, we distinguish between the cases $p \geq 2$ and $1 < p < 2$ and we use the following elementary inequalities (see Simons [98, formula (2.2)]): for all $\xi,\, \zeta \in \mathbb{R}^N$

$$|\xi - \zeta|^p \leq \begin{cases} c(|\xi|^{p-2}\xi - |\zeta|^{p-2}\zeta)(\xi - \zeta) & \text{if } 1 < p \geq 2; \\ c\langle|\xi|^{p-2}\xi - |\eta|^{p-2}\eta, \xi - \eta\rangle^{p/2} \left(|\xi|^p + |\eta|^p\right)^{(2-p)/2} & \text{if } 1 < p < 2, \end{cases} \tag{3.84}$$

where c is a positive constant.

Case 1: $p \geq 2$. By (3.84) and (3.83), we immediately conclude that

$$\|u_n - u^*\|_{a,b}^p \leq c\langle I'(u_n^*) - I'(u^*), u_n - u^*\rangle = o(1) \qquad \text{as } n \to \infty.$$

Case 2: $1 < p < 2$. Since by convexity for all $\gamma \geq 1$

$$(v + w)^\gamma \leq 2^{\gamma-1}(v^\gamma + w^\gamma) \qquad \text{for all } v, w \in \mathbb{R}_0^+, \tag{3.85}$$

we have for $\gamma = 2/p$,

$$\|u_n - u^*\|_{a,b}^2 \leq 2^{(2-p)/p} \left[\left(\int_\Omega a(x)|D(u_n - u^*)|^p\, dx \right)^{2/p} + \left(\int_\Gamma b(x)|u_n - u^*|^p\, d\sigma \right)^{2/p} \right].$$

Thus, in order to conclude that (3.81) holds, it is enough to show that

$$\int_\Omega a(x)|D(u_n - u^*)|^p\, dx \to 0 \qquad \text{and} \qquad \int_\Gamma b(x)|u_n - u^*|^p\, d\sigma \to 0$$

as $n \to \infty$. Indeed, combining (3.84) and (3.85), we have

$$
\begin{aligned}
&\int_\Omega a(x)|D(u_n - u^*)|^p\,dx \\
&\quad \le c\int_\Omega a(x)\left\{(|Du_n|^{p-2}Du_n - |Du^*|^{p-2}Du^*)\cdot D(u_n-u^*)\right\}^{p/2} \\
&\qquad \times (|Du_n|^p + |Du^*|^p)^{(2-p)/2}\,dx \\
&\quad \le c\left(\int_\Omega a(x)(|Du_n|^{p-2}Du_n - |Du^*|^{p-2}Du^*)\cdot D(u_n-u^*)\,dx\right)^{p/2} \\
&\qquad \times \left(\|u_n\|_{a,b}^p + \|u^*\|_{a,b}^p\right)^{(2-p)/2} \\
&\quad \le c\left(\int_\Omega a(x)(|Du_n|^{p-2}Du_n - |Du^*|^{p-2}Du^*)\cdot D(u_n-u^*)\,dx\right)^{p/2} \\
&\qquad \times \left(\|u_n\|_{a,b}^{(2-p)p/2} + \|u^*\|_{a,b}^{(2-p)p/2}\right) \\
&\quad \le C_1\left(\int_\Omega a(x)(|Du_n|^{p-2}Du_n - |Du^*|^{p-2}Du^*)\cdot D(u_n-u^*)\,dx\right)^{p/2},
\end{aligned}
$$

where $C_1 = 2c\left(\lambda^{q/(q-r)}A\right)^{(2-p)/2}$ by (3.78), and C_1 is independent of n by (3.76). Similar arguments yield

$$
\int_\Gamma b(x)(u_n - u^*)^p\,d\sigma \le C_2\left(\int_\Gamma b(x)\left[u_n^{p-1} - (u^*)^{p-1}\right](u_n - u^*)\,dx\right)^{p/2},
$$

with an appropriate positive constant C_2 independent of n. Combining the above two inequalities with (3.83) we conclude that $\|u_n - u^*\|_{a,b} = o(1)$ as $n \to \infty$, that is (3.81) holds and u^* is a nontrivial nonnegative weak solution of problem (3.71) corresponding to $\lambda = \lambda^*$.

Next, by Moser iteration, we deduce that u satisfies *(a)*, since $u \in W^{1,p}_{\mathrm{loc}}(\Omega)$. Using the main assumptions on the coefficient $a = a(x)$, an application of [20, Corollary on p. 830] due to DiBenedetto shows that the weak solution u satisfies also property *(b)*. Finally, *(c)* follows immediately by the strong maximum principle since u is a C^1 nonnegative weak solution of the differential inequality $\operatorname{div}(a(x)|Du|^{p-2}Du) - |u|^{q-2}u \le 0$ in Ω, with $q > p$. Property *(d)* follows by using Theorem 1 of Serrin [96]. □

3.5 Glossary

Biological Pattern Formation: It deals with the visible orderly outcomes of self-organization and the common principles behind similar patterns in nature.

Coercive Functional: It is a function that "grows rapidly" at the extremes of the space on which it is defined.

Fréchet Derivative: It is a derivative defined on Banach spaces, which is commonly used to generalize the derivative of a real-valued function of a single real variable to the case of a vector-valued function of several real variables. The notion is named after the French mathematician Maurice Fréchet (1878–1973).

Gâteaux Derivative: It is a generalization of the concept of directional derivative in differential calculus. The notion is named after the French mathematician René Gâteaux (1889–1914).

Indefinite Potential: Sign-changing weight function.

Lagrange Multipliers Theorem: It is a method in mathematical optimization for finding the local maxima and minima of a function subject to equality constraints. The result is named after the French mathematician Joseph Louis Lagrange (1736–1813).

Lower Semi-continuous Functional: It is a real-valued function f such that, roughly speaking, its values for arguments near a point x_0 are greater than $f(x_0)$.

Mathematical Optimization: It is the selection of the best element (with regards to certain criteria) from some set of available alternatives.

Set Function: It is a function whose input is a set. The output is usually a number. Often the input is a set of real numbers, a set of points in Euclidean space, or a set of points in some measure space.

Upper Semi-continuous Functional: It is a real-valued function f such that, roughly speaking, its values for arguments near a point x_0 are less than $f(x_0)$.

Chapter 4

Nonlinear Problems in Orlicz-Sobolev Spaces

4.1 Existence and multiplicity of solutions 118
  4.1.1 Some properties of Orlicz-Sobolev spaces 122
  4.1.2 Absorption case: existence of infinitely many solutions . 125
  4.1.3 Desorption case: existence of solutions for large parameters 132
4.2 A continuous spectrum for nonhomogeneous operators 134
  4.2.1 Case of small positive parameters 136
  4.2.2 Existence of low-energy solutions 140
4.3 Nonlinear eigenvalue problems with indefinite potential 145
  4.3.1 Concentration properties of the spectrum 149
  4.3.2 Optimization property for the Rayleigh quotient 156
  4.3.3 More about the Rayleigh quotient on spheres 158
4.4 Multiple solutions in Orlicz-Sobolev spaces 161
  4.4.1 More about Orlicz-Sobolev spaces 163
  4.4.2 Multiplicity versus the three critical points theorem ... 166
4.5 Neumann problems in Orlicz-Sobolev spaces 171
  4.5.1 More about generalized Orlicz-Sobolev spaces 174
  4.5.2 Conditions for concentration of the spectrum 179
4.6 Glossary 191

> Great minds discuss ideas,
> average minds discuss events,
> small minds discuss people.
>
> Eleanor Roosevelt (1884–1962)

The Orlicz-Sobolev spaces play a significant role in many fields of mathematics, such as approximation theory, partial differential equations, calculus of variations, nonlinear potential theory, theory of quasi-conformal mappings, differential geometry, geometric function theory, and probability theory. These spaces consists of functions that have weak derivatives and satisfy certain integrability conditions. The study of nonlinear elliptic equations involving quasilinear *homogeneous* type operators is based on the theory of Sobolev spaces $W^{m,p}(\Omega)$ and its goal is to find weak solutions. In the case of *nonhomogeneous* differential operators, the natural setting for this approach are Orlicz-Sobolev spaces.

In this chapter, we study several nonlinear eigenvalue problems involving nonhomogeneous differential operators. This analysis is carried out in the general setting corresponding to Orlicz-Sobolev spaces. We consider both Dirichlet and Neumann boundary conditions and we are mainly interested in the following two research topics: (i) the existence of a continuous spectrum; (ii) sufficient conditions for the existence of multiple solutions. The spectral analysis developed in this chapter is mainly based on variational methods and related energy estimates. The content of this chapter is related with the papers [85, 50, 53, 55, 58, 67, 72, 15].

4.1 Existence and multiplicity of solutions

This section is devoted to the study of weak solutions of problems of the type

$$\begin{cases} -\mathrm{div}(a(|\nabla u(x)|)\nabla u(x)) = f(u(x)), & \text{for } x \in \Omega \\ u(x) = 0, & \text{for } x \in \partial\Omega \end{cases} \tag{4.1}$$

where $\Omega \subset \mathbb{R}^N$ ($N \geq 3$) is a bounded domain with smooth boundary.

The case where $a(t) = t^{p-2}$ ($p > 1$, $t \geq 0$) is fairly well understood and a great variety of existence results are available. We shall focus on the case where $a : [0, \infty) \to \mathbb{R}$ is defined by $a(t) = \log(1 + t^q) \cdot t^p$, where p, $q > 1$. We treat separately the cases where either $f(t) = -\lambda|t|^{p-2}t + |t|^{r-2}t$ or $f(t) = \lambda|t|^{p-2}t - |t|^{r-2}t$, where $r < (Np - N + p)/(N - p)$ and λ is a positive parameter.

In Chapter 1 we introduced the Orlicz and Orlicz-Sobolev function spaces. In the sequel we study further properties of these spaces.

Define

$$\phi(t) := \log(1 + |t|^q) \cdot |t|^{p-2}t, \qquad \text{for all } t \in \mathbb{R}$$

and

$$\Phi(t) := \int_0^t \phi(s), \qquad \text{for all } t \in \mathbb{R}.$$

A straightforward computation yields

$$\Phi(t) = \frac{1}{p}\log(1 + |t|^q) \cdot |t|^p - \frac{q}{p}\int_0^{|t|} \frac{s^{p+q-1}}{1 + s^q}\, ds,$$

for all $t \in \mathbb{R}$. We point out that ϕ is an odd, increasing homeomorphism of $\mathbb{R}$ into $\mathbb{R}$, while Φ is a convex even function on $\mathbb{R}$ and an increasing function from $\mathbb{R}_+$ to $\mathbb{R}_+$.

Set

$$\Phi^\star(t) := \int_0^t \phi^{-1}(s)\, ds, \qquad \text{for all } t \in \mathbb{R}.$$

Define the Orlicz class

$$K_\Phi(\Omega) := \{u : \Omega \to \mathbb{R};\ \text{measurable function and } \int_\Omega \Phi(|u(x)|)\, dx < \infty\}$$

and the Orlicz space

$$L_\Phi(\Omega) := \text{ the linear hull of } K_\Phi(\Omega).$$

The space $L_\Phi(\Omega)$ is a Banach space endowed with the Luxemburg norm

$$\|u\|_\Phi := \inf\left\{k > 0;\ \int_\Omega \Phi\left(\frac{u(x)}{k}\right)\, dx \leq 1\right\}$$

or the equivalent norm (the Orlicz norm)

$$\|u\|_{(\Phi)} := \sup\left\{\left|\int_\Omega uvdx\right|;\ v \in K_{\overline{\Phi}}(\Omega),\ \int_\Omega \overline{\Phi}(|v|)dx \leq 1\right\},$$

where $\overline{\Phi}$ denotes the conjugate Young function of Φ, that is,

$$\overline{\Phi}(t) = \sup\{ts - \Phi(s);\ s \in \mathbb{R}\}.$$

According to Adams [2, p. 234], we have $L_\Phi(\Omega) = K_\Phi(\Omega)$.

Next, we denote by $W^1L_\Phi(\Omega)$ the Orlicz-Sobolev space defined by

$$W^1L_\Phi(\Omega) := \left\{u \in L_\Phi(\Omega);\ \frac{\partial u}{\partial x_i} \in L_\Phi(\Omega),\ i = 1, \ldots, N\right\}.$$

This is a Banach space with respect to the norm

$$\|u\|_{1,\Phi} := \|u\|_\Phi + \||\nabla u|\|_\Phi.$$

We also define the Orlicz-Sobolev space $W_0^1L_\Phi(\Omega)$ as the closure of $C_0^\infty(\Omega)$ in $W^1L_\Phi(\Omega)$. By the Poincaré inequality, we can consider on $W_0^1L_\Phi(\Omega)$ the equivalent norm

$$\|u\| := \||\nabla u|\|_\Phi.$$

The space $W_0^1L_\Phi(\Omega)$ is a reflexive Banach space.

We first study the boundary value problem

$$\begin{cases} -\text{div}(\log(1 + |\nabla u(x)|^q)|\nabla u(x)|^{p-2}\nabla u(x)) = \\ -\lambda|u(x)|^{p-2}u(x) + |u(x)|^{r-2}u(x), \quad \text{if } x \in \Omega \\ u(x) = 0, \quad \text{if } x \in \partial\Omega. \end{cases} \tag{4.2}$$

We say that $u \in W_0^1 L_\Phi(\Omega)$ is a *weak solution* of problem (4.2) if

$$\int_\Omega \log(1+|\nabla u(x)|^q)|\nabla u(x)|^{p-2}\nabla u \nabla v \, dx + \lambda \int_\Omega |u(x)|^{p-2}u(x)v(x)\, dx - \int_\Omega |u(x)|^{r-2}u(x)v(x)\, dx = 0$$

for all $v \in W_0^1 L_\Phi(\Omega)$.

We establish the following multiplicity property.

Theorem 23 *Assume that $p,\ q > 1$, $p+q < N$, $p+q < r$ and $r < (Np - N+p)/(N-p)$. Then for every $\lambda > 0$ problem (4.2) has infinitely many weak solutions.*

Next, we consider the problem

$$\begin{cases} -\mathrm{div}(\log(1+|\nabla u(x)|^q)|\nabla u(x)|^{p-2}\nabla u(x)) = \\ \lambda|u(x)|^{p-2}u(x) - |u(x)|^{r-2}u(x), \quad \text{if } x \in \Omega \\ u(x) = 0, \quad \text{if } x \in \partial\Omega. \end{cases} \tag{4.3}$$

We say that $u \in W_0^1 L_\Phi(\Omega)$ is a *weak solution* of problem (4.3) if

$$\int_\Omega \log(1+|\nabla u(x)|^q)|\nabla u(x)|^{p-2}\nabla u \nabla v \, dx - \lambda \int_\Omega |u(x)|^{p-2}u(x)v(x)\, dx + \int_\Omega |u(x)|^{r-2}u(x)v(x)\, dx = 0$$

for all $v \in W_0^1 L_\Phi(\Omega)$.

Theorem 24 *Assume that the hypotheses of Theorem 23 are fulfilled. Then there exists $\lambda_\star > 0$ such that for any $\lambda \geq \lambda_\star$, problem (4.3) has a nontrivial weak solution.*

A careful analysis of the proofs shows that Theorems 23 and 24 remain valid for more general classes of differential operators. Indeed, we can replace $\mathrm{div}(\log(1+|\nabla u(x)|^q)|\nabla u(x)|^{p-2}\nabla u(x))$ by $\mathrm{div}(a(|\nabla u(x)|)\nabla u(x))$, where $a(t)$ is such that the assumption

$$1 < \liminf_{t\to\infty} \frac{t\phi(t)}{\Phi(t)} \leq \sup_{t>0} \frac{t\phi(t)}{\Phi(t)} < \infty$$

is fulfilled. Some examples of potentials $a(t)$ satisfying this hypothesis are provided in Table 4.1.

We remark that if $a(t) = 1$ and $f(u) = -\lambda u + |u|^{r-2}u$, then problem (4.2) becomes

$$\begin{cases} -\Delta u = -\lambda u + |u|^{r-2}u, & \text{in } \Omega \\ u = 0, & \text{on } \partial\Omega\,. \end{cases} \tag{4.4}$$

This problem has been studied by Ambrosetti and Rabinowitz [4] for the

TABLE 4.1: Admissible potentials

Potential	**Values of exponents**
$a(t) = \|t\|^{\alpha-1}$	$\alpha > 0$
$a(t) = \|t\|^{\alpha}/\log(1+\|t\|^{\beta})$	$0 < \beta < \alpha$

case when $2 < r < 2^* = 2N/(N-2)$. Using the mountain pass theorem combined with the fact that the operator $-\Delta + \lambda I$ ($\lambda > 0$) is coercive in $H_0^1(\Omega)$, Ambrosetti and Rabinowitz showed that problem (4.4) has a positive solution for any $\lambda > 0$. Theorem 23 establishes the existence of *infinitely many solutions* (not necessarily positive) for a related class of boundary value problems, but involving another differential operator in the class of Orlicz-Sobolev spaces.

The strong difference between the results of Theorems 23 and 24 can be understood by the following elementary arguments. Indeed, consider the corresponding problems

$$\begin{cases} -\Delta u = -\lambda u + u^{r-1}, & \text{in } \Omega \\ u = 0, & \text{on } \partial\Omega \\ u > 0, & \text{in } \Omega \end{cases} \tag{4.5}$$

and

$$\begin{cases} -\Delta u = \lambda u - u^{r-1}, & \text{in } \Omega \\ u = 0, & \text{on } \partial\Omega \\ u > 0, & \text{in } \Omega\,. \end{cases} \tag{4.6}$$

As we have seen, the mountain pass theorem implies that problem (4.5) has at least one solution for any $\lambda > 0$, provided that $2 < r < 2^* = 2N/(N-2)$. Problem (4.6) corresponds to the case studied in Theorem 24. A simple multiplication with the first eigenfunction $\varphi_1 > 0$ in relation (4.6) implies

$$\lambda_1 \int_\Omega u\varphi_1 dx = \lambda \int_\Omega u\varphi_1 dx - \int_\Omega u^{r-1}\varphi_1 dx\,.$$

Thus, a necessary condition that problem (4.6) has a solution is that λ is sufficiently large. Indeed, under the assumptions of Theorem 23, the nonlinear term $f_1(u) := -\lambda|u|^{p-2}u + |u|^{r-2}u$ satisfies the Ambrosetti-Rabinowitz condition $0 \le r\int_0^t f_1(s)ds \le tf_1(t)$, for all $t \ge 0$. This condition fails if $f_2(u) := \lambda|u|^{p-2}u - |u|^{r-2}u$. However in this case, we show that the corresponding energy functional is coercive and lower semicontinuous.

Problems (4.2) and (4.3) have been formulated for the nonhomogeneous differential operator $\operatorname{div}(a(|\nabla u|)\nabla u)$, where the potential $a(\xi)$ is given by

$$a(t) := t^{p-2}\log(1+t^q), \qquad t \ge 0.$$

Problems of this type arise in nonlinear plasticity. Under suitable assumptions, related results can be formulated for the potentials

$$a(t) := (1+t^2)^\gamma - 1, \qquad t \geq 0,\, \gamma > \frac{1}{2}$$

and

$$a(t) := \int_0^t s^{1-\alpha}(\sinh^{-1} s)^\beta ds, \qquad t \geq 0,\, \alpha,\, \beta > 0.$$

The corresponding nonlinear partial differential equations describe phenomena in nonlinear elasticity, resp. generalized Newtonian fluids.

4.1.1 Some properties of Orlicz-Sobolev spaces

In many applications of Orlicz-Sobolev spaces to boundary value problems for nonlinear partial differential equations, the compactness of the embeddings plays a central role. Compact embedding theorems for Sobolev or Orlicz-Sobolev spaces are also intimately connected with the problem of discreteness of spectra of Schrödinger operators.

While the Banach spaces $W^1L_\Phi(\Omega)$ and $W_0^1L_\Phi(\Omega)$ can be defined by fairly general convex properties of Φ, these function spaces depend on the growth rate of Φ at infinity.

Define the Orlicz-Sobolev conjugate $\Phi_\star$ of Φ by

$$\Phi_\star^{-1}(t) := \int_0^t \frac{\Phi^{-1}(s)}{s^{\frac{N+1}{N}}}\, ds.$$

Proposition 5 *Assume that the hypotheses of Theorems 23 or 24 are fulfilled. Then the following properties hold:*

(a) $\displaystyle\lim_{t\to 0}\int_t^1 \frac{\Phi^{-1}(s)}{s^{\frac{N+1}{N}}}\, ds < \infty;$

(b) $\displaystyle\lim_{t\to \infty}\int_1^t \frac{\Phi^{-1}(s)}{s^{\frac{N+1}{N}}}\, ds = \infty;$

(c) $\displaystyle\lim_{t\to\infty}\frac{|t|^{\gamma+1}}{\Phi_\star(kt)} = 0$, *for all* $k > 0$ *and all* $1 \leq \gamma < \dfrac{Np - N + p}{N - p}$.

Proof. (a) By L'Hôpital's rule we have

$$\begin{aligned}\lim_{t\searrow 0}\frac{\Phi(t)}{t^{p+q}} &= \lim_{t\searrow 0}\frac{\phi(t)}{(p+q)t^{p+q-1}}\\ &= \frac{1}{p+q}\lim_{t\searrow 0}\frac{\log(1+t^q)}{t^q} = \frac{1}{p+q}\lim_{t\searrow 0}\frac{\frac{qt^{q-1}}{1+t^q}}{qt^{q-1}} = \frac{1}{p+q}.\end{aligned}$$

We deduce that Φ is equivalent to t^{p+q} near zero. Using this fact and the remarks in Adams [2, p. 248], we infer that (a) holds if and only if

$$\lim_{t\to 0}\int_t^1 \frac{s^{\frac{1}{p+q}}}{s^{\frac{N+1}{N}}}\, ds < \infty,$$

or

$$p+q<N.$$

The last inequality holds since the hypotheses of Theorems 23 or 24 are fulfilled.

(b) By the change of variable $s=\Phi(\tau)$ we obtain

$$\int_1^t \frac{\Phi^{-1}(s)}{s^{\frac{N+1}{N}}}\,ds=\int_{\Phi^{-1}(1)}^{\Phi^{-1}(t)}\frac{\tau\phi(\tau)}{\Phi(\tau)}(\Phi(\tau))^{-1/N}\,d\tau. \tag{4.7}$$

By elementary computation we deduce that

$$0\le \lim_{\tau\to\infty}\frac{\displaystyle\int_0^\tau \frac{s^{p+q-1}}{1+s^q}\,ds}{\tau^p\log(1+\tau^q)}\le \lim_{\tau\to\infty}\frac{\displaystyle\int_0^\tau \frac{s^{p+q-1}}{s^q}\,ds}{\tau^p\log(1+\tau^q)}=\lim_{\tau\to\infty}\frac{\frac{1}{p}\tau^p}{\tau^p\log(1+\tau^q)}=0.$$

Thus

$$\lim_{\tau\to\infty}\frac{\displaystyle\int_0^\tau \frac{s^{p+q-1}}{1+s^q}\,ds}{\tau^p\log(1+\tau^q)}=0. \tag{4.8}$$

The first consequence of the above relation is that

$$\lim_{t\to\infty}\frac{\Phi(t)}{t^p\log(1+t^q)}=\frac{1}{p}. \tag{4.9}$$

On the other hand, by (4.8),

$$\begin{aligned}\lim_{\tau\to\infty}\frac{\tau\phi(\tau)}{\Phi(\tau)} &= \lim_{\tau\to\infty}\frac{\tau^p\log(1+\tau^q)}{\frac{1}{p}\tau^p\log(1+\tau^q)-\frac{q}{p}\displaystyle\int_0^\tau\frac{s^{p+q-1}}{1+s^q}\,ds}\\ &= p\lim_{\tau\to\infty}\left(1-q\cdot\frac{\displaystyle\int_0^\tau\frac{s^{p+q-1}}{1+s^q}\,ds}{\tau^p\log(1+\tau^q)}\right)^{-1}=p\end{aligned} \tag{4.10}$$

and

$$\lim_{t\to\infty}\Phi(t)=\lim_{t\to\infty}\frac{1}{p}t^p\log(1+t^q)\left[1-q\cdot\frac{\displaystyle\int_0^{|t|}\frac{s^{p+q-1}}{1+s^q}\,ds}{t^p\log(1+t^q)}\right]=\infty. \tag{4.11}$$

Relations (4.7), (4.10) and (4.11) yield

$$\lim_{t\to\infty}\int_1^t\frac{\Phi^{-1}(s)}{s^{\frac{N+1}{N}}}\,ds=\infty\,.$$

Equivalently, we can write

$$\int_{\Phi^{-1}(1)}^{\infty} \frac{d\tau}{[\Phi(\tau)]^{1/N}} = \infty$$

or by (4.9),

$$\int_{\Phi^{-1}(1)}^{\infty} \frac{d\tau}{\tau^{p/N}[\log(1+\tau^q)]^{1/N}} = \infty. \tag{4.12}$$

Since

$$\log(1+\theta) \le \theta, \quad \forall\, \theta > 0$$

we can deduce that

$$\frac{1}{\tau^{p/N}[\log(1+\tau^q)]^{1/N}} \ge \frac{1}{\tau^{(p+q)/N}}, \quad \forall\, \tau > 0.$$

Since $p+q<N$, we find

$$\int_{\Phi^{-1}(1)}^{\infty} \tau^{-(p+q)/N}\, d\tau = \infty$$

and thus relation (4.12) holds. We conclude that

$$\lim_{t\to\infty} \int_1^t \frac{\Phi^{-1}(s)}{s^{\frac{N+1}{N}}}\, ds = \infty.$$

(c) Let γ be fixed such that $1 \le \gamma < (Np-N+p)/(N-p)$. By Adams [2, p. 231], we have

$$\lim_{t\to\infty} \frac{|t|^{\gamma+1}}{\Phi_\star(kt)} = 0, \quad \forall\, k > 0$$

if and only if

$$\lim_{t\to\infty} \frac{\Phi_\star^{-1}(t)}{t^{1/(\gamma+1)}} = 0. \tag{4.13}$$

Using again L'Hôpital's rule we deduce that

$$\limsup_{t\to\infty} \frac{\Phi_\star^{-1}(t)}{t^{1/(\gamma+1)}} \le (\gamma+1) \limsup_{t\to\infty} \frac{\Phi^{-1}(t)}{t^{\frac{1}{\gamma+1}+\frac{1}{N}}}.$$

Setting $\tau = \Phi(t)$ we obtain

$$\limsup_{t\to\infty} \frac{\Phi_\star^{-1}(t)}{t^{1/(\gamma+1)}} \le (\gamma+1) \limsup_{\tau\to\infty} \frac{\tau}{[\Phi(\tau)]^{\frac{1}{\gamma+1}+\frac{1}{N}}}.$$

Since $\gamma < (Np-N+p)/(N-p)$ we have

$$p > \frac{N(\gamma+1)}{N+\gamma+1}.$$

Using the above inequality and (4.8) we obtain

$$\limsup_{\tau\to\infty} \frac{\tau^{\frac{N(\gamma+1)}{N+\gamma+1}}}{\Phi(\tau)} = 0.$$

We conclude that (c) holds. □

Proposition 5 enables us to apply Theorem 8.33 from Adams [2] in order to obtain that $W_0^1L_\Phi(\Omega)$ is compactly embedded in $L^{\gamma+1}(\Omega)$, provided that $1 \le \gamma < (Np - N + p)/(N - p)$. This is one of the versions of the Rellich-Kondrachov theorem for Orlicz-Sobolev spaces.

In what follows an important role will be played by

$$p^0 := \sup_{t>0} \frac{t\phi(t)}{\Phi(t)}.$$

It follows from Example 2 in [19, p. 243] that

$$p^0 = p + q.$$

4.1.2 Absorption case: existence of infinitely many solutions

The key argument in the proof of Theorem 23 is the following $\mathbb{Z}_2$-symmetric version of the mountain pass theorem.

Symmetric Mountain Pass Lemma. *Let X be an infinite dimensional real Banach space and let $I \in C^1(X, \mathbb{R})$ be even, satisfying the Palais-Smale condition (that is, any sequence $\{x_n\} \subset X$ such that $\{I(x_n)\}$ is bounded and $I'(x_n) \to 0$ in $X^\star$, has a convergent subsequence) and $I(0) = 0$. Suppose that*

(I1) there exist constants ρ, $b > 0$ such that $I(x) \ge b$ if $\|x\| = \rho$;

(I2) for each finite dimensional subspace $X_1 \subset X$, the set $\{x \in X_1;\ I(x) \ge 0\}$ is bounded.

Then I has an unbounded sequence of critical values.

Let E denote the Orlicz-Sobolev space $W_0^1L_\Phi(\Omega)$. Let $\lambda > 0$ be arbitrary but fixed.

The energy functional $J_\lambda : E \to \mathbb{R}$ associated to problem (4.2) is defined by

$$J_\lambda(u) := \int_\Omega \Phi(|\nabla u(x)|)\, dx + \frac{\lambda}{p}\int_\Omega |u(x)|^p\, dx - \frac{1}{r}\int_\Omega |u(x)|^r\, dx.$$

Due to the compact embedding of $W_0^1L_\Phi(\Omega)$ into $L^{\gamma+1}(\Omega)$, J_λ is well defined.

Let $J_{\lambda,1}$, $J_{\lambda,2} : E \to \mathbb{R}$ denote the functionals

$$J_{\lambda,1}(u) := \int_\Omega \Phi(|\nabla u(x)|)\, dx \quad \text{and} \quad J_{\lambda,2}(u) := \frac{\lambda}{p}\int_\Omega |u(x)|^p\, dx - \frac{1}{r}\int_\Omega |u(x)|^r\, dx.$$

Therefore

$$J_\lambda(u) = J_{\lambda,1}(u) + J_{\lambda,2}(u), \quad \forall\, u \in E.$$

Then $J_{\lambda,1}$ is a C^1 functional, with the Gâteaux derivative given by

$$\langle J'_{\lambda,1}(u), v\rangle = \int_\Omega \log(1+|\nabla u(x)|^q)|\nabla u(x)|^{p-2}\nabla u(x)\nabla v(x)\, dx\,,$$

for all u, $v \in E$.

We also have that $J_{\lambda,2}$ is of class C^1 and

$$\langle J'_{\lambda,2}(u), v\rangle = \lambda \int_\Omega |u(x)|^{p-2}u(x)v(x)\, dx - \int_\Omega |u(x)|^{r-2}u(x)v(x)\, dx\,,$$

for all u, $v \in E$.

We deduce that $J_\lambda \in C^1(E,\mathbb{R})$ and

$$\begin{aligned}\langle J'_\lambda(u), v\rangle &= \int_\Omega \log(1+|\nabla u(x)|^q)|\nabla u(x)|^{p-2}\nabla u(x)\nabla v(x)\, dx \\ &+ \lambda \int_\Omega |u(x)|^{p-2}u(x)v(x)\, dx - \int_\Omega |u(x)|^{r-2}u(x)v(x)\, dx\end{aligned}$$

for all u, $v \in E$. Thus, the weak solutions of problem (4.2) coincide with the critical points of J_λ.

Lemma 25 *There exist $\eta > 0$ and $\alpha > 0$ such that $J_\lambda(u) \geq \alpha > 0$ for any $u \in E$ with $\|u\| = \eta$.*

Proof. We first show that

$$\Phi(t) \geq \tau^{p^0}\Phi(t/\tau), \quad \forall\, t > 0 \text{ and } \tau \in (0,1]\,. \tag{4.14}$$

Indeed, since

$$p^0 = \sup_{t>0} \frac{t\phi(t)}{\Phi(t)}$$

we have

$$\frac{t\phi(t)}{\Phi(t)} \leq p^0, \quad \forall\, t > 0.$$

Let $\tau \in (0,1]$ be fixed. We have

$$\log(\Phi(t/\tau)) - \log(\Phi(t)) = \int_t^{t/\tau} \frac{\phi(s)}{\Phi(s)}\, ds \leq \int_t^{t/\tau} \frac{p^0}{s}\, ds = \log(\tau^{-p^0})$$

and it follows that (4.14) holds.

Fix $u \in E$ with $\|u\| < 1$ and $\xi \in (0, \|u\|)$. Using relation (4.14) we have

$$\int_\Omega \Phi(|\nabla u(x)|)\, dx \geq \xi^{p^0} \int_\Omega \Phi\left(\frac{|\nabla u(x)|}{\xi}\right)\, dx. \tag{4.15}$$

Defining $v(x) = |\nabla u(x)|/\xi$, for all $x \in \Omega$, we have $\|v\|_\Phi = \|u\|/\xi > 1$. Since $\Phi(t) \leq \frac{t\phi(t)}{p}$, for all $t \in \mathbb{R}$, we deduce by Lemma C.9 from [19] that

$$\int_\Omega \Phi(v(x))\, dx \geq \|v\|_\Phi^p > 1. \tag{4.16}$$

Relations (4.15) and (4.16) show that

$$\int_\Omega \Phi(|\nabla u(x)|)\, dx \geq \xi^{p^0}.$$

Letting $\xi \nearrow \|u\|$ in the above inequality we obtain

$$\int_\Omega \Phi(|\nabla u(x)|)\, dx \geq \|u\|^{p^0}, \quad \forall\, u \in E \text{ with } \|u\| < 1. \tag{4.17}$$

On the other hand, since E is continuously embedded in $L^r(\Omega)$, it follows that there exists a positive constant $C_1 > 0$ such that

$$\int_\Omega |u(x)|^r\, dx \leq C_1 \cdot \|u\|^r, \quad \forall\, u \in E. \tag{4.18}$$

Using relations (4.17) and (4.18), we deduce that for all $u \in E$ with $\|u\| \leq 1$ we have

$$\begin{aligned} J_\lambda(u) &\geq \int_\Omega \Phi(|\nabla u(u)|)\, dx - \frac{1}{r}\int_\Omega |u(x)|^r\, dx \\ &\geq \|u\|^{p^0} - \frac{C_1}{r} \cdot \|u\|^r \\ &= \left(1 - \frac{C_1}{r} \cdot \|u\|^{r-p^0}\right) \|u\|^{p^0}. \end{aligned}$$

Using the hypotheses of Theorem 23, we have $p^0 = p + q < r$. We conclude that Lemma 25 holds. □

Lemma 26 *Assume that E_1 is a finite dimensional subspace of E. Then the set $S = \{u \in E_1;\ J_\lambda(u) \geq 0\}$ is bounded.*

Proof. With the same arguments as those used in the proof of relation (4.14) we have

$$\frac{\Phi(\sigma t)}{\Phi(t)} \leq \sigma^{p^0}, \quad \forall\, t > 0 \text{ and } \sigma > 1. \tag{4.19}$$

Then, for all $u \in E$ with $\|u\| > 1$, relation (4.19) implies

$$\begin{aligned} \int_\Omega \Phi(|\nabla u(x)|)\, dx &= \int_\Omega \Phi\left(\|u\| \frac{|\nabla u(x)|}{\|u\|}\right) dx \\ &\leq \|u\|^{p^0} \int_\Omega \Phi\left(\frac{|\nabla u(x)|}{\|u\|}\right) dx \\ &\leq \|u\|^{p^0}. \end{aligned} \tag{4.20}$$

On the other hand, since E is continuously embedded in $L^p(\Omega)$, it follows that there exists a positive constant $C_2 > 0$ such that

$$\int_\Omega |u(x)|^p \, dx \leq C_2 \cdot \|u\|^p, \quad \forall \, u \in E. \tag{4.21}$$

Relations (4.20) and (4.21) yield

$$J_\lambda(u) \leq \|u\|^{p^0} + \frac{\lambda}{p} \cdot C_2 \cdot \|u\|^p - \frac{1}{r} \int_\Omega |u(x)|^r \, dx \,, \tag{4.22}$$

for all $u \in E$ with $\|u\| > 1$.

We point out that the functional $|\cdot|_r : E \to \mathbb{R}$ defined by

$$|u|_r = \left(\int_\Omega |u(x)|^r \, dx \right)^{1/r}$$

is a norm in E. In the finite dimensional subspace E_1 the norms $|.|_r$ and $\|.\|$ are equivalent, so there exists a positive constant $C_3 = C_3(E_1)$ such that

$$\|u\| \leq C_3 \cdot |u|_r, \quad \forall \, u \in E_1.$$

The above remark and relation (4.22) imply

$$J_\lambda(u) \leq \|u\|^{p^0} + \frac{\lambda}{p} \cdot C_2 \cdot \|u\|^p - \frac{1}{r} \cdot C_3^{-1} \cdot \|u\|^r \,,$$

for all $u \in E_1$ with $\|u\| > 1$.

Hence

$$\|u\|^{p^0} + \frac{\lambda}{p} \cdot C_2 \cdot \|u\|^p - \frac{1}{r} \cdot C_3^{-1} \cdot \|u\|^r \geq 0 \,, \tag{4.23}$$

for all $u \in S$ with $\|u\| > 1$. Since by our hypotheses we have $r > p^0 > p$, the above relation implies that S is bounded in E. □

Lemma 27 *Assume that $\{u_n\} \subset E$ is a sequence which satisfies the properties*

$$|J_\lambda(u_n)| < M \tag{4.24}$$

$$J'_\lambda(u_n) \to 0 \quad \text{as } n \to \infty \,, \tag{4.25}$$

where M is a positive constant. Then $\{u_n\}$ possesses a convergent subsequence.

Proof. We first show that $\{u_n\}$ is bounded in E. Assume to the contrary. Then passing eventually to a subsequence, still denoted by $\{u_n\}$, we may assume that $\|u_n\| \to \infty$ as $n \to \infty$. Thus we may suppose that $\|u_n\| > 1$ for any integer n.

By (4.25) we deduce that there exists $N_1 > 0$ such that for any $n > N_1$ we have

$$\|J'_\lambda(u_n)\| \leq 1.$$

On the other hand, for any $n > N_1$ fixed, the application

$$E \ni v \to \langle J'_\lambda(u_n), v\rangle$$

is linear and continuous. Therefore

$$|\langle J'_\lambda(u_n), v\rangle| \leq \|J'_\lambda(u_n)\| \cdot \|v\| \leq \|v\|, \quad \forall v \in E,\ n > N_1.$$

Setting $v = u_n$ we have

$$-\|u_n\| \leq \int_\Omega \log(1 + |\nabla u_n(u)|^q)|\nabla u_n(x)|^p\, dx + \lambda \int_\Omega |u_n(x)|^p\, dx - \int_\Omega |u_n(x)|^r\, dx \leq \|u_n\|,$$

for all $n > N_1$. We obtain

$$-\|u_n\| - \int_\Omega \log(1 + |\nabla u_n(u)|^q)|\nabla u_n(x)|^p\, dx - \lambda \int_\Omega |u_n(x)|^p\, dx \leq - \int_\Omega |u_n(x)|^r\, dx, \tag{4.26}$$

for any $n > N_1$.

If $\|u_n\| > 1$, then relations (4.24) and (4.26) imply

$$\begin{aligned} M > J_\lambda(u_n) &= \int_\Omega \Phi(|\nabla u_n(x)|)\, dx + \frac{\lambda}{p}\int_\Omega |u_n(x)|^p\, dx - \frac{1}{r}\int_\Omega |u_n(x)|^r\, dx \\ &\geq \int_\Omega \Phi(|\nabla u_n(x)|)\, dx + \lambda \cdot \left(\frac{1}{p} - \frac{1}{r}\right) \cdot \int_\Omega |u_n(x)|^p\, dx - \\ &\quad \frac{1}{r} \cdot \int_\Omega \log(1 + |\nabla u_n(u)|^q)|\nabla u_n(x)|^p\, dx - \frac{1}{r} \cdot \|u_n\| \\ &= \int_\Omega \Phi(|\nabla u_n(x)|)\, dx - \frac{1}{r} \cdot \int_\Omega \phi(|\nabla u_n(x)|)|\nabla u_n(x)|\, dx + \\ &\quad \lambda \cdot \left(\frac{1}{p} - \frac{1}{r}\right) \cdot \int_\Omega |u_n(x)|^p\, dx - \frac{1}{r} \cdot \|u_n\|. \end{aligned}$$

Since

$$p^0 \geq \frac{t\phi(t)}{\Phi(t)}, \quad \forall\, t > 0$$

we find

$$\int_\Omega \Phi(|\nabla u_n(x)|)\, dx - \frac{1}{r} \cdot \int_\Omega \phi(|\nabla u_n(x)|)|\nabla u_n(x)|\, dx \geq \left(1 - \frac{p^0}{r}\right) \int_\Omega \Phi(|\nabla u_n(x)|)\, dx.$$

Using the above relations we deduce that for any $n > N_1$ such that $\|u_n\| > 1$ we have

$$M > \left(1 - \frac{p^0}{r}\right) \cdot \int_\Omega \Phi(|\nabla u_n(x)|)\, dx - \frac{1}{r} \cdot \|u_n\|. \tag{4.27}$$

Since $\Phi(t) \leq (t\phi(t))/p$ for all $t \in \mathbb{R}$ we deduce by Lemma C.9 in [19] that

$$\int_\Omega \Phi(|\nabla u_n(x)|)\, dx \geq \|u_n\|^p, \tag{4.28}$$

for all $n > N_1$ with $\|u_n\| > 1$.

Relations (4.27) and (4.28) imply

$$M > \left(1 - \frac{p^0}{r}\right) \cdot \|u_n\|^p - \frac{1}{r} \cdot \|u_n\|,$$

for all $n > N_1$ with $\|u_n\| > 1$. Since $p^0 < r$, letting $n \to \infty$ we obtain a contradiction. It follows that $\{u_n\}$ is bounded in E.

Since $\{u_n\}$ is bounded in E, we deduce that there exists a subsequence, still denoted by $\{u_n\}$, and $u_0 \in E$ such that $\{u_n\}$ weakly converges to u_0 in E. Since E is compactly embedded in $L^p(\Omega)$ and $L^r(\Omega)$ it follows that $\{u_n\}$ strongly converges to u_0 in $L^p(\Omega)$ and $L^r(\Omega)$. Hence

$$\lim_{n\to\infty} J_{\lambda,2}(u_n) = J_{\lambda,2}(u_0) \quad \text{and} \quad \lim_{n\to\infty} J'_{\lambda,2}(u_n) = J'_{\lambda,2}(u_0). \tag{4.29}$$

Since

$$J_{\lambda,1}(u) = J_\lambda(u) - J_{\lambda,2}(u), \quad \forall\, u \in E\,,$$

relations (4.29) and (4.25) imply

$$\lim_{n\to\infty} J'_{\lambda,1}(u_n) = -J'_{\lambda,2}(u_0), \quad \text{in } E^\star. \tag{4.30}$$

Using the fact that Φ is convex and thus $J_{\lambda,1}$ is also convex, we have

$$J_{\lambda,1}(u_n) \leq J_{\lambda,1}(u_0) + \langle J'_{\lambda,1}(u_n), u_n - u_0\rangle.$$

Passing to the limit as $n \to \infty$ and using (4.30) we deduce that

$$\limsup_{n\to\infty} J_{\lambda,1}(u_n) \leq J_{\lambda,1}(u_0). \tag{4.31}$$

Using again the fact that $J_{\lambda,1}$ is convex, it follows that $J_{\lambda,1}$ is weakly lower semicontinuous and hence

$$\liminf_{n\to\infty} J_{\lambda,1}(u_n) \geq J_{\lambda,1}(u_0). \tag{4.32}$$

By (4.31) and (4.32) we find

$$\lim_{n\to\infty} J_{\lambda,1}(u_n) = J_{\lambda,1}(u_0)$$

or

$$\lim_{n\to\infty} \int_\Omega \Phi(|\nabla u_n(x)|)\, dx = \int_\Omega \Phi(|\nabla u_0(x)|)\, dx\,. \tag{4.33}$$

Since Φ is increasing and convex, it follows that

$$\begin{aligned}\Phi\left(\frac{1}{2}|\nabla u_n(x)-\nabla u_0(x)|\right) &\leq \Phi\left(\frac{1}{2}\left(|\nabla u_n(x)|+|\nabla u_0(x)|\right)\right)\\ &\leq \frac{\Phi(|\nabla u_n(x)|)+\Phi(|\nabla u_0(x)|)}{2},\end{aligned}$$

for all $x\in\Omega$ and all n. Integrating the above inequalities over Ω we find

$$0\leq\int_\Omega\Phi\left(\frac{1}{2}|\nabla(u_n-u_0)(x)|\right)\,dx\leq\frac{\int_\Omega\Phi(|\nabla u_n(x)|)\,dx+\int_\Omega\Phi(|\nabla u_0(x)|)\,dx}{2},$$

for all n.

We point out that Lemma C.9 from [19] implies

$$\int_\Omega\Phi(|\nabla u_n(x)|)\,dx\leq\|u_n\|^p<1,\quad\text{provided that }\|u_n\|<1\,,$$

while relation (4.19) yields

$$\int_\Omega\Phi(|\nabla u_n(x)|)\,dx\leq\|u_n\|^{p^0},\quad\text{provided that }\|u_n\|>1.$$

Since $\{u_n\}$ is bounded in E, the above inequalities prove the existence of a positive constant K_1 such that

$$\int_\Omega\Phi(|\nabla u_n(x)|)\,dx\leq K_1,$$

for all n. So there exists a positive constant K_2 such that

$$0\leq\int_\Omega\Phi\left(\frac{1}{2}|\nabla(u_n-u_0)(x)|\right)\,dx\leq K_2,\tag{4.34}$$

for all n.

On the other hand, since $\{u_n\}$ weakly converges to u_0 in E, we have

$$\int_\Omega\frac{\partial u_n}{\partial x_i}v\,dx\to\int_\Omega\frac{\partial u_0}{\partial x_i}v\,dx,\quad\forall\,v\in L_{\Phi^\star}(\Omega),\ \forall\,i=1,\ldots,N.$$

In particular this holds for all $v\in L^\infty(\Omega)$. Hence $\{\frac{\partial u_n}{\partial x_i}\}$ weakly converges to $\frac{\partial u_0}{\partial x_i}$ in $L^1(\Omega)$ for all $i=1,\ldots,N$. Thus we deduce that

$$\nabla u_n(x)\to\nabla u_0(x)\quad\text{a.e. }x\in\Omega.\tag{4.35}$$

Relations (4.34) and (4.35) and the Lebesgue dominated convergence theorem imply

$$\lim_{n\to\infty}\int_\Omega\Phi\left(\frac{1}{2}|\nabla(u_n-u_0)(x)|\right)\,dx=0.$$

Thus by Adams [2, p. 236],

$$\lim_{n\to\infty}\left\|\frac{1}{2}(u_n-u_0)\right\|=0$$

and hence

$$\lim_{n\to\infty}\|(u_n-u_0)\|=0.$$

The proof of Lemma 27 is complete. □

We now return to the proof of Theorem 23. We first observe that J_λ is even and $J_\lambda(0)=0$. Lemma 27 implies that J_λ satisfies the Palais-Smale condition. On the other hand, Lemmas 25 and 26 show that conditions (I1) and (I2) in the symmetric mountain pass lemma are satisfied. We conclude that problem (4.2) has infinitely many weak solutions in E. □

We point out that the Orlicz-Sobolev space E cannot be replaced by a classical Sobolev space since in this case condition (I1) in the symmetric mountain pass lemma is not satisfied.

4.1.3 Desorption case: existence of solutions for large parameters

Fix $\lambda>0$. Let $I_\lambda : E\to\mathbb{R}$ be defined by

$$I_\lambda(u):=\int_\Omega \Phi(|\nabla u(x)|)\,dx-\frac{\lambda}{p}\int_\Omega |u(x)|^p\,dx+\frac{1}{r}\int_\Omega |u(x)|^r\,dx.$$

Then I_λ is well-defined and $I_\lambda\in C^1(E,\mathbb{R})$ with the Gâteaux derivative given by

$$\begin{aligned}\langle I_\lambda'(u),v\rangle &= \int_\Omega \log(1+|\nabla u(x)|^q)|\nabla u(x)|^{p-2}\nabla u(x)\nabla v(x)\,dx\\ &- \lambda\int_\Omega |u(x)|^{p-2}u(x)v(x)\,dx+\int_\Omega |u(x)|^{r-2}u(x)v(x)\,dx\,,\end{aligned}$$

for all $u,\,v\in E$.

In what follows, we show that I_λ possesses a nontrivial global minimum point in E.

Lemma 28 *The functional I_λ is coercive on E.*

Proof. We first show that for any $b,\,d>0$ and all $0<k<l$, the following inequality holds:

$$b\cdot t^k-d\cdot t^l\le b\cdot\left(\frac{b}{d}\right)^{k/(l-k)}\,,\quad \forall\, t\ge 0. \tag{4.36}$$

Indeed, since the function

$$[0,\infty)\ni t\to t^\theta$$

is increasing for any $\theta > 0$, it follows that

$$b - d \cdot t^{l-k} < 0, \quad \forall\, t > \left(\frac{b}{d}\right)^{1/(l-k)},$$

and

$$t^k \cdot (b - d \cdot t^{l-k}) \leq b \cdot t^k < b \cdot \left(\frac{b}{d}\right)^{k/(l-k)}, \quad \forall\, t \in \left[0, \left(\frac{b}{d}\right)^{1/(l-k)}\right].$$

The above two inequalities show that (4.36) holds.

Using (4.36) we deduce that for any $x \in \Omega$ and $u \in E$ we have

$$\frac{\lambda}{p} \cdot |u(x)|^p - \frac{1}{r} \cdot |u(x)|^r \leq \frac{\lambda}{p} \cdot \left[\frac{\lambda \cdot r}{p}\right]^{(p/(r-p))} = D_1,$$

where D_1 is a positive constant independent of u and x. Integrating the above inequality we find

$$\frac{\lambda}{p} \int_\Omega |u(x)|^p \, dx - \frac{1}{r} \int_\Omega |u(x)|^r \, dx \leq D_2, \quad \forall\, u \in E, \tag{4.37}$$

where D_2 is a positive constant independent of u.

Using inequalities (4.28) and (4.37) we obtain that for any $u \in E$ with $\|u\| > 1$ we have

$$I_\lambda(u) \geq \|u\|^p - D_2.$$

Thus I_λ is coercive and the proof of Lemma 28 is complete. □

In what follows, we provide the proof of Theorem 24. First, we prove that I_λ is weakly lower semicontinuous on E. Indeed, using the definitions of $J_{\lambda,1}$ and $J_{\lambda,2}$ introduced in the previous section we get

$$I_\lambda(u) = J_{\lambda,1}(u) - J_{\lambda,2}(u), \quad \forall\, u \in E.$$

Since Φ is convex it is clear that $J_{\lambda,1}$ is convex and thus weakly lower semicontinuous on E. As we have already remarked, the functional $J_{\lambda,2}$ is also lower semicontinuous on E. Thus, we obtain that I_λ is weakly lower semicontinuous on E.

By Lemma 28 we deduce that I_λ is coercive on E. Thus there exists $u_\lambda \in E$ a global minimizer of I_λ and thus a weak solution of problem (4.3).

We show that u_λ is not trivial for large enough λ. Indeed, letting $t_0 > 1$ be a fixed real and Ω_1 be an open subset of Ω with $|\Omega_1| > 0$, we deduce that there exists $u_1 \in C_0^\infty(\Omega) \subset E$ such that $u_1(x) = t_0$ for any $x \in \overline{\Omega}_1$ and $0 \leq u_1(x) \leq t_0$ in $\Omega \setminus \Omega_1$. We have

$$\begin{aligned} I_\lambda(u_1) &= \int_\Omega \Phi(|\nabla u_1(x)|)\, dx - \frac{\lambda}{p} \int_\Omega |u_1(x)|^p \, dx + \frac{1}{r} \int_\Omega |u_1(x)|^r \, dx \\ &\leq L - \frac{\lambda}{p} \int_{\Omega_1} |u_1(x)|^p \, dx \\ &\leq L - \frac{\lambda}{p} \cdot t_0^p \cdot |\Omega_1|, \end{aligned}$$

where L is a positive constant. Thus, there exists $\lambda_\star > 0$ such that $I_\lambda(u_1) < 0$ for any $\lambda \in [\lambda_\star, \infty)$. It follows that $I_\lambda(u_\lambda) < 0$ for any $\lambda \geq \lambda_\star$ and thus u_λ is a nontrivial weak solution of problem (4.3) for large enough λ. The proof of Theorem 24 is complete. □

4.2 A continuous spectrum for nonhomogeneous operators

Let Ω be a bounded domain in $\mathbb{R}^N$ $(N \geq 3)$ with smooth boundary $\partial\Omega$. In this section, we are concerned with the following eigenvalue problem:

$$\begin{cases} -\operatorname{div}(a(|\nabla u|)\nabla u) = \lambda |u|^{q(x)-2}u, & \text{for} \quad x \in \Omega \\ u = 0, & \text{for} \quad x \in \partial\Omega\,. \end{cases} \tag{4.38}$$

Our first purpose is to establish a sufficient condition for the existence of a *continuous spectrum* that concentrates near the origin. Next, we consider the case when *every* $\lambda > 0$ is an eigenvalue of problem (4.38). In such a case, we prove the existence of a sequence of *low-energy solutions*.

We assume that the function $a : (0, \infty) \to \mathbb{R}$ is such that the mapping $\phi : \mathbb{R} \to \mathbb{R}$ defined by

$$\phi(t) = \begin{cases} a(|t|)t, & \text{for} \quad t \neq 0 \\ 0, & \text{for} \quad t = 0\,, \end{cases}$$

is an odd, increasing homeomorphism from $\mathbb{R}$ onto $\mathbb{R}$. We also suppose that $\lambda > 0$ and $q : \overline{\Omega} \to (0, \infty)$ is a continuous function.

Define

$$\Phi(t) = \int_0^t \phi(s)\, ds, \quad \Phi^\star(t) = \int_0^t \phi^{-1}(s)\, ds, \qquad \text{for all } t \in \mathbb{R}.$$

We recall that Φ is a *Young function*, that is, $\Phi(0) = 0$, Φ is convex, and $\lim_{x\to\infty} \Phi(x) = +\infty$. Furthermore, since $\Phi(x) = 0$ if and only if $x = 0$, $\lim_{x\to 0} \Phi(x)/x = 0$, and $\lim_{x\to\infty} \Phi(x)/x = +\infty$, then Φ is called an N–*function*. The function $\Phi^\star$ is called the *complementary* function of Φ and it satisfies

$$\Phi^\star(t) = \sup\{st - \Phi(s);\ s \geq 0\}, \qquad \text{for all } t \geq 0\,.$$

We observe that $\Phi^\star$ is also an N–function and the following Young-type inequality holds:

$$st \leq \Phi(s) + \Phi^\star(t), \qquad \text{for all } s, t \geq 0\,.$$

The Orlicz space $L_\Phi(\Omega)$ defined by the N–function Φ is the space of measurable functions $u : \Omega \to \mathbb{R}$ such that

$$\|u\|_{L_\Phi} := \sup\left\{ \int_\Omega uv\, dx;\ \int_\Omega \Phi^\star(|v|)\, dx \leq 1 \right\} < \infty\,.$$

Then $(L_\Phi(\Omega), \|\cdot\|_{L_\Phi})$ is a Banach space whose norm is equivalent to the Luxemburg norm

$$\|u\|_\Phi := \inf\left\{k > 0;\ \int_\Omega \Phi\left(\frac{u(x)}{k}\right)\,dx \le 1\right\}.$$

We denote by $W_0^1 L_\Phi(\Omega)$ the corresponding Orlicz-Sobolev space for problem (4.38), equipped with the norm

$$\|u\| = \||\nabla u|\|_\Phi.$$

In this section, we assume that Φ and $\Phi^\star$ satisfy the Δ_2-condition (at infinity), namely

$$1 < \liminf_{t\to\infty} \frac{t\phi(t)}{\Phi(t)} \le \limsup_{t>0} \frac{t\phi(t)}{\Phi(t)} < \infty.$$

Then $L_\Phi(\Omega)$ and $W_0^1 L_\Phi(\Omega)$ are reflexive Banach spaces.

The Orlicz-Sobolev conjugate of Φ is defined by

$$\Phi_\star^{-1}(t) = \int_0^t \frac{\Phi^{-1}(s)}{s^{(N+1)/N}}\,ds.$$

We assume that

$$\lim_{t\to 0}\int_t^1 \frac{\Phi^{-1}(s)}{s^{(N+1)/N}}\,ds < \infty, \quad \text{and} \quad \lim_{t\to\infty}\int_1^t \frac{\Phi^{-1}(s)}{s^{(N+1)/N}}\,ds = \infty. \tag{4.39}$$

Finally, we define

$$p_0 := \inf_{t>0} \frac{t\phi(t)}{\Phi(t)} \quad \text{and} \quad p^0 := \sup_{t>0} \frac{t\phi(t)}{\Phi(t)}.$$

We say that $\lambda \in \mathbb{R}$ is an eigenvalue of problem (4.38) if there exists $u \in W_0^1 L_\Phi(\Omega) \setminus \{0\}$ such that

$$\int_\Omega a(|\nabla u|)\nabla u \nabla v\,dx - \lambda \int_\Omega |u|^{q(x)-2} uv\,dx = 0,$$

for all $v \in W_0^1 L_\Phi(\Omega)$. We point out that if λ is an eigenvalue of problem (4.38) then the corresponding $u \in W_0^1 L_\Phi(\Omega)\setminus\{0\}$ is a weak solution of (4.38), called an eigenvector of equation (4.38) corresponding to the eigenvalue λ.

The first main result in this section shows that, under certain circumstances, any positive and sufficiently small λ is an eigenvalue of (4.38).

Theorem 25 *Assume that relation* (4.39) *is fulfilled and furthermore that*

$$1 < \inf_{x\in\Omega} q(x) < p_0\,, \tag{4.40}$$

and

$$\lim_{t\to\infty} \frac{|t|^{q^+}}{\Phi_\star(kt)} = 0, \quad \text{for all } k > 0. \tag{4.41}$$

Then there exists $\lambda^\star > 0$ *such that any* $\lambda \in (0, \lambda^\star)$ *is an eigenvalue of problem* (4.38).

The above result implies

$$\inf_{u \in W_0^1 L_\Phi(\Omega)\setminus\{0\}} \frac{\displaystyle\int_\Omega \Phi(|\nabla u|)\, dx}{\displaystyle\int_\Omega |u|^{q(x)}\, dx} = 0.$$

The second result of this section asserts that, in certain cases, the set of eigenvalues may coincide with the *entire* positive semi-axis.

Theorem 26 *Assume that relations* (4.39) *and* (4.41) *are fulfilled and furthermore*

$$\sup_{x\in\Omega} q(x) < p_0. \tag{4.42}$$

Then every $\lambda > 0$ *is an eigenvalue of problem* (4.38). *Moreover, for any* $\lambda > 0$ *there exists a sequence of eigenvectors* $\{u_n\} \subset E$ *such that* $\lim_{n\to\infty} u_n = 0$ *in* $W_0^1 L_\Phi(\Omega)$.

Remark 4 *Relations* (4.39) *and* (4.41) *enable us to apply Theorem 8.33 from Adams [2] in order to obtain that* $W_0^1 L_\Phi(\Omega)$ *is compactly embedded in* $L^{q^+}(\Omega)$. *This fact combined with the continuous embedding of* $L^{q^+}(\Omega)$ *in* $L^{q(x)}(\Omega)$ *ensures that* $W_0^1 L_\Phi(\Omega)$ *is compactly embedded in* $L^{q(x)}(\Omega)$.

Remark 5 *The assertion of Theorems 25 and 26 still remains valid if we replace hypothesis* (4.41) *in Theorems 25 and 26 by the following assumption:*

$$N < p_0 < \liminf_{t\to\infty} \frac{\log(\Phi(t))}{\log(t)}. \tag{4.43}$$

Indeed, using Lemma D.2 from [19], it follows that $W_0^1 L_\Phi(\Omega)$ *is continuously embedded in* $W_0^{1,p_0}(\Omega)$. *On the other hand, since we assume* $p_0 > N$, *we deduce that* $W_0^{1,p_0}(\Omega)$ *is compactly embedded in* $C(\overline{\Omega})$. *Thus we obtain that* $W_0^1 L_\Phi(\Omega)$ *is compactly embedded in* $C(\overline{\Omega})$. *Since* Ω *is bounded it follows that* $W_0^1 L_\Phi(\Omega)$ *is continuously embedded in* $L^{q(x)}(\Omega)$.

4.2.1 Case of small positive parameters

Let E denote the Orlicz-Sobolev space $W_0^1 L_\Phi(\Omega)$.

For any $\lambda > 0$ the energy functional $J_\lambda : E \to \mathbb{R}$ corresponding to problem (4.38) is defined by

$$J_\lambda(u) = \int_\Omega \Phi(|\nabla u|)\, dx - \lambda \int_\Omega \frac{1}{q(x)} |u|^{q(x)}\, dx.$$

Then $J_\lambda \in C^1(E, \mathbb{R})$ and

$$\langle J'_\lambda(u), v\rangle = \int_\Omega a(|\nabla u|)\nabla u \nabla v\, dx - \lambda \int_\Omega |u|^{q(x)-2} uv\, dx,$$

for all $u,\ v \in E$. Thus the weak solutions of problem (4.38) coincide with the critical points of J_λ. If such a weak solution exists and is nontrivial then the corresponding λ is an eigenvalue of problem (4.38).

The following result asserts that J_λ satisfies one of the geometric hypotheses (existence of a "mountain") of the mountain pass theorem, provided that λ is small enough.

Lemma 29 *There exists $\lambda^\star > 0$ such that for any $\lambda \in (0, \lambda^\star)$ there are ρ and $\alpha > 0$ such that $J_\lambda(u) \geq \alpha > 0$ for any $u \in E$ with $\|u\| = \rho$.*

Proof. By the definition of p^0 and since $\frac{\mathrm{d}}{\mathrm{d}\tau}\left(\tau^{p_0}\Phi(t/\tau)\right) \geq 0$ we obtain

$$\Phi(t) \geq \tau^{p^0}\Phi(t/\tau), \quad \forall\, t > 0 \text{ and } \tau \in (0,1]\,.$$

Therefore

$$\int_\Omega \Phi(|\nabla u(x)|)\, dx \geq \|u\|^{p^0}, \quad \forall\, u \in E \text{ with } \|u\| < 1. \tag{4.44}$$

On the other hand, since E is continuously embedded in $L^{q(x)}(\Omega)$, there exists a positive constant c_1 such that

$$|u|_{q(x)} \leq c_1 \|u\|, \quad \forall\, u \in E. \tag{4.45}$$

We fix $\rho \in (0,1)$ such that $\rho < 1/c_1$. Then relation (4.45) implies

$$|u|_{q(x)} < 1, \quad \forall\, u \in E, \text{ with } \|u\| = \rho. \tag{4.46}$$

Furthermore, we have

$$\int_\Omega |u|^{q(x)}\, dx \leq |u|^{q^-}_{q(x)}, \quad \forall\, u \in E, \text{ with } \|u\| = \rho. \tag{4.47}$$

Relations (4.45) and (4.47) imply

$$\int_\Omega |u|^{q(x)}\, dx \leq c_1^{q^-} \|u\|^{q^-}, \quad \forall\, u \in E, \text{ with } \|u\| = \rho. \tag{4.48}$$

Taking into account relations (4.44) and (4.48), we deduce that for any $u \in E$ with $\|u\| = \rho$ the following inequalities hold:

$$J_\lambda(u) \geq \|u\|^{p^0} - \frac{\lambda}{q^-} \int_\Omega |u|^{q(x)}\, dx = \rho^{q^-} \left(\rho^{p^0 - q^-} - \frac{\lambda}{q^-} c_1^{q^-} \right).$$

We point out that by relation (4.40) and the definition of p^0 we have $q^- < l \leq p^0$. By the above inequality we observe that if we define

$$\lambda^\star = \frac{\rho^{p^0 - q^-}}{2} \cdot \frac{q^-}{c_1^{q^-}} \tag{4.49}$$

then for any $\lambda \in (0, \lambda^\star)$ and any $u \in E$ with $\|u\| = \rho$ there exists $\alpha = \frac{\rho^{p^0}}{2} > 0$ such that

$$J_\lambda(u) \geq \alpha > 0.$$

The proof of Lemma 29 is complete. □

The next property shows that J_λ has a "valley" near the origin, hence the second geometric hypothesis of the mountain pass theorem is *not* fulfilled.

Lemma 30 *There exists $\phi \in E$ such that $\phi \geq 0$, $\varphi \neq 0$ and $J_\lambda(t\phi) < 0$, for small enough $t > 0$.*

Proof. Assumption (4.40) implies that $q^- < p_0$. Let $\epsilon_0 > 0$ be such that $q^- + \epsilon_0 < p_0$. On the other hand, since $q \in C(\overline{\Omega})$ it follows that there exists an open set $\Omega_0 \subset \Omega$ such that $|q(x) - q^-| < \epsilon_0$ for all $x \in \Omega_0$. Thus we conclude that $q(x) \leq q^- + \epsilon_0 < p_0$ for all $x \in \Omega_0$.

Let $\phi \in C_0^\infty(\Omega)$ be such that $\text{supp}(\phi) \supset \overline{\Omega}_0$, $\phi(x) = 1$ for all $x \in \overline{\Omega}_0$ and $0 \leq \phi \leq 1$ in Ω.

We point out that there exists $t_0 \in (0,1)$ such that for any $t \in (0, t_0)$ we have

$$\|t|\nabla\phi|\|_\Phi = t\|\phi\| < 1.$$

Taking into account all the above information and using Lemma C.9 from [19] we have

$$\begin{aligned}
J_\lambda(t\phi) &= \int_\Omega \Phi(t|\nabla\phi(x)|)\, dx - \lambda \int_\Omega \frac{t^{q(x)}}{q(x)} |\phi|^{q(x)}\, dx \\
&\leq \int_\Omega \Phi(t|\nabla\phi(x)|)\, dx - \frac{\lambda}{q^+} \int_\Omega t^{q(x)} |\phi|^{q(x)}\, dx \\
&\leq \int_\Omega \Phi(t|\nabla\phi(x)|)\, dx - \frac{\lambda}{q^+} \int_{\Omega_0} t^{q(x)} |\phi|^{q(x)}\, dx \\
&\leq t^{p_0} \|\phi\|^{p_0} - \frac{\lambda \cdot t^{q^- + \epsilon_0}}{q^+} |\Omega_0|,
\end{aligned}$$

for any $t \in (0,1)$, where $|\Omega_0|$ denotes the Lebesgue measure of Ω_0. Therefore

$$J_\lambda(t\phi) < 0$$

for $t < \delta^{1/(p_0 - q^- - \epsilon_0)}$, where

$$0 < \delta < \min\left\{ t_0, \frac{\frac{\lambda}{q^+}|\Omega_0|}{\|\phi\|^{p_0}} \right\}.$$

The proof of Lemma 30 is complete. □

Next, we give the proof of Theorem 25.

Let $\lambda^\star > 0$ be defined as in (4.49) and $\lambda \in (0, \lambda^\star)$. By Lemma 29 it follows that on the boundary of the ball in E centered at the origin and of radius ρ, denoted by $B_\rho(0)$, we have

$$\inf_{\partial B_\rho(0)} J_\lambda > 0. \tag{4.50}$$

On the other hand, by Lemma 30, there exists $\phi \in E$ such that $J_\lambda(t\phi) < 0$ for all small enough $t > 0$. Moreover, relations (4.44) and (4.48) imply that for any $u \in B_\rho(0)$ we have

$$J_\lambda(u) \geq \|u\|^{p^0} - \frac{\lambda}{q^-} c_1^{q^-} \|u\|^{q^-}.$$

It follows that

$$-\infty < \underline{c} := \inf_{B_\rho(0)} J_\lambda < 0.$$

We now let $0 < \epsilon < \inf_{\partial B_\rho(0)} J_\lambda - \inf_{B_\rho(0)} J_\lambda$. Applying Ekeland's variational principle to the functional $J_\lambda : \overline{B_\rho(0)} \to \mathbb{R}$, we find $u_\epsilon \in \overline{B_\rho(0)}$ such that

$$\begin{aligned} J_\lambda(u_\epsilon) &< \inf_{\overline{B_\rho(0)}} J_\lambda + \epsilon \\ J_\lambda(u_\epsilon) &< J_\lambda(u) + \epsilon \cdot \|u - u_\epsilon\|, \quad u \neq u_\epsilon. \end{aligned}$$

Since

$$J_\lambda(u_\epsilon) \leq \inf_{\overline{B_\rho(0)}} J_\lambda + \epsilon \leq \inf_{B_\rho(0)} J_\lambda + \epsilon < \inf_{\partial B_\rho(0)} J_\lambda,$$

we deduce that $u_\epsilon \in B_\rho(0)$. Now, we define $I_\lambda : \overline{B_\rho(0)} \to \mathbb{R}$ by $I_\lambda(u) = J_\lambda(u) + \epsilon \cdot \|u - u_\epsilon\|$. It is clear that u_ϵ is a minimum point of I_λ and thus

$$\frac{I_\lambda(u_\epsilon + t \cdot v) - I_\lambda(u_\epsilon)}{t} \geq 0$$

for small $t > 0$ and any $v \in B_1(0)$. The above relation yields

$$\frac{J_\lambda(u_\epsilon + t \cdot v) - J_\lambda(u_\epsilon)}{t} + \epsilon \cdot \|v\| \geq 0.$$

Letting $t \to 0$ it follows that $\langle J_\lambda'(u_\epsilon), v\rangle + \epsilon \cdot \|v\| > 0$ and we infer that $\|J_\lambda'(u_\epsilon)\| \leq \epsilon$.

We deduce that there exists a sequence $\{w_n\} \subset B_\rho(0)$ such that

$$J_\lambda(w_n) \to \underline{c} \quad \text{and} \quad J_\lambda'(w_n) \to 0. \tag{4.51}$$

The sequence $\{w_n\}$ is bounded in E. Thus there exists $w \in E$ such that, up to a subsequence, $\{w_n\}$ weakly converges to w in E. By Remark 5, the function space E is compactly embeddded in $L^{q(x)}(\Omega)$, hence $\{w_n\}$ strongly converges to w in $L^{q(x)}(\Omega)$. Using the Hölder inequality, we obtain

$$\lim_{n\to\infty}\int_\Omega |w_n|^{q(x)}\,dx = \int_\Omega |w|^{q(x)}\,dx \quad\text{and}\quad \lim_{n\to\infty}\int_\Omega |w_n|^{q(x)-2}w_n v\,dx = \int_\Omega |w|^{q(x)-2}wv\,dx$$

for any $v \in E$.

We conclude that w is a nontrivial weak solution of problem (4.38) and thus any $\lambda \in (0,\lambda^\star)$ is an eigenvalue of problem (4.38). Next, we deduce that $\{w_n\}$ strongly converges to w in E. So by (4.51),

$$J_\lambda(w) = \underline{c} < 0 \quad\text{and}\quad J'_\lambda(w) = 0.$$

The proof of Theorem 25 is complete. □

4.2.2 Existence of low-energy solutions

The proof of Theorem 26 relies on a symmetric version of the mountain pass lemma, cf. Kajikia [40].

Definition 1 *Let X be a real Banach space. We say that a subset A of X is symmetric if $u \in A$ implies $-u \in A$. For a closed symmetric set A which does not contain the origin, we define the genus $\gamma(A)$ of A as the smallest integer k such that there exists an odd continuous mapping from A to $\mathbb{R}^k \setminus \{0\}$. If there does not exist such an integer k, we define $\gamma(A) = +\infty$. Moreover, we set $\gamma(\emptyset) = 0$. Finally, we denote by Γ_k the family*

$$\Gamma_k = \{A \subset X;\ 0 \notin A \text{ and } \gamma(A) \geq k\}.$$

We now state the symmetric mountain pass lemma of Kajikia [40].

Theorem 27 *Assume that X is an infinite dimensional Banach space and that $\Lambda \in C^1(X,\mathbb{R})$ satisfies conditions (A1) and (A2) below.*

(A1) $\Lambda(u)$ is even, bounded from below, $\Lambda(0) = 0$ and $\Lambda(u)$ satisfies the Palais-Smale condition (that is, any sequence $\{u_n\}$ in X such that $\{\Lambda(u_n)\}$ is bounded and $\Lambda'(u_n) \to 0$ in $X^\star$ as $n \to \infty$, has a convergent subsequence);

(A2) For each $k \in \mathbb{N}$, there exists $A_k \in \Gamma_k$ such that $\sup_{u\in A_k}\Lambda(u) < 0$.

Under the above assumptions, either (i) or (ii) below holds.

(i) There exists a sequence $\{u_n\}$ such that $\Lambda'(u_n) = 0$, $\Lambda(u_n) < 0$ and $\{u_n\}$ converges to zero;

(ii) There exist sequences $\{u_n\}$ and $\{v_n\}$ such that $\Lambda'(u_n) = 0$, $\Lambda(u_n) = 0$, $u_n \neq 0$, $\lim_{n\to\infty} u_n = 0$, $\Lambda'(v_n) = 0$, $\Lambda(v_n) = 0$, and v_n converges to a non-zero limit.

In order to apply Theorem 27 to the functional J_λ, we prove two auxiliary results.

Lemma 31 *The functional* J_λ *satisfies condition (A1) from Theorem 27.*

Proof. We have $J_\lambda(u) = J_\lambda(-u)$ for any $u \in E$ and $J_\lambda(0) = 0$. On the other hand, by relation (4.44) we have

$$\int_\Omega \Phi(|\nabla u(x)|)\, dx \geq \|u\|^{p^0}, \quad \forall\, u \in E \text{ with } \|u\| < 1,$$

while by Lemma C.9 from [19] we have

$$\int_\Omega \Phi(|\nabla u(x)|)\, dx \geq \|u\|^{p_0}, \quad \forall\, u \in E \text{ with } \|u\| > 1.$$

Therefore

$$\int_\Omega \Phi(|\nabla u(x)|)\, dx \geq \alpha(\|u\|), \quad \forall\, u \in E, \tag{4.52}$$

where $\alpha : [0, \infty) \to \mathbb{R}$, $\alpha(t) = t^{p^0}$ if $t < 0$ and $\alpha(t) = t^{p_0}$ if $t > 1$.

By Remark 4, the space E is continuously embedded in $L^{q^\pm}(\Omega)$. Thus there exist positive constants d_1 and d_2 such that

$$\int_\Omega |u|^{q^+}\, dx \leq d_1 \|u\|^{q^+}, \quad \int_\Omega |u|^{q^-}\, dx \leq d_2 \|u\|^{q^-}, \quad \forall\, u \in E. \tag{4.53}$$

Combining relations (4.52) and (4.53) we obtain

$$J_\lambda(u) \geq \alpha(\|u\|) - \frac{d_1 \lambda}{q^-}\|u\|^{q^+} - \frac{d_2 \lambda}{q^-}\|u\|^{q^-}, \quad \forall\, u \in E.$$

Since by relation (4.37) we have $q^+ < p_0$, the above relation shows that J_λ is bounded from below.

Next, we show that J_λ satisfies the Palais-Smale condition. Let $\{u_n\}$ be a sequence in E such that $\{J_\lambda(u_n)\}$ is bounded and $J'(u_n) \to 0$ in E', as $n \to \infty$. We show that $\{u_n\}$ is bounded in E. Assume to the contrary. Then, passing eventually to a subsequence, still denoted by $\{u_n\}$, we may assume that $\|u_n\| \to \infty$ as $n \to \infty$. Thus we may consider that $\|u_n\| > 1$ for any integer n.

By our assumptions, there is a positive constant M such that for all large enough n we have

$$
\begin{aligned}
M+1+\|u_n\| &\geq J_\lambda(u_n)-\frac{1}{q^-}\langle J'(u_n),u_n\rangle \\
&= \int_\Omega \Phi(|\nabla u_n|)\,dx-\lambda\int_\Omega \frac{1}{q(x)}|u_n|^{q(x)}\,dx \\
&- \frac{1}{q^-}\cdot\int_\Omega \phi(|\nabla u_n(x)|)|\nabla u_n(x)|\,dx \\
&+ \frac{\lambda}{q^-}\int_\Omega |u_n|^{q(x)}\,dx \\
&\geq \int_\Omega \Phi(|\nabla u_n|)\,dx-\frac{1}{q^-}\cdot\int_\Omega \phi(|\nabla u_n(x)|)|\nabla u_n(x)|\,dx \\
&\geq \left(1-\frac{p^0}{q^-}\right)\int_\Omega \Phi(|\nabla u_n|)\,dx \\
&\geq \left(1-\frac{p^0}{q^-}\right)\|u_n\|^{p_0}.
\end{aligned}
$$

Since $p_0>1$, letting $n\to\infty$ we obtain a contradiction. It follows that $\{u_n\}$ is bounded in E. Similar arguments as those used at the end of the proof of Theorem 25 imply that, up to a subsequence, $\{u_n\}$ strongly converges in E.

The proof of Lemma 31 is complete. □

Lemma 32 *The functional J_λ satisfies condition (A2) from Theorem 27.*

Proof. We construct a sequence of subsets $A_k\in\Gamma_k$ such that $\sup_{u\in A_k}J_\lambda(u)<0$, for each $k\in\mathbb{N}$.

Let $x_1\in\Omega$ and $r_1>0$ be such that $\overline{B_{r_1}(x_1)}\subset\Omega$ and $|\overline{B_{r_1}(x_1)}|<|\Omega|/2$. Consider $\theta_1\in C_0^\infty(\Omega)$ be a function with $\mathrm{supp}(\theta_1)=\overline{B_{r_1}(x_1)}$. Define $\Omega_1=\Omega\setminus\overline{B_{r_1}(x_1)}$.

Next, let $x_2\in\Omega$ and $r_2>0$ be such that $\overline{B_{r_2}(x_2)}\subset\Omega_1$ and $|\overline{B_{r_2}(x_2)}|<|\Omega_1|/2$. Consider $\theta_2\in C_0^\infty(\Omega)$ be a function with $\mathrm{supp}(\theta_2)=\overline{B_{r_2}(x_2)}$.

Continuing the process described above we can construct inductively a sequence of functions $\theta_1,\theta_2,\ldots,\theta_k\in C_0^\infty(\Omega)$ such that $\mathrm{supp}(\theta_i)\neq\mathrm{supp}(\theta_j)$ if $i\neq j$ and $|\mathrm{supp}(\theta_i)|>0$ for any $i,\ j\in\{1,\ldots,k\}$.

We define the following finite dimensional subspace of E:

$$F=\mathrm{span}\{\theta_1,\theta_2,\ldots,\theta_k\}.$$

Clearly, $\dim F=k$ and $\int_\Omega|\theta|^{q(x)}\,dx>0$, for all $\theta\in F\setminus\{0\}$. We denote by S_1 the unit sphere in E, that is, $S_1=\{u\in E;\ \|u\|=1\}$. For any number $t\in(0,1)$ we define the set

$$A_k(t)=t\cdot(S_1\cap F).$$

Since for any bounded symmetric neighborhood ω of the origin in $\mathbb{R}^k$, we have $\gamma(\partial\omega) = k$, it follows that $\gamma(A_k(t)) = k$ for any $t \in (0,1)$.

Finally, we show that for each integer k there exists $t_k \in (0,1)$ such that

$$\sup_{u \in A_k(t_k)} J_\lambda(u) < 0.$$

For any $t \in (0,1)$ we have

$$\begin{aligned}
\sup_{u \in A_k(t)} J_\lambda(u) &\leq \sup_{\theta \in S_1 \cap F} J_\lambda(t\theta) \\
&= \sup_{\theta \in S_1 \cap F} \left\{ \int_\Omega \Phi(t|\nabla\theta|)\,dx - \lambda \int_\Omega \frac{1}{q(x)} t^{q(x)} |\theta|^{q(x)}\,dx \right\} \\
&\leq \sup_{\theta \in S_1 \cap F} \left\{ t^{p_0} \int_\Omega \Phi(|\nabla\theta|)\,dx - \frac{\lambda t^{q^+}}{q^+} \int_\Omega |\theta|^{q(x)}\,dx \right\} \\
&= \sup_{\theta \in S_1 \cap F} \left\{ t^{p_0} \left(1 - \frac{\lambda}{q^+} \cdot \frac{1}{t^{p_0 - q^+}} \cdot \int_\Omega |\theta|^{q(x)}\,dx \right) \right\}.
\end{aligned}$$

Since $S_1 \cap F$ is compact we have $m = \min_{\theta \in S_1 \cap F} \int_\Omega |\theta|^{q(x)}\,dx > 0$. Combining that fact with the information given by relation (4.42), that is $p_0 > q^+$, we deduce that we can choose small enough $t_k \in (0,1)$ such that

$$1 - \frac{\lambda}{q^+} \cdot \frac{1}{t^{p_0 - q^+}} \cdot m < 0.$$

The above relations yield

$$\sup_{u \in A_k(t_k)} J_\lambda(u) < 0.$$

The proof of Lemma 32 is complete. □

Using Lemmas 31 and 32, we can apply Theorem 27 to the functional J_λ. So there exists a sequence $\{u_n\} \subset E$ such that $J'(u_n) = 0$, for each n, $J_\lambda(u_n) \leq 0$ and $\{u_n\}$ converges to zero in E.

The proof of Theorem 26 is complete. □

We illustrate these abstract results with two examples of boundary value problems with variable exponents.

We first consider the following nonlinear problem

$$\begin{cases} -\operatorname{div}(\log(1 + |\nabla u|^r)|\nabla u|^{p-2}\nabla u) = \lambda |u|^{q(x)-2}u, & \text{for } x \in \Omega \\ u = 0, & \text{for } x \in \partial\Omega, \end{cases} \tag{4.54}$$

where p and r are real numbers such that $1 < p$, r, $N > p + r$ and $q(x)$ is a continuous function on $\overline{\Omega}$ such that $1 < q(x)$ for all $x \in \overline{\Omega}$ and furthermore

$$\inf_\Omega q(x) < p \quad \text{and} \quad \sup_\Omega q(x) < \frac{Np}{N-p}.$$

In this case we have

$$\phi(t) = \log(1 + |t|^r) \cdot |t|^{p-2}t, \qquad \text{for all } t \in \mathbb{R}$$

and

$$\Phi(t) = \int_0^t \phi(s), \qquad \text{for all } t \in \mathbb{R}.$$

Then ϕ is an odd, increasing homeomorphism of $\mathbb{R}$ onto $\mathbb{R}$, while Φ is convex and even on $\mathbb{R}$ and an increasing function from $\mathbb{R}_+$ to $\mathbb{R}_+$. We have

$$p_0 = p \quad \text{and} \quad p^0 = p + r$$

and thus relation (4.40) in Theorem 25 is satisfied. On the other hand, relations (4.39) and (4.41) are fulfilled. Thus, by Theorem 25, there exists $\lambda^\star > 0$ such that any $\lambda \in (0, \lambda^\star)$ is an eigenvalue for problem (4.54).

Next, we consider the problem

$$\begin{cases} -\text{div}\left(\dfrac{|\nabla u|^{p-2}\nabla u}{\log(1 + |\nabla u|)}\right) = \lambda |u|^{q(x)-2}u, & \text{for} \quad x \in \Omega \\ u = 0, & \text{for} \quad x \in \partial\Omega\,, \end{cases} \tag{4.55}$$

where p is a real number such that $p > N + 1$ and $q \in C(\overline{\Omega})$ satisfies $1 < q(x) < p - 1$ for any $x \in \overline{\Omega}$. In this case we have that

$$\phi(t) = \frac{|t|^{p-2}}{\log(1 + |t|)}t$$

and that

$$\Phi(t) = \int_0^t \phi(s)\, ds,$$

is an increasing continuous function from $\mathbb{R}^+$ to $\mathbb{R}^+$, with $\Phi(0) = 0$ and such that the function $\Phi(\sqrt{t})$ is convex. From Example 3 on p. 243 in [19] we infer that

$$p_0 = p - 1 < p^0 = p = \liminf_{t\to\infty} \frac{\log(\Phi(t))}{\log(t)}.$$

Thus, conditions (4.39), (4.42) and (4.43) from Theorem 26 and Remark 5 are verified. We deduce that every $\lambda > 0$ is an eigenvalue of problem (4.55). Moreover, for each $\lambda > 0$ there exists a sequence of eigenvectors $\{u_n\}$ such that $\lim_{n\to\infty} u_n = 0$ in $W_0^1 L_\Phi(\Omega)$.

4.3 Nonlinear eigenvalue problems with indefinite potential

The energy of the mind is the essence of life.

Aristotle (384–322 B.C.)

Let Ω be a bounded domain in $\mathbb{R}^N$ ($N \geq 3$) with smooth boundary $\partial\Omega$. Assume that $a_i : (0, \infty) \to \mathbb{R}$, $i = 1, 2$, are functions such that the mappings $\phi_i : \mathbb{R} \to \mathbb{R}$, $i = 1, 2$, defined by

$$\phi_i(t) = \begin{cases} a_i(|t|)t, & \text{for} \quad t \neq 0 \\ 0, & \text{for} \quad t = 0\,, \end{cases}$$

are odd, increasing homeomorphisms from $\mathbb{R}$ onto $\mathbb{R}$, λ is a real number, $V(x)$ is a potential and q_1, q_2, $m : \overline{\Omega} \to (1, \infty)$ are continuous functions.

In this section, we are concerned with the study of the eigenvalue problem

$$\begin{cases} -\mathrm{div}((a_1(|\nabla u|) + a_2(|\nabla u|))\nabla u) + V(x)|u|^{m(x)-2}u & \text{if} \quad x \in \Omega \\ \quad = \lambda(|u|^{q_1(x)-2} + |u|^{q_2(x)-2})u, & \\ u = 0, & \text{if} \quad x \in \partial\Omega\,. \end{cases} \tag{4.56}$$

Set

$$\Phi_i(t) = \int_0^t \phi_i(s)\, ds, \quad (\Phi_i)^\star(t) = \int_0^t (\phi_i)^{-1}(s)\, ds, \qquad \text{for all } t \in \mathbb{R},\ i = 1, 2\,.$$

We observe that Φ_i, $i = 1, 2$, are *Young functions*, that is, $\Phi_i(0) = 0$, Φ_i are convex, and $\lim_{x\to\infty} \Phi_i(x) = +\infty$. Furthermore, since $\Phi_i(x) = 0$ if and only if $x = 0$, $\lim_{x\to 0} \Phi_i(x)/x = 0$, and $\lim_{x\to\infty} \Phi_i(x)/x = +\infty$, then Φ_i are called *N–functions.* The functions $(\Phi_i)^\star$, $i = 1, 2$, are called the *complementary* functions of Φ_i, $i = 1, 2$, and they satisfy

$$(\Phi_i)^\star(t) = \sup\{st - \Phi_i(s);\ s \geq 0\}, \qquad \text{for all } t \geq 0\,.$$

We also observe that $(\Phi_i)^\star$, $i = 1, 2$, are N–functions and the Young inequality holds

$$st \leq \Phi_i(s) + (\Phi_i)^\star(t), \qquad \text{for all } s, t \geq 0\,.$$

For an easier manipulation of the Orlicz and Orlicz-Sobolev spaces $L_{\Phi_i}(\Omega)$, $W^1 L_{\Phi_i}(\Omega)$ and $W_0^1 L_{\Phi_i}(\Omega)$, we define

$$(\phi_i)_0 := \inf_{t>0} \frac{t\phi_i(t)}{\Phi_i(t)} \quad \text{and} \quad (\phi_i)^0 := \sup_{t>0} \frac{t\phi_i(t)}{\Phi_i(t)},\ i \in \{1, 2\}\,.$$

In this section, we assume that for all $i \in \{1, 2\}$ we have

$$1 < (\phi_i)_0 \leq \frac{t\phi_i(t)}{\Phi_i(t)} \leq (\phi_i)^0 < \infty, \quad \forall\, t \geq 0\,. \tag{4.57}$$

TABLE 4.2: Homeomorphisms and distinguished values

Homeomorphism	$(\phi)_0$ **and** $(\phi)^0$
$\phi(t) = p\lvert t\rvert^{p-2}t \ (p>1,\ t\in\mathbb{R})$	$(\phi)_0 = (\phi)^0 = p$
$\phi(t) = \log(1+\lvert t\rvert^s)\lvert t\rvert^{p-2}t \ (p,s>1,\ t\in\mathbb{R})$	$(\phi)_0 = p,\ (\phi)^0 = p+s$
$\phi(t) = \dfrac{\lvert t\rvert^{p-2}t}{\log(1+\lvert t\rvert)} \ (t\neq 0);\ \phi(0)=0 \ (p>2)$	$(\phi)_0 = p-1,\ (\phi)^0 = p$

The above relation implies that each Φ_i, $i \in \{1,2\}$, satisfies the Δ_2-condition, that is,

$$\Phi_i(2t) \le K\Phi_i(t), \quad \forall\, t \ge 0\,, \tag{4.58}$$

where K is a positive constant.

We also recall that the following relations hold for $i \in \{1,2\}$:

$$\|u\|_i^{(\phi_i)^0} \le \int_\Omega \Phi_i(|\nabla u|)\,dx \le \|u\|_i^{(\phi_i)_0}, \quad \forall\, u \in W_0^1L_{\Phi_i}(\Omega),\ \|u\|_i < 1,, \tag{4.59}$$

$$\|u\|_i^{(\phi_i)_0} \le \int_\Omega \Phi_i(|\nabla u|)\,dx \le \|u\|_i^{(\phi_i)^0}, \quad \forall\, u \in W_0^1L_{\Phi_i}(\Omega),\ \|u\|_i > 1\,. \tag{4.60}$$

Furthermore, in this section we assume that for each $i \in \{1,2\}$ the function Φ_i satisfies the following condition

$$\text{the function } [0,\infty) \ni t \longmapsto \Phi_i(\sqrt{t}) \text{ is convex}\,. \tag{4.61}$$

Conditions (4.58) and (4.61) assure that for each $i \in \{1,2\}$ the Orlicz spaces $L_{\Phi_i}(\Omega)$ are uniformly convex spaces and thus reflexive Banach spaces. This fact implies that also the Orlicz-Sobolev spaces $W_0^1L_{\Phi_i}(\Omega)$, $i \in \{1,2\}$, are reflexive Banach spaces.

In Table 4.2 we provide examples of functions $\phi : \mathbb{R} \to \mathbb{R}$ which are odd, increasing homeomorphisms from $\mathbb{R}$ onto $\mathbb{R}$ and satisfy conditions (4.57) and (4.61).

Now we turn back to problem (4.56). We study this problem when q_1, q_2, $m : \overline{\Omega} \to (1,\infty)$ are continuous functions satisfying the following assumptions:

$$1 < (\phi_2)_0 \le (\phi_2)^0 < q_2^- \le q_2^+ \le m^- \le m^+ \le q_1^- \le q_1^+ < (\phi_1)_0 \le (\phi_1)^0 < N\,, \tag{4.62}$$

$$q_1^+ < [(\phi_2)_0]^\star := \frac{N(\phi_2)_0}{N-(\phi_2)_0}, \quad \forall\, x \in \overline{\Omega}\,, \tag{4.63}$$

and the potential $V : \Omega \to \mathbb{R}$ satisfies

$$V \in L^{r(x)}(\Omega), \ \text{ with } r(x) \in C(\overline{\Omega}) \text{ and } r(x) > \frac{N}{m^-}\ \forall\, x \in \overline{\Omega}\,. \tag{4.64}$$

Condition (4.62), which describes the competition between the growth rates involved in equation (4.56), actually assures a balance between them and thus it represents the *key* of the present study. Such a balance is essential since we are working on a non-homogeneous (eigenvalue) problem for which a minimization technique based on the Lagrange multipliers theorem cannot be applied in order to find (principal) eigenvalues (unlike in the case of homogeneous operators). Thus in the case of nonlinear non-homogeneous eigenvalue problems the classical theory used in the homogeneous case does not work entirely, but some of its ideas can still be useful and some particular results can still be obtained in certain aspects while in other aspects entirely new phenomena can occur. To focus on our case, condition (4.62) together with conditions (4.63) and (4.64) imply

$$\lim_{\|u\|_1\to 0}\frac{\displaystyle\int_\Omega \Phi_1(|\nabla u|)\,dx+\int_\Omega \Phi_2(|\nabla u|)\,dx+\int_\Omega \frac{V(x)}{m(x)}|u|^{m(x)}\,dx}{\displaystyle\int_\Omega \frac{1}{q_1(x)}|u|^{q_1(x)}\,dx+\int_\Omega \frac{1}{q_2(x)}|u|^{q_2(x)}\,dx}=\infty$$

and

$$\lim_{\|u\|_1\to \infty}\frac{\displaystyle\int_\Omega \Phi_1(|\nabla u|)\,dx+\int_\Omega \Phi_2(|\nabla u|)\,dx+\int_\Omega \frac{V(x)}{m(x)}|u|^{m(x)}\,dx}{\displaystyle\int_\Omega \frac{1}{q_1(x)}|u|^{q_1(x)}\,dx+\int_\Omega \frac{1}{q_2(x)}|u|^{q_2(x)}\,dx}=\infty\,.$$

In other words, the absence of homogeneity is balanced by the behavior (actually, the blow-up) of the Rayleigh quotient associated to problem (4.56) in the origin and at infinity. The consequences of the above remarks is that the infimum of the Rayleigh quotient associated to problem (4.56) is a real number, that is,

$$\inf_{u\in W_0^1L_{\Phi_1}(\Omega)\setminus\{0\}}\frac{\displaystyle\int_\Omega \Phi_1(|\nabla u|)\,dx+\int_\Omega \Phi_2(|\nabla u|)\,dx+\int_\Omega \frac{V(x)}{m(x)}|u|^{m(x)}\,dx}{\displaystyle\int_\Omega \frac{1}{q_1(x)}|u|^{q_1(x)}\,dx+\int_\Omega \frac{1}{q_2(x)}|u|^{q_2(x)}\,dx}\in\mathbb{R}\,, \tag{4.65}$$

and it will be attained for a function $u_0 \in W_0^{1,p_1(x)}(\Omega)\setminus\{0\}$. Moreover, the value in (4.65) represents an eigenvalue of problem (4.56) with the corresponding eigenfunction u_0. On the other hand, we can prove that any λ superior to the value given by relation (4.65) is also an eigenvalue of problem (4.56). Thus we conclude that problem (4.56) possesses a continuous family of eigenvalues.

Related with the above ideas we will also discuss the *optimization* of the eigenvalues described by relation (4.65) with respect to the potential V, provided that V belongs to a bounded, closed and convex subset of $L^{r(x)}(\Omega)$ (where $r(x)$ is given by relation (4.64)). By optimization, we understand the existence of some potentials $V_\star$ and $V^\star$ such that the eigenvalue described in relation (4.65) is minimal or maximal with respect to the set to which V belongs. The results that we obtain in the context of optimization of eigenvalues are motivated by the above advances in this field in the case of homogeneous (linear or nonlinear) eigenvalue problems.

It follows by relation (4.62) that $W_0^1L_{\Phi_1}(\Omega)$ is continuously embedded in $W_0^1L_{\Phi_2}(\Omega)$. Therefore the natural function space suitable to analyze problem (4.56) is $W_0^1L_{\Phi_1}(\Omega)$.

We say that $\lambda \in \mathbb{R}$ is an *eigenvalue* of problem (4.56) if there exists $u \in W_0^1L_{\Phi_1}(\Omega) \setminus \{0\}$ such that

$$\int_\Omega (a_1(|\nabla u|) + a_2(|\nabla u|))\nabla u \nabla v\, dx \quad + \int_\Omega V(x)|u|^{m(x)-2}uv\, dx \\ -\lambda \int_\Omega (|u|^{q_1(x)-2} + |u|^{q_2(x)-2})uv\, dx = 0\,,$$

for all $v \in W_0^1L_{\Phi_1}(\Omega)$.

We point out that if λ is an eigenvalue of problem (4.56) then the corresponding *eigenfunction* $u \in W_0^1L_{\Phi_1}(\Omega) \setminus \{0\}$ is a *weak solution* of problem (4.56).

For each *indefinite* potential $V \in L^{r(x)}(\Omega)$, we define the following *Rayleigh quotients*:

$$A(V) := \inf_{u \in W_0^1L_{\Phi_1}(\Omega)\setminus\{0\}} \frac{\displaystyle\int_\Omega \left(\Phi_1(|\nabla u|) + \Phi_2(|\nabla u|) + \frac{V(x)}{m(x)}|u|^{m(x)}\right) dx}{\displaystyle\int_\Omega \left(\frac{1}{q_1(x)}|u|^{q_1(x)} + \frac{1}{q_2(x)}|u|^{q_2(x)}\right) dx}$$

and

$$B(V) := \inf_{u \in W_0^1L_{\Phi_1}(\Omega)\setminus\{0\}} \frac{\displaystyle\int_\Omega \left(a_1(|\nabla u|)|\nabla u|^2 + a_2(|\nabla u|)|\nabla u|^2 + V(x)|u|^{m(x)}\right) dx}{\displaystyle\int_\Omega \left(|u|^{q_1(x)} + |u|^{q_2(x)}\right) dx}\,.$$

Thus we can define functions $A,\ B : L^{r(x)}(\Omega) \to \mathbb{R}$.

The first result of this section is the following theorem.

Theorem 28 *Assume that conditions* (4.62), (4.63) *and* (4.64) *are fulfilled. Then $A(V)$ is an eigenvalue of problem* (4.56). *Moreover, there exists an eigenfunction $u_V \in W_0^1L_{\Phi_1}(\Omega) \setminus \{0\}$, corresponding to the eigenvalue $A(V)$, such that*

$$A(V) = \frac{\displaystyle\int_\Omega \left(\Phi_1(|\nabla u_V|) + \Phi_2(|\nabla u_V|) + \frac{V(x)}{m(x)}|u_V|^{m(x)}\right) dx}{\displaystyle\int_\Omega \left(\frac{1}{q_1(x)}|u_V|^{q_1(x)} + \frac{1}{q_2(x)}|u_V|^{q_2(x)}\right) dx}\,.$$

Furthermore, $B(V) \le A(V)$, each $\lambda \in (A(V), \infty)$ is an eigenvalue of problem (4.56), *while no $\lambda \in (-\infty, B(V))$ is an eigenvalue of problem* (4.56).

Next, we show that on each convex, bounded and closed subset of $L^{r(x)}(\Omega)$ the function A defined above is bounded from below and attains its minimum. The result is the following:

Theorem 29 *Assume that conditions* (4.62), (4.63) *and* (4.64) *are fulfilled. Assume that S is a convex, bounded and closed subset of $L^{r(x)}(\Omega)$. Then there exists $V_\star \in S$ which minimizes $A(V)$ on S, that is,*

$$A(V_\star) = \inf_{V \in S} A(V)\,.$$

Finally, we focus our attention on the special case when the set S from Theorem 29 is a ball in $L^{r(x)}(\Omega)$. Thus, we will denote each closed ball centered at the origin of radius R in $L^{r(x)}(\Omega)$ by $\overline{B}_R(0)$, that is,

$$\overline{B}_R(0) := \{u \in L^{r(x)}(\Omega);\ \ |u|_{r(x)} \le R\}\,.$$

By Theorem 29, we can define the function $A_\star : [0, \infty) \to \mathbb{R}$ by

$$A_\star(R) = \min_{V \in \overline{B}_R(0)} A(V)\,.$$

The result establishes some qualitative properties of $A_\star$.

Theorem 30 *a) The function $A_\star$ is not constant and decreases monotonically.*
b) The function $A_\star$ is continuous.

Results similar to Theorems 29 and 30 can be obtained if we notice that on each convex, bounded and closed subset of $L^{r(x)}(\Omega)$ the function A defined in Theorem 28 is also bounded from above and attains its maximum. We also note that we can define a function $A^\star : [0, \infty) \to \mathbb{R}$ by

$$A^\star(R) = \max_{V \in \overline{B}_R(0)} A(V)\,,$$

which has similar properties as $A_\star$.

4.3.1 Concentration properties of the spectrum

Let X denote the generalized Sobolev space $W_0^1 L_{\Phi_1}(\Omega)$. Relation (4.62), Lemma 8.12(b) from Adams [2] and the Rellich-Kondrachov theorem imply that

$$\begin{aligned} W_0^1 L_{\Phi_1}(\Omega) &\subset W_0^1 L_{\Phi_2}(\Omega) \subset W_0^{1,(\phi_2)_0}(\Omega) \hookrightarrow L^{q_1^+}(\Omega) \\ &\subset L^{q_1(x)}(\Omega) \subset L^{m(x)}(\Omega) \subset L^{q_2(x)}(\Omega)\,, \end{aligned} \tag{4.66}$$

where we denoted by $\subset$ a *continuous* embedding while by $\hookrightarrow$ we denoted a *compact* embedding.

Define the functionals J_V, $I : X \to \mathbb{R}$ by

$$J_V(u) = \int_\Omega \Phi_1(|\nabla u|)\, dx + \int_\Omega \Phi_2(|\nabla u|)\, dx + \int_\Omega \frac{V(x)}{m(x)}|u|^{m(x)}\, dx\,,$$

$$I(u) = \int_\Omega \frac{1}{q_1(x)}|u|^{q_1(x)}\, dx + \int_\Omega \frac{1}{q_2(x)}|u|^{q_2(x)}\, dx\,.$$

Relation (4.66) assures that the functionals defined above are well-defined. We notice that for any V satisfying condition (4.64) we have

$$J_V(u) = J_0(u) + \int_\Omega \frac{V(x)}{m(x)}|u|^{m(x)}\, dx, \quad \forall\, u \in X\,,$$

where J_0 is obtained in the case when $V = 0$ in Ω.

We have J_V, $I \in C^1(X, \mathbb{R})$ and for all u, $v \in X$,

$$\langle J_V'(u), v\rangle = \int_\Omega (a_1(|\nabla u|) + a_2(|\nabla u|))\nabla u \nabla v\, dx + \int_\Omega V(x)|u|^{m(x)-2}uv\, dx\,,$$

$$\langle I'(u), v\rangle = \int_\Omega |u|^{q_1(x)-2}uv\, dx + \int_\Omega |u|^{q_2(x)-2}uv\, dx\,.$$

In order to prove Theorem 28, we first establish some auxiliary results.

Lemma 33 *Assume conditions* (4.62), (4.63) *and* (4.64) *are fulfilled. Then for each* $\epsilon > 0$ *there exists* $C_\epsilon > 0$ *such that*

$$\left|\int_\Omega \frac{V(x)}{m(x)}|u|^{m(x)}\, dx\right| \le \epsilon \int_\Omega (\Phi_1(|\nabla u|) + \Phi_2(|\nabla u|))\, dx + C_\epsilon |V|_{r(x)} \int_\Omega (|u|^{m^-} + |u|^{m^+})\, dx\,,$$

for all $u \in X$.

Proof. First, we point out that since $r(x) > r^-$ on $\overline{\Omega}$ it follows that $L^{r(x)}(\Omega) \subset L^{r^-}(\Omega)$. On the other hand, since $r(x) > \frac{N}{m^-}$ for each $x \in \overline{\Omega}$ it follows that $r^- > \frac{N}{m^-}$. Thus, we infer that $V \in L^{r^-}(\Omega)$ and $r^- > \frac{N}{m^-}$.

Now, let $\epsilon > 0$ be fixed. We claim that there exists $D_\epsilon > 0$ such that

$$\int_\Omega |V(x)|\cdot|u|^{m^-}\, dx \le \epsilon \int_\Omega |\nabla u|^{m^-}\, dx + D_\epsilon |V|_{r^-} \int_\Omega |u|^{m^-}\, dx, \ \forall\, u \in W_0^{1,m^-}(\Omega)\,. \tag{4.67}$$

In order to establish (4.67), we show that for each $s \in (1, \frac{Nm^-}{N-m^-})$ there exists $D'_\epsilon > 0$ such that

$$|v|_s \le \epsilon|\, |\nabla v|\, |_{m^-} + D'_\epsilon |v|_{m^-}, \quad \forall\, u \in W_0^{1,m^-}(\Omega)\,. \tag{4.68}$$

Indeed, suppose to the contrary that relation (4.68) does not hold for each

$\epsilon > 0$. Then there exists $\epsilon_0 > 0$ and a sequence $(v_n) \subset W_0^{1,m^-}(\Omega)$ such that $|v_n|_s = 1$ and

$$\epsilon_0 |\, |\nabla v_n|\, |_{m^-} + n|v_n|_{m^-} < 1, \quad \forall\, n .$$

The sequence (v_n) is bounded in $W_0^{1,m^-}(\Omega)$ and $|v_n|_{m^-} \to 0$. Thus we deduce that passing eventually to a subsequence we can assume that v_n weakly converges to a function v in $W_0^{1,m^-}(\Omega)$ and actually $v = 0$. Since $s \in (1, \frac{Nm^-}{N-m^-})$ it follows by the Rellich-Kondrachov theorem that $W_0^{1,m^-}(\Omega)$ is compactly embedded in $L^s(\Omega)$ and thus v_n converges to 0 in $L^s(\Omega)$. On the other hand, since $|v_n|_s = 1$ for each n we deduce that $|v|_s = 1$ and this is a contradiction. We have thus obtained that relation (4.68) holds.

Next, we point out that since $r^- > \frac{N}{m^-}$ then $m^- \cdot r^{-\prime} < \frac{Nm^-}{N-m^-}$, where $r^{-\prime} = \frac{r^-}{r^- - 1}$. Thus, by the Hölder inequality we have

$$\int_\Omega |V(x)| \cdot |u|^{m^-}\, dx \leq |V|_{r^-} \cdot |u|^{m^-}_{m^- \cdot r^{-\prime}}, \quad \forall\, u \in W_0^{1,m^-}(\Omega) .$$

Combining the last inequality with relation (4.68), we infer that relation (4.67) holds.

Similar arguments as those used in the proof of relation (4.67), combined with the fact that since $r^- > \frac{N}{m^-}$ we also have $r^- > \frac{N}{m^+}$, imply that there exists D_ϵ'' such that for all $u \in W_0^{1,m^+}(\Omega)$

$$\int_\Omega |V(x)| \cdot |u|^{m^+}\, dx \leq \epsilon \int_\Omega |\nabla u|^{m^+}\, dx + D_\epsilon'' |V|_{r^-} \int_\Omega |u|^{m^+}\, dx . \qquad (4.69)$$

Using relation (4.62), we deduce that $m^- \leq m^+ < (\phi_1)_0$ and hence $W_0^{1,(\phi_1)_0}(\Omega) \subset W_0^{1,m^\pm}(\Omega)$. On the other hand, we have $W_0^1 L_{\Phi_1}(\Omega) \subset W_0^{1,(\phi_1)_0}(\Omega)$. The above facts imply that relations (4.67) and (4.69) hold for any $u \in X$. Moreover, on the right-hand sides of inequalities (4.67) and (4.69), we can take $|V|_{r(x)}$ instead of $|V|_{r^-}$ since $L^{r(x)}(\Omega)$ is continuously embedded in $L^{r^-}(\Omega)$ by virtue of the Hölder inequality.

Finally, we point out that since by (4.62) we have $(\phi_2)^0 < m^- \leq m(x) \leq m^+ < (\phi_1)_0$ for each $x \in \overline{\Omega}$, we deduce that

$$\left| \int_\Omega \frac{V(x)}{m(x)} |u|^{m(x)}\, dx \right| \leq \frac{1}{m^-} \int_\Omega |V(x)| \cdot (|u|^{m^-} + |u|^{m^+})\, dx, \quad \forall\, u \in X \quad (4.70)$$

and

$$\int_\Omega (|\nabla u|^{m^-} + |\nabla u|^{m^+})\, dx \leq \int_\Omega \left(|\nabla u|^{(\phi_2)^0}\, dx + |\nabla u|^{(\phi_1)_0} \right) dx, \quad \forall\, u \in X . \qquad (4.71)$$

Relations (4.67), (4.69), (4.70), (4.71), (4.62) and (4.66) imply that Lemma 33 is true. □

Lemma 34 *The following relations hold:*

$$\lim_{\|u\|_1\to\infty}\frac{J_V(u)}{I(u)}=\infty \tag{4.72}$$

and

$$\lim_{\|u\|_1\to 0}\frac{J_V(u)}{I(u)}=\infty. \tag{4.73}$$

Proof. First, we point out that by relation (4.62), $q_2(x)<m^\pm<q_1(x)$ for any $x\in\overline{\Omega}$. Therefore

$$|u(x)|^{m^-}+|u(x)|^{m^+}\leq 2(|u(x)|^{q_1(x)}+|u(x)|^{q_2(x)}),\quad \forall\, x\in\overline{\Omega}\text{ and }\forall\, u\in X\,.$$

Integrating the above inequality we infer that

$$\frac{\displaystyle\int_\Omega(|u|^{m^-}+|u|^{m^+})\,dx}{\displaystyle\int_\Omega(|u|^{q_1(x)}+|u|^{q_2(x)})\,dx}\leq 2,\quad \forall\, u\in X\,. \tag{4.74}$$

Using Lemma 33 we find that for $\epsilon\in(0,1)$, there exists $C_\epsilon>0$ such that

$$\frac{J_V(u)}{I(u)}\geq\frac{(1-\epsilon)\displaystyle\int_\Omega(\Phi_1(|\nabla u|)+\Phi_2(|\nabla u|))\,dx-C_\epsilon|V|_{r(x)}\int_\Omega(|u|^{m^-}+|u|^{m^+})\,dx}{\displaystyle\frac{1}{q_2^-}\int_\Omega(|u|^{q_1(x)}+|u|^{q_2(x)})\,dx},$$

for any $u\in X$.

From the above inequality and relation (4.74) we deduce that there exist positive constants β and γ such that

$$\frac{J_V(u)}{I(u)}\geq\frac{\beta\displaystyle\int_\Omega(\Phi_1(|\nabla u|)+\Phi_2(|\nabla u|))\,dx}{\displaystyle\int_\Omega(|u|^{q_1(x)}+|u|^{q_2(x)})\,dx}-\gamma|V|_{r(x)},\quad\forall\, u\in X\,. \tag{4.75}$$

For any $u\in X$ with $\|u\|_1>1$ relation (4.75) implies

$$\frac{J_V(u)}{I(u)}\geq\frac{\beta\displaystyle\int_\Omega\Phi_1(|\nabla u|)\,dx}{|u|_{q_1^-}^{q_1^-}+|u|_{q_1^+}^{q_1^+}+|u|_{q_2^-}^{q_2^-}+|u|_{q_2^+}^{q_2^+}}-\gamma|V|_{r(x)},\quad\forall\, u\in X\text{ with }\|u\|_1>1\,.$$

Now, taking into account the continuous embedding of X in $L^{q_i^\pm}(\Omega)$ for $i=1,2$ (given by relations (4.62) and (4.66)), we deduce the existence of a constant $\delta>0$ such that

$$\frac{J_V(u)}{I(u)}\geq\frac{\delta\|u\|_1^{(\phi_1)_0}}{\|u\|_1^{q_1^-}+\|u\|_1^{q_1^+}+\|u\|_1^{q_2^-}+\|u\|_1^{q_2^+}}-\gamma|V|_{r(x)},\quad\forall\, u\in X\text{ with }\|u\|_1>1\,.$$

Since $(\phi_1)_0 > q_1^+ \geq q_1^- \geq q_2^+ \geq q_2^-$, passing to the limit as $\|u\|_1 \to \infty$ in the above inequality we deduce that relation (4.72) holds.

Relation (4.66) shows that the space $W_0^1 L_{\Phi_1}(\Omega)$ is continuously embedded in $W_0^1 L_{\Phi_2}(\Omega)$. Thus, if $\|u\|_1 \to 0$, then $\|u\|_2 \to 0$.

The above remarks enable us to affirm that for any $u \in X$ with small enough $\|u\|_1 < 1$, we have $\|u\|_2 < 1$.

Using again relation (4.66), we deduce that $W_0^1 L_{\Phi_2}(\Omega)$ is continuously embedded in $L^{q_i^{\pm}}(\Omega)$ with $i = 1, 2$. It follows that there exist positive constants d_{i1} and d_{i2} with $i = 1, 2$ such that

$$\|u\|_2 \geq d_{i1} \cdot |u|_{q_i^+}, \quad \forall\, u \in W_0^1 L_{\Phi_2}(\Omega) \text{ and } i = 1, 2 \tag{4.76}$$

and

$$\|u\|_2 \geq d_{i2} \cdot |u|_{q_i^-}, \quad \forall\, u \in W_0^1 L_{\Phi_2}(\Omega) \text{ and } i = 1, 2\,. \tag{4.77}$$

Thus for any $u \in X$ with small enough $\|u\|_1 < 1$, relation (4.75) implies

$$\frac{J_V(u)}{I(u)} \geq \frac{\beta \displaystyle\int_\Omega \Phi_2(|\nabla u|)\, dx}{|u|_{q_1^-}^{q_1^-} + |u|_{q_1^+}^{q_1^+} + |u|_{q_2^-}^{q_2^-} + |u|_{q_2^+}^{q_2^+}} - \gamma |V|_{r(x)}\,.$$

Next, relations (4.76) and (4.77) imply that there exists a constant $\xi > 0$ such that

$$\frac{J_V(u)}{I(u)} \geq \frac{\xi \|u\|_2^{(\phi_2)^0}}{\|u\|_2^{q_1^-} + \|u\|_2^{q_1^+} + \|u\|_2^{q_2^-} + \|u\|_2^{q_2^+}} - \gamma |V|_{r(x)}\,,$$

for any $u \in X$ with small enough $\|u\|_1 < 1$. Since $(\phi_2)^0 < q_2^- \leq q_2^+ \leq q_1^- \leq q_1^+$, passing to the limit as $\|u\|_1 \to 0$ (and thus, $\|u\|_2 \to 0$) in the above inequality we deduce that relation (4.73) holds. The proof of Lemma 34 is complete. □

We point out that using relation (4.75) we deduce that for V satisfying (4.64), the quotient $\frac{J_V(u)}{I(u)}$ is bounded from below for $u \in X \setminus \{0\}$, that is, $A(V)$ is a real number. Similarly, $B(V)$ is also a real number.

Lemma 35 *There exists $u \in X \setminus \{0\}$ such that $\frac{J_V(u)}{I(u)} = A(V)$.*

Proof. Let $(u_n) \subset X \setminus \{0\}$ be a minimizing sequence for $A(V)$, that is,

$$\lim_{n\to\infty} \frac{J_V(u_n)}{I(u_n)} = A(V)\,. \tag{4.78}$$

By relation (4.72), the sequence $\{u_n\}$ is bounded in X. Since X is reflexive, it follows that there exists $u \in X$ such that, up to a subsequence, (u_n) weakly converges to u in X. On the other hand, the functional J_0 (obtained for $V = 0$ on Ω) is weakly lower semi-continuous. Thus we find

$$\liminf_{n\to\infty} J_0(u_n) \geq J_0(u)\,. \tag{4.79}$$

By the compact embedding theorem for Sobolev spaces and assumptions (4.62), (4.63) and (4.64), it follows that X is compactly embedded in $L^{\sigma(x)}(\Omega)$ (where $\sigma(x) = m(x) \cdot r(x)/(r(x)-1)$) and $L^{q_i(x)}(\Omega)$ with $i = 1, 2$. Thus (u_n) strongly converges in $L^{\sigma(x)}(\Omega)$ and $L^{q_i(x)}(\Omega)$ with $i = 1, 2$. Then by relations (4.66) combined with the Hölder inequality, it follows that

$$\lim_{n\to\infty} I(u_n) = I(u) \tag{4.80}$$

and

$$\lim_{n\to\infty} \int_\Omega V(x)|u_n|^{m(x)}\, dx = \int_\Omega V(x)|u|^{m(x)}\, dx\,. \tag{4.81}$$

Relations (4.79), (4.80) and (4.81) imply that if $u \not\equiv 0$ then

$$\frac{J_V(u)}{I(u)} = A(V)\,.$$

Thus, in order to conclude that the lemma holds, it is enough to show that u is not trivial. Assume to the contrary. Then u_n weakly converges to 0 in X and strongly in $L^{s(x)}(\Omega)$ for any $s(x) \in C(\overline{\Omega})$ with $1 < s(x) < \frac{N(\phi_1)_0}{N-(\phi_1)_0}$ on $\overline{\Omega}$. In other words, we have

$$\lim_{n\to\infty} I(u_n) = 0\,, \tag{4.82}$$

and

$$\lim_{n\to\infty} \int_\Omega V(x)|u_n|^{m(x)}\, dx = 0\,. \tag{4.83}$$

Letting $\epsilon \in (0, |A(V)|)$ be fixed by relation (4.78), we deduce that for large enough n we have

$$|J_V(u_n) - A(V)I(u_n)| < \epsilon I(u_n)\,,$$

or

$$(|A(V)| - \epsilon)I(u_n) < J_V(u_n) < (|A(V)| + \epsilon)I(u_n)\,.$$

Passing to the limit in the above inequalities and invoking relation (4.82), we find

$$\lim_{n\to\infty} J_V(u_n) = 0\,.$$

Next, by relation (4.83), we obtain

$$\lim_{n\to\infty} J_0(u_n) = 0\,.$$

This fact implies that u_n actually strongly converges to 0 in X, that is, $\lim_{n\to\infty} \|u_n\|_1 = 0$. From this information and relation (4.73) we deduce that

$$\lim_{n\to\infty} \frac{J_V(u_n)}{I(u_n)} = \infty,$$

which is a contradiction. Thus $u \not\equiv 0$. The proof of Lemma 35 is complete. □

By Lemma 35 we conclude that there exists $u \in X \setminus \{0\}$ such that

$$\frac{J_V(u)}{I(u)} = A(V) = \inf_{w \in X \setminus \{0\}} \frac{J_V(w)}{I(w)} . \tag{4.84}$$

Then we have for any $w \in X$

$$\frac{d}{d\epsilon} \frac{J_V(u+\epsilon w)}{I(u+\epsilon w)} |_{\epsilon=0} = 0 .$$

A simple computation yields

$$\langle J_V'(u), w\rangle I(u) - J_V(u)\langle I'(u), w\rangle = 0 , \tag{4.85}$$

for all $w \in X$. Relation (4.85) combined with the fact that $J_V(u) = A(V) \cdot I(u)$ and $I(u) \neq 0$ implies that $A(V)$ is an eigenvalue of problem (4.56).

Next, we show that any $\lambda \in (A(V), \infty)$ is an eigenvalue of problem (4.56). Let $\lambda \in (A(V), \infty)$ be arbitrary but fixed. Define $T_{V,\lambda} : X \to \mathbb{R}$ by

$$T_{V,\lambda}(u) = J_V(u) - \lambda I(u) .$$

Clearly, $T_{V,\lambda} \in C^1(X, \mathbb{R})$ with

$$\langle T_{V,\lambda}'(u), v\rangle = \langle J_V'(u), v\rangle - \lambda\langle I'(u), v\rangle, \quad \forall\, u \in X.$$

Thus λ is an eigenvalue of problem (4.56) if and only if there exists a critical point $u_\lambda \in X \setminus \{0\}$ of $T_{V,\lambda}$.

With similar arguments as in the proof of relation (4.72) we can show that $T_{V,\lambda}$ is coercive, that is, $\lim_{\|u\| \to \infty} T_{V,\lambda}(u) = \infty$. On the other hand, the functional $T_{V,\lambda}$ is weakly lower semi-continuous. Thus there exists $u_\lambda \in X$ a global minimum point of $T_{V,\lambda}$, hence a critical point of $T_{V,\lambda}$. It is enough to show that u_λ is not trivial. Indeed, since $A(V) = \inf_{u \in X \setminus \{0\}} \frac{J_V(u)}{I(u)}$ and $\lambda > A(V)$ it follows that there exists $v_\lambda \in X$ such that

$$J_V(v_\lambda) < \lambda I(v_\lambda) ,$$

or

$$T_{V,\lambda}(v_\lambda) < 0 .$$

Thus

$$\inf_X T_{V,\lambda} < 0$$

and we conclude that u_λ is a nontrivial critical point of $T_{V,\lambda}$, or λ is an eigenvalue of problem (4.56).

Finally, we prove that no $\lambda < B(V)$ is an eigenvalue of problem (4.56). We assume to the contrary that there exists an eigenvalue $\lambda < B(V)$ of problem (4.56). It follows that there exists $u_\lambda \in X \setminus \{0\}$ such that

$$\langle J_V'(u_\lambda), u_\lambda\rangle = \lambda\langle I'(u_\lambda), u_\lambda\rangle .$$

Since $u_\lambda \neq 0$ we have $\langle I'(u_\lambda), u_\lambda\rangle > 0$. Using this fact and the definition of $B(V)$ it follows that

$$\langle J_V'(u_\lambda), u_\lambda\rangle = \lambda\langle I'(u_\lambda), u_\lambda\rangle < B(V)\langle I'(u_\lambda), u_\lambda\rangle \leq \langle J_V'(u_\lambda), u_\lambda\rangle\,.$$

Obviously, this is a contradiction. We deduce that no $\lambda \in (-\infty, B(V))$ is an eigenvalue of problem (4.56). Furthermore, we have $A(V) \geq B(V)$. The proof of Theorem 28 is complete. □

Remark 6 *In the case when $V = 0$ in Ω, we deduce by direct computation that $A(0) > 0$.*

4.3.2 Optimization property for the Rayleigh quotient

We shall now give the proof of Theorem 29.

Let S be a convex, bounded and closed subset of $L^{r(x)}(\Omega)$ and

$$A_\star := \inf_{V\in S} A(V)\,.$$

Relation (4.75) assures that $A_\star$ is finite.

Let $(V_n) \subset S$ be a minimizing sequence for $A_\star$, that is,

$$A(V_n) \to A_\star, \quad \text{as } n \to \infty\,.$$

Obviously, (V_n) is a bounded sequence and thus, there exists $V_\star \in L^{r(x)}(\Omega)$ such that V_n weakly converges to $V_\star$ in $L^{r(x)}(\Omega)$. Moreover, since S is convex and closed it is also weakly closed, hence $V_\star \in S$.

Next, we show that $A(V_\star) = A_\star$. Indeed, by Theorem 28, for each positive integer n there exists $u_n \in X \setminus \{0\}$ such that

$$\frac{J_{V_n}(u_n)}{I(u_n)} = A(V_n)\,. \tag{4.86}$$

Since $(A(V_n))$ is a bounded sequence, we have by relation (4.75)

$$\frac{J_{V_n}(u_n)}{I(u_n)} \geq \beta\frac{J_0(u_n)}{I(u_n)} - C, \quad \text{for any } n\,,$$

where C is a positive constant. We infer that (u_n) is bounded in X and it cannot contain a subsequence converging to 0 (otherwise we obtain a contradiction by applying Lemma 34). Thus there exists $u_0 \in X \setminus \{0\}$ such that (u_n) weakly converges to u_0 in X. Using relation (4.63) (and thus, $W_0^1L_{\Phi_1}(\Omega) \subset W_0^{1,(\phi_1)_0}(\Omega)$) and the Rellich-Kondrachov theorem we deduce that (u_n) strongly converges to u_0 in $L^{s(x)}(\Omega)$ for any $s(x) \in C(\overline{\Omega})$ satisfying $1 < s(x) < \frac{N(\phi_1)_0}{N-(\phi_1)_0}$ for any $x \in \overline{\Omega}$. In particular, using conditions (4.62), (4.63) and (4.64), we deduce that (u_n) converges to u_0 in $L^{m(x)}(\Omega)$ and in

$L^{m(x)\cdot r'(x)}(\Omega)$ where $r'(x) = \frac{r(x)}{r(x)-1}$. Using this information, the Hölder inequality and the facts that $V_\star \in L^{r(x)}(\Omega)$ and (V_n) is bounded in $L^{r(x)}(\Omega)$, we conclude

$$\lim_{n\to\infty} \int_\Omega \frac{V_\star(x)}{m(x)} |u_n|^{m(x)}\, dx = \int_\Omega \frac{V_\star(x)}{m(x)} |u_0|^{m(x)}\, dx \tag{4.87}$$

and

$$\lim_{n\to\infty} \int_\Omega \left(\frac{V_n(x)}{m(x)} |u_n|^{m(x)} - \frac{V_n(x)}{m(x)} |u_0|^{m(x)} \right) dx = 0\,. \tag{4.88}$$

On the other hand, since (V_n) weakly converges to $V_\star$ in $L^{r(x)}(\Omega)$ and $u_0 \in L^{m(x)\cdot r'(x)}(\Omega)$, where $r'(x) = \frac{r(x)}{r(x)-1}$, we deduce that

$$\lim_{n\to\infty} \int_\Omega \frac{V_n(x)}{m(x)} |u_0|^{m(x)}\, dx = \int_\Omega \frac{V_\star(x)}{m(x)} |u_0|^{m(x)}\, dx\,. \tag{4.89}$$

Combining the equality

$$\begin{aligned}
&\int_\Omega \frac{V_\star(x)}{m(x)} |u_n|^{m(x)}\, dx - \int_\Omega \frac{V_n(x)}{m(x)} |u_n|^{m(x)}\, dx = \\
&\int_\Omega \frac{V_\star(x)}{m(x)} |u_n|^{m(x)}\, dx - \int_\Omega \frac{V_\star(x)}{m(x)} |u_0|^{m(x)}\, dx + \\
&\int_\Omega \frac{V_\star(x)}{m(x)} |u_0|^{m(x)}\, dx - \int_\Omega \frac{V_n(x)}{m(x)} |u_0|^{m(x)}\, dx + \\
&\int_\Omega \frac{V_n(x)}{m(x)} |u_0|^{m(x)}\, dx - \int_\Omega \frac{V_n(x)}{m(x)} |u_n|^{m(x)}\, dx\,,
\end{aligned}$$

with relations (4.87), (4.88) and (4.89) we obtain

$$\lim_{n\to\infty} \int_\Omega \left(\frac{V_\star(x)}{m(x)} |u_n|^{m(x)} - \frac{V_n(x)}{m(x)} |u_n|^{m(x)} \right) dx = 0\,. \tag{4.90}$$

Since

$$A(V_\star) = \inf_{u\in X\setminus\{0\}} \frac{J_{V_\star}(u)}{I(u)}\,,$$

it follows that

$$A(V_\star) \le \frac{J_{V_\star}(u_n)}{I(u_n)}\,.$$

Combining the above inequality and equality (4.86) we obtain

$$A(V_\star) \le \frac{J_{V_\star}(u_n) - J_{V_n}(u_n)}{I(u_n)} + A(V_n)\,.$$

Taking into account relation (4.90), the fact that $I(u_n)$ is bounded and does not converge to 0 and $(A(V_n))$ converges to $A_\star$, then passing to the limit as $n \to \infty$ in the last inequality, we infer that

$$A(V_\star) \le A_\star \,.$$

But using the definition of $A_\star$ and the fact that $V_\star \in S$ we conclude that actually

$$A(V_\star) = A_\star \,.$$

The proof of Theorem 29 is complete. □

4.3.3 More about the Rayleigh quotient on spheres

In this section we provide the proof of Theorem 30. This result is stated on *spheres* and establishes that the minimum of the Rayleigh quotient with respect to all admissible indefinite potentials is nonconstant, continuous and nonincreasing.

We first argue that the function $A_\star$ is nonconstant. Indeed, by Remark 6, we have $A_\star(0) = A(0) > 0$. On the other hand, we have

$$\lambda_m := \inf_{u \in X \setminus \{0\}} \frac{\displaystyle\int_\Omega \Phi_1(|\nabla u|)\, dx + \int_\Omega \Phi_2(|\nabla u|)\, dx}{\displaystyle\int_\Omega \frac{1}{m(x)} |u|^{m(x)}\, dx} > 0 \,.$$

Moreover, there exists $u_m \in X \setminus \{0\}$ such that

$$\lambda_m = \frac{\displaystyle\int_\Omega \Phi_1(|\nabla u_m|)\, dx + \int_\Omega \Phi_2(|\nabla u_m|)\, dx}{\displaystyle\int_\Omega \frac{1}{m(x)} |u_m|^{m(x)}\, dx} \,.$$

Thus, taking $V_m(x) = -\lambda_m$ for all $x \in \Omega$ it is clear that $V_m \in L^\infty(\Omega) \subset L^{r(x)}(\Omega)$ and

$$\frac{J_{V_m}(u_m)}{I(u_m)} = 0 \,.$$

It follows that

$$A(V_m) \le 0 \,,$$

and we find

$$A_\star(\lambda_m) \le 0 \,.$$

We conclude that $A_\star$ is nonconstant. Furthermore, the function $A_\star$ also takes on negative values. Indeed, we just notice that for all $\lambda > \lambda_m$, there exists $u_\lambda \in X \setminus \{0\}$ such that taking $V_\lambda = -\lambda$ for all $x \in \Omega$ we have

$$\frac{J_{V_\lambda}(u_\lambda)}{I(u_\lambda)} < 0 \,.$$

Next, we show that $A_\star$ decreases monotonically. Indeed, if $0 \leq R_1 < R_2$ then $\overline{B}_{R_1}(0) \subset \overline{B}_{R_2}(0)$. Thus the definition of $A_\star$ implies that $A_\star(R_1) \geq A_\star(R_2)$.

Finally, we show that the function $A_\star$ is continuous. Let $R > 0$ and $t \in (0, R)$ be fixed. We verify that $\lim_{t\searrow 0} A_\star(R+t) = \lim_{t\searrow 0} A_\star(R-t) = A_\star(R)$.

First, we prove that $\lim_{t\searrow 0} A_\star(R+t) = A_\star(R)$. By Theorem 30 a) we have

$$A_\star(R) \geq A_\star(R+t)\,.$$

Moreover, by Theorem 29 it follows that there exists $V_{R+t} \in \overline{B}_{R+t}(0)$ (i.e. $|V_{R+t}|_{r(x)} \leq R+t$) such that

$$A(V_{R+t}) = A_\star(R+t)\,.$$

Taking now $V_{R,t} := \frac{R}{R+t} V_{R+t}$ we have

$$|V_{R,t}|_{r(x)} = \frac{R}{R+t}|V_{R+t}|_{r(x)} \leq R\,,$$

or $V_{R,t} \in \overline{B}_R(0)$. Therefore, we obviously have $A(V_{R,t}) \geq A_\star(R)$.

On the other hand, by Theorem 28, there exists $u_t \in X \setminus \{0\}$ such that

$$A(V_{R+t}) = \frac{J_{V_{R+t}}(u_t)}{I(u_t)}\,.$$

Therefore

$$\begin{aligned}
A_\star(R+t) = A(V_{R+t}) &= \frac{J_{V_{R+t}}(u_t)}{I(u_t)} \\
&= \frac{J_{\frac{R+t}{R}\cdot V_{R,t}}(u_t)}{I(u_t)} \\
&= \frac{R+t}{R}\cdot\frac{J_{V_{R,t}}(u_t)}{I(u_t)} - \frac{t}{R}\cdot\frac{J_0(u_t)}{I(u_t)} \\
&\geq \frac{R+t}{R}\cdot A_\star(R) - \frac{t}{R}\cdot\frac{J_0(u_t)}{I(u_t)}\,.
\end{aligned}$$

On the other hand, by relation (4.75) we have that for each $t \in (0, R)$

$$\begin{aligned}
A_\star(R) \geq A_\star(R+t) = A(V_{R+t}) &= \frac{J_{V_{R+t}}(u_t)}{I(u_t)} \\
&\geq \beta_1\cdot\frac{J_0(u_t)}{I(u_t)} - \gamma\cdot|V_{R+t}|_{r(x)} \\
&= \beta_1\cdot\frac{J_0(u_t)}{I(u_t)} - \gamma\cdot 2R\,,
\end{aligned}$$

where $\beta_1 > 0$ and $\gamma > 0$ are real constants.

Combining the last two inequalities we deduce that

$$A_\star(R) \geq A_\star(R+t) \geq \frac{R+t}{R} \cdot A_\star(R) - \frac{t}{R} \cdot \frac{A_\star(R) + \gamma \cdot 2R}{\beta_1},$$

for each $t \in (0, R)$.

We conclude that

$$\lim_{t \searrow 0} A_\star(R+t) = A_\star(R).$$

Next we argue that $\lim_{t \searrow 0} A_\star(R-t) = A_\star(R)$.

Obviously,

$$A_\star(R) \leq A_\star(R-t), \quad \forall\, t \in (0, R).$$

By Theorem 29, there exists $V_R \in \overline{B}_R(0)$ such that

$$A_\star(R) = A(V_R).$$

Moreover, by Theorem 28, there exists $u_0 \in X \setminus \{0\}$ such that

$$A(V_R) = \frac{J_{V_R}(u_0)}{I(u_0)}.$$

Define now

$$V_t := \frac{R-t}{R} V_R, \quad \forall\, t \in (0, R).$$

Then $V_t \in \overline{B}_{R-t}(0)$. It follows that

$$\frac{J_{V_t}(u_0)}{I(u_0)} \geq A_\star(R-t), \quad \forall\, t \in (0, R).$$

Thus for all $t \in (0, R)$

$$\begin{aligned} A_\star(R) = A(V_R) = \frac{J_{V_R}(u_0)}{I(u_0)} &= \frac{J_{\frac{R}{R-t}V_t}(u_0)}{I(u_0)} \\ &= \frac{J_{V_t}(u_0)}{I(u_0)} + \frac{t}{R-t} \cdot \frac{\displaystyle\int_\Omega \frac{V_t(x)}{m(x)} |u_0|^{m(x)}\, dx}{I(u_0)} \\ &\geq A_\star(R-t) + \frac{t}{R} \cdot \frac{\displaystyle\int_\Omega \frac{V_R(x)}{m(x)} |u_0|^{m(x)}\, dx}{I(u_0)}. \end{aligned}$$

We infer that

$$\lim_{t \searrow 0} A_\star(R-t) = A_\star(R).$$

It follows that the function $A_\star$ is continuous. The proof of Theorem 30 is complete. □

By Theorem 30 a), the function $A_\star$ decreases monotonically. We notice that in the special case when $q_1(x) = m(x) = q_2(x) = q$ for each $x \in \overline{\Omega}$, where $q > 1$ is a real number, for which conditions (4.62), (4.63) and (4.64) are fulfilled, the above result can be improved, in the sense that we can show that $A_\star$ is actually strictly decreasing in $[0, \infty)$. Indeed, letting $0 \leq R_1 < R_2$ be given, by Theorem 29, there exists $V_1 \in \overline{B}_{R_1}(0)$ such that

$$A(V_1) = A_\star(R_1).$$

Then for each real number $t \in (0, R_2 - R_1)$ we have $V_1 - t \in \overline{B}_{R_2}(0)$ since $|V_1 - t|_{r(x)} \leq |V_1|_{r(x)} + t \leq R_2$. Next, by Theorem 28 there exists $u_1 \in X \setminus \{0\}$ such that

$$A(V_1) = \frac{J_{V_1}(u_1)}{I(u_1)}.$$

Taking into account all the above remarks we infer

$$A_\star(R_1) - \frac{t}{2} = A(V_1) - \frac{t}{2} = \frac{J_{V_1}(u_1)}{I(u_1)} - \frac{t}{2} = \frac{J_{V_1 - t}(u_1)}{I(u_1)} \geq A(V_1 - t) \geq A_\star(R_2),$$

or

$$A_\star(R_1) > A_\star(R_2).$$

Finally, we point out that by Theorem 30 b) we have

$$A_\star(R) = \inf_{s \leq R} A_\star(s) \quad \text{and} \quad A_\star(R) = \sup_{s \geq R} A_\star(s).$$

4.4 Multiple solutions in Orlicz-Sobolev spaces

In this section, we consider the problem

$$\begin{cases} -\mathrm{div}(a(|\nabla u(x)|)\nabla u(x)) + a(|u(x)|)u(x) = \lambda\, f(x, u(x)), & \text{for} \quad x \in \Omega, \\ \dfrac{\partial u}{\partial \nu}(x) = 0, & \text{for} \quad x \in \partial\Omega, \end{cases} \tag{4.91}$$

where Ω is a bounded domain in $\mathbb{R}^N$ ($N \geq 3$) with smooth boundary $\partial\Omega$, ν is the outer unit normal to $\partial\Omega$, while $f : \Omega \times \mathbb{R} \to \mathbb{R}$ is a Carathéodory function, and λ is a positive parameter. Throughout this section we assume that the function $a : (0, \infty) \to \mathbb{R}$ is such that the mapping $\phi : \mathbb{R} \to \mathbb{R}$ defined by

$$\phi(t) = \begin{cases} a(|t|)t, & \text{for} \quad t \neq 0 \\ 0, & \text{for} \quad t = 0, \end{cases} \tag{4.92}$$

is an odd, strictly increasing homeomorphism from $\mathbb{R}$ onto $\mathbb{R}$.

Set

$$\Phi(t) = \int_0^t \phi(s)\, ds, \quad \Phi^\star(t) = \int_0^t \phi^{-1}(s)\, ds \qquad \text{for all } t \in \mathbb{R}. \tag{4.93}$$

We observe that Φ is a *Young function*, that is, $\Phi(0) = 0$, Φ is convex, and $\lim_{t\to\infty} \Phi(t) = +\infty$. Furthermore, since $\Phi(t) = 0$ if and only if $t = 0$, $\lim_{t\to 0} \Phi(t)/t = 0$, and $\lim_{t\to\infty} \Phi(t)/t = +\infty$, then Φ is called an *N–function*. The function $\Phi^\star$ is called the *complementary* function of Φ and it satisfies

$$\Phi^\star(t) = \sup\{st - \Phi(s);\ s \geq 0\}, \qquad \text{for all } t \geq 0\,.$$

We observe that $\Phi^\star$ is also an N–function and the following Young-type inequality holds:

$$st \leq \Phi(s) + \Phi^\star(t), \qquad \text{for all } s, t \geq 0\,.$$

We assume that

$$1 < \liminf_{t\to\infty} \frac{t\phi(t)}{\Phi(t)} \leq \sup_{t>0} \frac{t\phi(t)}{\Phi(t)} < \infty. \tag{Φ_0}$$

Due to assumption (Φ_0), we may define the numbers

$$p_0 := \inf_{t>0} \frac{t\phi(t)}{\Phi(t)} \quad \text{and} \quad p^0 := \sup_{t>0} \frac{t\phi(t)}{\Phi(t)}. \tag{4.94}$$

We assume that the nonlinear term $f : \Omega \times \mathbb{R} \to \mathbb{R}$ satisfies the following hypotheses:

(f0) there exist $c_0 > 0$ and $0 < s < p_0 - 1$ such that $|f(x,t)| \leq c_0(1 + |t|^s)$ for every $(x,t) \in \Omega \times \mathbb{R}$;

(f1) there exists $b \in \mathbb{R}$ such that $B_F = \int_\Omega F(x,b)dx > 0$, where $F(x,t) = \int_0^t f(x,w)dw$, $t \in \mathbb{R}$;

(f2) there exists $\delta > 0$ such that $f(x,t)t \leq 0$ for every $x \in \Omega$ and $t \in [-\delta, \delta]$.

Roughly speaking, by (f0), the growth of $f(x,\cdot)$ is $(p_0 - 1)$-*sublinear at infinity*. In this setting, the presence of the eigenvalue $\lambda > 0$ in problem (4.91) is indispensable. Indeed, if we analyze even the simplest case $a(t) = 1$ that corresponds to the Laplace equation, and we assume that $f(x,\cdot)$ is uniformly Lipschitz with Lipschitz constant $L > 0$ (uniformly for $x \in \Omega$), then problem (4.91) has only the *trivial* weak solution whenever $\lambda < L^{-1}$. Moreover, (f2) implies in particular that $f(x,0) = 0$ for every $x \in \Omega$; thus $u = 0$ can always be considered a solution of problem (4.91).

Finally, we assume that

$$N < p_0 < \liminf_{t\to\infty} \frac{\log(\Phi(t))}{\log(t)}. \tag{Φ_1}$$

We say that $u \in W^1L_\Phi(\Omega)$ is a *weak solution* for problem (4.91) if

$$\int_\Omega a(|\nabla u|)\nabla u \nabla v\, dx + \int_\Omega a(|u|)uv\, dx - \lambda \int_\Omega f(x,u)v\, dx = 0, \quad \forall v \in W^1L_\Phi(\Omega).$$

The main result in this section establishes the following multiplicity property.

Theorem 31 *Assume that conditions* (Φ_0) *and* (Φ_1) *hold and that the function* $[0,\infty) \ni t \to \Phi(\sqrt{t})$ *is convex. Let* $f : \Omega \times \mathbb{R} \to \mathbb{R}$ *be a Carathéodory function which satisfies hypotheses* (f0), (f1) *and* (f2).

Then there exist a nonempty open interval $\Lambda \subset \left(0, 2\Phi(b)|\Omega|B_F^{-1}\right)$ *and* $\mu > 0$ *such that for any* $\lambda \in \Lambda$ *problem* (4.91) *has at least two nontrivial weak solutions whose norms are less than* μ.

Moreover, we give an estimate for the interval $\Lambda \subset (0,\infty)$ where problem (4.91) has at least two nontrivial weak solutions.

We point out that it is possible for the technical assumption, namely that the function $[0,\infty) \ni t \to \Phi(\sqrt{t})$ is convex, not to be a necessary condition. Actually, this will be used in the proof of Theorem 31 in order to obtain a Clarkson-type inequality for the function Φ, that is,

$$\begin{gathered}\frac{1}{2}\left[\int_\Omega \Phi(|\nabla u|)\,dx + \int_\Omega \Phi(|\nabla v|)\,dx\right] \geq \\ \int_\Omega \Phi\left(\left|\frac{\nabla u + \nabla v}{2}\right|\right)\,dx + \int_\Omega \Phi\left(\left|\frac{\nabla u - \nabla v}{2}\right|\right)\,dx\,,\end{gathered} \tag{4.95}$$

for any $u, v \in W^1L_\Phi(\Omega)$. Inequality (4.95) extends the *classical* Clarkson inequality, obtained for the homogeneous function $\Phi(t) = t^p$ with $p \geq 2$. Unfortunately, at present it is unknown if an inequality of type (4.95) can be stated for a class of functions which do not satisfy the requirement that $t \longmapsto \Phi(\sqrt{t})$ is convex.

4.4.1 More about Orlicz-Sobolev spaces

By our hypotheses, the Orlicz-Sobolev space $W^1L_\Phi(\Omega)$ is continuously embedded in $W^{1,p_0}(\Omega)$. On the other hand, since $p_0 > N$ then $W^{1,p_0}(\Omega)$ is compactly embedded in $C(\overline{\Omega})$. Thus $W^1L_\Phi(\Omega)$ is compactly embedded in $C(\overline{\Omega})$. Defining $\|u\|_\infty = \sup_{x\in\overline{\Omega}} |u(x)|$ we find a positive constant $c > 0$ such that

$$\|u\|_\infty \leq c\,\|u\|_{1,\Phi}, \quad \forall\, u \in W^1L_\Phi(\Omega).$$

We now establish some useful properties of the norms on Orlicz-Sobolev spaces.

Lemma 36 *On* $W^1L_\Phi(\Omega)$ *the norms*

$$\|u\|_{1,\Phi} = \||\nabla u|\|_\Phi + \|u\|_\Phi,$$

$$\|u\|_{2,\Phi} = \max\{\||\nabla u|\|_\Phi, \|u\|_\Phi\},$$

$$\|u\| = \inf\left\{\mu > 0;\ \int_\Omega \left[\Phi\left(\frac{|u(x)|}{\mu}\right) + \Phi\left(\frac{|\nabla u(x)|}{\mu}\right)\right]\,dx \leq 1\right\},$$

are equivalent. More precisely, for every $u \in W^1L_\Phi(\Omega)$ *we have*

$$\|u\| \leq 2\|u\|_{2,\Phi} \leq 2\|u\|_{1,\Phi} \leq 4\|u\|.$$

Proof. We first point out that $\|\cdot\|_{1,\Phi}$ and $\|\cdot\|_{2,\Phi}$ are equivalent, since

$$\|u\|_{2,\Phi} \leq \|u\|_{1,\Phi} \leq 2\|u\|_{2,\Phi}, \quad \forall\, u \in W^1L_\Phi(\Omega). \tag{4.96}$$

In the sequel, we assume that $u \neq 0$. We note that

$$\int_\Omega \Phi\left(\frac{|u(x)|}{\|u\|_\Phi}\right) dx \leq 1 \quad \text{and} \quad \int_\Omega \Phi\left(\frac{|\nabla u(x)|}{\||\nabla u|\|_\Phi}\right) dx \leq 1, \tag{4.97}$$

and

$$\int_\Omega \left[\Phi\left(\frac{|u(x)|}{\|u\|}\right) + \Phi\left(\frac{|\nabla u(x)|}{\|u\|}\right)\right] dx \leq 1. \tag{4.98}$$

By relation (4.98) we obtain

$$\int_\Omega \Phi\left(\frac{|u(x)|}{\|u\|}\right) dx \leq 1 \quad \text{and} \quad \int_\Omega \Phi\left(\frac{|\nabla u(x)|}{\|u\|}\right) dx \leq 1.$$

Using the definition of $\|\cdot\|_\Phi$, we obtain

$$\|u\|_{1,\Phi} = \||\nabla u|\|_\Phi + \|u\|_\Phi \leq 2\|u\|, \quad \forall\, u \in W^1L_\Phi(\Omega). \tag{4.99}$$

On the other hand, since

$$\Phi(t) \leq \frac{t\phi(t)}{p_0}, \quad \forall\, t > 0,$$

with $p_0 > N$, by Lemma C.4 (ii) from [19] we deduce that

$$\Phi(2t) \geq 2\Phi(t), \quad \forall\, t > 0.$$

Thus we obtain for all $u \in W^1L_\Phi(\Omega)$ and all $x \in \Omega$

$$2\Phi\left(\frac{|u(x)|}{2\|u\|_{2,\Phi}}\right) \leq \Phi\left(\frac{|u(x)|}{\|u\|_{2,\Phi}}\right) \quad \text{and} \quad 2\Phi\left(\frac{|\nabla u(x)|}{2\|u\|_{2,\Phi}}\right) \leq \Phi\left(\frac{|\nabla u(x)|}{\|u\|_{2,\Phi}}\right).$$

It follows that

$$\begin{gathered}\int_\Omega \left[\Phi\left(\frac{|u(x)|}{2\|u\|_{2,\Phi}}\right) + \Phi\left(\frac{|\nabla u(x)|}{2\|u\|_{2,\Phi}}\right)\right] dx \leq \\ \frac{1}{2}\left\{\int_\Omega \left[\Phi\left(\frac{|u(x)|}{\|u\|_{2,\Phi}}\right) + \Phi\left(\frac{|\nabla u(x)|}{\|u\|_{2,\Phi}}\right)\right] dx\right\}.\end{gathered} \tag{4.100}$$

But, since

$$\|u\|_{2,\Phi} \geq \|u\|_\Phi \quad \text{and} \quad \|u\|_{2,\Phi} \geq \||\nabla u|\|_\Phi, \quad \forall\, u \in W^1L_\Phi(\Omega),$$

we get for all $u \in W^1L_\Phi(\Omega)$ and all $x \in \Omega$

$$\frac{|u(x)|}{\|u\|_\Phi} \geq \frac{|u(x)|}{\|u\|_{2,\Phi}} \quad \text{and} \quad \frac{|\nabla u(x)|}{\||\nabla u|\|_\Phi} \geq \frac{|\nabla u(x)|}{\|u\|_{2,\Phi}}. \tag{4.101}$$

Taking into account that Φ is increasing on $[0,\infty)$, we obtain by (4.100), (4.101) and (4.97)

$$\int_\Omega \left[\Phi\left(\frac{|u(x)|}{2\|u\|_{2,\Phi}}\right)+\Phi\left(\frac{|\nabla u(x)|}{2\|u\|_{2,\Phi}}\right)\right]\,dx \leq$$
$$\frac{1}{2}\left\{\int_\Omega \left[\Phi\left(\frac{|u(x)|}{\|u\|_\Phi}\right)+\Phi\left(\frac{|\nabla u(x)|}{\||\nabla u|\|_\Phi}\right)\right]\,dx\right\}\leq 1,$$

for all $u\in W^1L_\Phi(\Omega)$. Thus we conclude that

$$\|u\|\leq 2\|u\|_{2,\Phi},\quad \forall\, u\in W^1L_\Phi(\Omega). \tag{4.102}$$

We deduce from relations (4.96), (4.99) and (4.102) that Lemma 36 holds. □

Lemma 37 *The following relations hold:*

$$\int_\Omega [\Phi(|u(x)|)+\Phi(|\nabla u(x)|)]\,dx\geq \|u\|^{p_0},\quad \forall\, u\in W^1L_\Phi(\Omega) \text{ with } \|u\|>1$$

and

$$\int_\Omega [\Phi(|u(x)|)+\Phi(|\nabla u(x)|)]\,dx\geq \|u\|^{p^0},\quad \forall\, u\in W^1L_\Phi(\Omega) \text{ with } \|u\|<1.$$

Proof. We first assume that $\|u\|>1$. Let $\beta\in(1,\|u\|)$. By Lemma C.4 (ii) from [19] we have

$$\int_\Omega [\Phi(|u(x)|)+\Phi(|\nabla u(x)|)]\,dx\geq$$
$$\beta^{p_0}\cdot\int_\Omega\left[\Phi\left(\frac{|u(x)|}{\beta}\right)+\Phi\left(\frac{|\nabla u(x)|}{\beta}\right)\right]\,dx\geq\beta^{p_0}.$$

Letting $\beta\nearrow\|u\|$ we find

$$\int_\Omega [\Phi(|u(x)|)+\Phi(|\nabla u(x)|)]\,dx\geq \|u\|^{p_0},\quad \forall\, u\in W^1L_\Phi(\Omega) \text{ with } \|u\|>1.$$

Next, we assume that $\|u\|<1$. Let $\xi\in(0,\|u\|)$. By the definition of p^0 (see (4.94)), we deduce that

$$\Phi(t)\geq\tau^{p^0}\Phi(t/\tau),\quad \forall\, t>0,\ \tau\in(0,1).$$

Therefore

$$\begin{aligned}&\int_\Omega [\Phi(|u(x)|)+\Phi(|\nabla u(x)|)]\,dx\geq\\ &\xi^{p^0}\cdot\int_\Omega\left[\Phi\left(\frac{|u(x)|}{\xi}\right)+\Phi\left(\frac{|\nabla u(x)|}{\xi}\right)\right]\,dx.\end{aligned} \tag{4.103}$$

Defining $v(x) = u(x)/\xi$, for all $x \in \Omega$, we have $\|v\| = \|u\|/\xi > 1$. Using the first inequality of this lemma we obtain

$$\int_\Omega [\Phi(|v(x)|) + \Phi(|\nabla v(x)|)]\, dx \geq \|v\|^{p_0} > 1. \tag{4.104}$$

Relations (4.103) and (4.104) yield

$$\int_\Omega [\Phi(|u(x)|) + \Phi(|\nabla u(x)|)]\, dx \geq \xi^{p^0}.$$

Letting $\xi \nearrow \|u\|$ in the above inequality we obtain for all $u \in W^1L_\Phi(\Omega)$ with $\|u\| < 1$

$$\int_\Omega [\Phi(|u(x)|) + \Phi(|\nabla u(x)|)]\, dx \geq \|u\|^{p^0}.$$

The proof of Lemma 37 is complete. □

4.4.2 Multiplicity versus the three critical points theorem

The key argument in the proof of Theorem 31 is the *three critical points* theorem due to Ricceri [94]. Ricceri's result goes back to an elementary property established by Pucci and Serrin [83], which asserts that if a functional of class C^1 defined on a real Banach space has two local minima, then it has a third critical point. This is an auxiliary result related to a problem of Rabinowitz [89], who raised the question whether critical points of mountain pass type must necessarily be saddle points. To the best of our knowledge, the first *three critical points* property was first found by Krasnoselskii [43]. He showed that if f is a coercive C^1 functional defined on a finite dimensional space having a nondegenerate critical point x_0 (that is, the *topological index* $\operatorname{ind} f'(x_0)(0)$ is different from zero) which is not a global minimum, then f admits a third critical point.

We now state the Ricceri multiplicity theorem.

Theorem 32 *Let E be a separable and reflexive real Banach space, and let $J, I : E \to \mathbb{R}$ be continuously Gâteaux differentiable functionals. Assume that there exist $u_0 \in E$ such that $J(u_0) = I(u_0) = 0$ and $J(u) \geq 0$ for every $u \in E$ and that there exist $u_1 \in E$, $r > 0$, such that*

(i) $r < J(u_1)$;

(ii) $\displaystyle\sup_{J(u)<r} I(u) < r\frac{I(u_1)}{J(u_1)}$.

Furthermore, put

$$\overline{a} = \frac{\zeta r}{r\frac{I(u_1)}{J(u_1)} - \sup_{J(u)<r} I(u)},$$

with $\zeta > 1$, and assume that the functional $J - \lambda I$ is sequentially weakly lower semicontinuous, satisfies the Palais-Smale condition and

(iii) $\lim_{\|u\|\to+\infty}(J(u)-\lambda I(u))=+\infty$, *for every* $\lambda\in[0,\overline{a}]$.

Then there exist a nonempty open interval $\Lambda\subset[0,\overline{a}]$ *and a number* $\mu>0$ *such that for each* $\lambda\in\Lambda$, *the equation* $J'(u)-\lambda I'(u)=0$ *admits at least three solutions in* E *having the norm less than* μ.

From now on, we assume that the hypotheses of Theorem 32 are fulfilled. Denote $E=W^1L_\Phi(\Omega)$ and define the functionals J, $I:E\to\mathbb{R}$ by

$$J(u)=\int_\Omega(\Phi(|\nabla u|)+\Phi(|u|))\,dx \quad\text{and}\quad I(u)=\int_\Omega F(x,u(x))\,dx.$$

Then J, $I\in C^1(E,\mathbb{R})$ and

$$\langle J'(u),v\rangle=\int_\Omega a(|\nabla u|)\nabla u\nabla v\,dx+\int_\Omega a(|u|)uv\,dx,$$

$$\langle I'(u),v\rangle=\int_\Omega f(x,u)v\,dx,$$

for any u, $v\in E$.

Let us observe that $u\in E$ is a weak solution of problem (4.91) if there exists $\lambda>0$ such that u is a critical point of the functional $J-\lambda I$. Therefore, we can seek for weak solutions of problem (4.91) by applying Theorem 32. In the sequel, we verify the hypotheses of Theorem 32. We first prove the following auxiliary result.

Lemma 38 $J':E\to E^\star$ *has a continuous inverse operator on* $E^\star$.

Proof. We use Theorem 26.A (d) from Zeidler [105], namely it is enough to verify that J' is coercive, hemicontinuous and uniformly monotone.

Indeed, since Φ is convex it follows that J is convex as well. Thus we have

$$J(u)\le\langle J'(u),u\rangle,\quad\forall\,u\in E.$$

By Lemma 37, for any $u\in E$ with $\|u\|>1$ we have

$$\frac{\langle J'(u),u\rangle}{\|u\|}\ge\frac{J(u)}{\|u\|}\ge\|u\|^{p_0-1}.$$

Thus,

$$\lim_{\|u\|\to\infty}\frac{\langle J'(u),u\rangle}{\|u\|}=\infty,$$

hence J' is coercive.

The fact that J' is hemicontinuous can be verified by standard arguments.

Finally, we show that J' is uniformly monotone. Indeed, since Φ is convex we have

$$\Phi(|\nabla u(x)|)\le$$
$$\Phi\left(\left|\frac{\nabla u(x)+\nabla v(x)}{2}\right|\right)+a(|\nabla u(x)|)\nabla u(x)\cdot\frac{\nabla u(x)-\nabla v(x)}{2},$$

and

$$\Phi(|\nabla v(x)|) \leq \Phi\left(\left|\frac{\nabla u(x) + \nabla v(x)}{2}\right|\right) + a(|\nabla v(x)|)\nabla v(x) \cdot \frac{\nabla v(x) - \nabla u(x)}{2},$$

for every $u, v \in E$ and $x \in \Omega$. Adding these relations and integrating we find

$$\frac{1}{2}\int_\Omega (a(|\nabla u|)\nabla u - a(|\nabla v|)\nabla v) \cdot (\nabla u - \nabla v)\, dx \geq \int_\Omega \Phi(|\nabla u|)\, dx + \int_\Omega \Phi(|\nabla v|)\, dx - 2\int_\Omega \Phi\left(\left|\frac{\nabla u + \nabla v}{2}\right|\right)\, dx, \tag{4.105}$$

for any $u,\ v \in E$.

On the other hand, since $\Phi : [0, \infty) \to \mathbb{R}$ is an increasing continuous function with $\Phi(0) = 0$, and $t \mapsto \Phi(\sqrt{t})$ is convex, we deduce that

$$\frac{1}{2}\left[\int_\Omega \Phi(|\nabla u|)\, dx + \int_\Omega \Phi(|\nabla v|)\, dx\right] \geq \int_\Omega \Phi\left(\left|\frac{\nabla u + \nabla v}{2}\right|\right)\, dx + \int_\Omega \Phi\left(\left|\frac{\nabla u - \nabla v}{2}\right|\right)\, dx, \tag{4.106}$$

for any $u,\ v \in E$.

Relations (4.105) and (4.106) yield

$$\int_\Omega (a(|\nabla u|)\nabla u - a(|\nabla v|)\nabla v) \cdot (\nabla u - \nabla v)\, dx \geq 4\int_\Omega \Phi\left(\left|\frac{\nabla u - \nabla v}{2}\right|\right)\, dx, \quad \forall u, v \in E. \tag{4.107}$$

Similarly, for all $u, v \in E$,

$$\int_\Omega (a(|u|)u - a(|v|)v) \cdot (u - v)\, dx \geq 4\int_\Omega \Phi\left(\left|\frac{u - v}{2}\right|\right)\, dx. \tag{4.108}$$

Relations (4.107) and (4.108) yield

$$\langle J'(u) - J'(v), u - v\rangle \geq 4J\left(\frac{u - v}{2}\right).$$

Define the function $\alpha : [0, \infty) \to [0, \infty)$ by

$$\alpha(t) = \frac{1}{2^{p^0 - 2}} \cdot \begin{cases} t^{p^0 - 1}, & \text{for} \quad t \leq 1 \\ t^{p_0 - 1}, & \text{for} \quad t \geq 1\,. \end{cases}$$

Then α is an increasing function with $\alpha(0) = 0$ and $\lim_{t\to\infty} \alpha(t) = \infty$. Thus, by Lemma 37, we deduce that

$$\langle J'(u) - J'(v), u - v\rangle \geq \alpha(\|u - v\|) \cdot \|u - v\|, \quad \forall\, u,\ v \in E,$$

hence J' is uniformly monotone, which concludes our proof. □

Now, we verify the hypotheses of Theorem 32 in three steps.

Step 1. For every $\lambda > 0$, the functional $J - \lambda I$ is coercive, that is, (iii) *holds.*

Indeed, by Lemma 37 we deduce that for any $u \in E$ with $\|u\| > 1$ we have $J(u) \geq \|u\|^{p_0}$. On the other hand, by (f0), there exists $c_1 > 0$ such that for all $u, v \in E$

$$\int_\Omega F(x, u(x))\, dx \leq c_1 (\int_\Omega (|u| + |u|^{s+1})\, dx \leq c_1 |\Omega| (\|u\|_\infty + \|u\|_\infty^{s+1}).$$

Since E is compactly embedded into $C(\overline{\Omega})$, by Lemma 36 we obtain $c_2 > 0$ such that

$$J(u) - \lambda I(u) \geq \|u\|^{p_0} - \lambda c_2 |\Omega| (\|u\| + \|u\|^{s+1}), \quad \forall\, u \in E.$$

Since $1 < s + 1 < p_0$ it follows that

$$\lim_{\|u\| \to \infty} (J(u) - \lambda I(u)) = \infty, \quad \forall\, \lambda > 0,$$

thus (iii) is verified.

Step 2. For every $\lambda > 0$, the functional $J - \lambda I$ is sequentially weakly lower semicontinuous and satisfies the Palais-Smale condition.

The fact that E is compactly embedded into $C(\overline{\Omega})$ implies that the operator $I' : E \to E^\star$ is compact. Consequently, the functional $I : X \to \mathbb{R}$ is sequentially weakly continuous. On the other hand, the convexity of $J : X \to \mathbb{R}$ implies the sequentially weakly lower semicontinuity of J. This proves the first part.

Combining Step 1, Lemma 38 and the fact that $I' : E \to E^\star$ is compact, we obtain that $J - \lambda I$ satisfies the Palais-Smale condition.

Step 3. Let $0 < r < \min\left\{1, (\frac{\delta}{2c})^{p^0}, \Phi(b)|\Omega|\right\}$ and $u_1(x) = b \in E$. Then (i) *and* (ii) *are verified.*

We first observe that $b \neq 0$ (this appears in (f1)). Therefore, $\Phi(b) = \Phi(-b) > 0$, that is, we can choose $r > 0$ as above. Now, we have

$$J(u_1) = \int_\Omega (\Phi(|\nabla u_1|) + \Phi(|u_1|))\, dx = \int_\Omega \Phi(|b|) = \Phi(b)|\Omega| > r,$$

hence (i) is verified.

Now, let $J(u) < r$. Then by Lemma 37 (and $r < 1$), we have $\|u\|^{p^0} \leq J(u) < r$. Therefore, $\|u\| \leq \frac{\delta}{2c}$. By Lemma 36, we have

$$|u(x)| \leq \|u\|_\infty \leq c\|u\|_{1,\Phi} \leq 2c\|u\| \leq \delta, \quad \forall x \in \Omega. \tag{4.109}$$

On the other hand, due to (f2), we obtain

$$F(x, t) = F(x, t) - F(x, 0) = f(x, \theta t)t = \frac{1}{\theta} f(x, \theta t)\theta t \leq 0 \quad (\text{with } \theta \in (0, 1))$$

for every $x \in \Omega$ and $t \in [-\delta, \delta]$. Consequently, for every $u \in E$, such that $J(u) < r$, we have $I(u) = \int_\Omega F(x, u(x))dx \leq 0$ (see (4.109)), thus

$$\sup_{J(u)<r} I(u) \leq 0. \tag{4.110}$$

However, by (f1), we have

$$r\frac{I(u_1)}{J(u_1)} = \frac{rB_F}{\Phi(b)|\Omega|} > 0,$$

which proves (ii).

Now, we return to the proof of Theorem 31. We have $I(0) = J(0) = 0$ and $J(u) \geq 0$ for every $u \in E$. Choosing $u_0 = 0$ and taking into account Steps 1-3, all the hypotheses of Theorem 32 can be verified. Denoting

$$\overline{a} = \frac{2r}{r\frac{I(u_1)}{J(u_1)} - \sup_{J(u)<r} I(u)},$$

there exist a nonempty open interval $\Lambda \subset [0, \overline{a}]$ and a number $\mu > 0$ such that for each $\lambda \in \Lambda$, the equation $J'(u) - \lambda I'(u) = 0$ admits at least three solutions in E (thus, at least two nontrivial weak solutions of problem (4.91)) having the norm less than μ. Moreover, due to (4.110), we have

$$\overline{a} \leq \frac{2r}{r\frac{I(u_1)}{J(u_1)}} = \frac{2J(u_1)}{I(u_1)} = \frac{2\Phi(b)|\Omega|}{B_F},$$

which completes the proof of Theorem 31. □

Example 9 *Consider the nonlinear problem*

$$\begin{cases} -\text{div}\left(\dfrac{|\nabla u|^{p-2}\nabla u}{\log(1+|\nabla u|)}\right) + \dfrac{|u|^{p-2}u}{\log(1+|u|)} = \lambda \ln\left(1 + (u-1)u_+^{q(x)}\right), & x \in \Omega, \\ \dfrac{\partial u}{\partial \nu} = 0, & x \in \partial\Omega, \end{cases} \tag{4.111}$$

where p is a real number such that $p > N + 1$ and $q \in C(\overline{\Omega})$ satisfies $2 < q(x) < p - 1$ for any $x \in \overline{\Omega}$ and $u_+ = \max(u, 0)$.

We define

$$\phi(t) = \frac{|t|^{p-2}}{\log(1+|t|)}t \quad \textit{for } t \neq 0, \quad \textit{and } \phi(0) = 0,$$

and

$$\Phi(t) = \int_0^t \phi(s)\, ds.$$

An easy computation shows that the function $[0,\infty) \ni t \mapsto \Phi(\sqrt{t})$ *is convex. Moreover, by [19, Example 3, p. 243] we have*

$$p_0 = p - 1 < p^0 = p = \liminf_{t\to\infty} \frac{\log(\Phi(t))}{\log(t)}.$$

Thus, conditions (Φ_0) *and* (Φ_1) *are verified.*

Now we define the function $f : \Omega \times \mathbb{R} \to \mathbb{R}$ *by*

$$f(x,t) = \ln\left(1 + (t-1)t_+^{q(x)}\right), \quad \forall\, x \in \Omega \text{ and } t \in \mathbb{R}.$$

Then $F : \Omega \times \mathbb{R} \to \mathbb{R}$ *is*

$$\begin{aligned} F(x,t) &= t\ln\left(1 + (t-1)t_+^{q(x)}\right) - (q(x)+1)t_+ \\ &+ \int_0^{t_+} \frac{q(x)+1-s^{q(x)}}{1+s^{q(x)+1}-s^{q(x)}}ds, \quad \forall\, x \in \Omega \text{ and } t \in \mathbb{R}. \end{aligned}$$

Then f is a Carathéodory function and (f0) is fulfilled by choosing $s = 1$. *Moreover, for large enough* $b > 0$, *(f1) is also verified. Finally, (f2) is verified for* $\delta = 1$. *Consequently, we can apply Theorem 31 and hence problem* (4.111) *has at least two nontrivial solutions for certain eigenvalues* $\lambda > 0$.

4.5 Neumann problems in Orlicz-Sobolev spaces

In this section, we consider the nonhomogeneous Neumann problem

$$\begin{cases} -\text{div}(a(x,|\nabla u(x)|)\nabla u(x)) + a(x,|u(x)|)u(x) = \lambda\, g(x,u(x)), & \text{for} \quad x \in \Omega \\ \dfrac{\partial u}{\partial \nu}(x) = 0, & \text{for} \quad x \in \partial\Omega\,, \end{cases} \tag{4.112}$$

where Ω is a bounded domain in $\mathbb{R}^N$ $(N \geq 3)$ with smooth boundary $\partial\Omega$, ν is the outward unit normal to $\partial\Omega$ and λ is a positive parameter. We assume that the function $a(x,t) : \overline{\Omega} \times \mathbb{R} \to \mathbb{R}$ in (4.112) is such that the mapping $\varphi(x,t) : \overline{\Omega} \times \mathbb{R} \to \mathbb{R}$,

$$\varphi(x,t) = \begin{cases} a(x,|t|)t, & \text{for} \quad t \neq 0 \\ 0, & \text{for} \quad t = 0\,, \end{cases}$$

satisfies the following condition

(φ) for all $x \in \Omega$, $\varphi(x,\cdot) : \mathbb{R} \to \mathbb{R}$ is an odd, increasing homeomorphism from $\mathbb{R}$ onto $\mathbb{R}$; and

$\Phi(x,t) : \overline{\Omega} \times \mathbb{R} \to \mathbb{R}$,

$$\Phi(x,t) = \int_0^t \varphi(x,s)\, ds, \quad \forall\, x \in \overline{\Omega},\ t \geq 0\,,$$

belongs to the *class* Φ, that is, Φ satisfies the following conditions

(Φ_1) for all $x \in \Omega$, $\Phi(x,\cdot) : [0,\infty) \to \mathbb{R}$ is a nondecreasing continuous function, with $\Phi(x,0) = 0$ and $\Phi(x,t) > 0$ whenever $t > 0$; $\lim_{t\to\infty} \Phi(x,t) = \infty$;

(Φ_2) for every $t \geq 0$, $\Phi(\cdot,t) : \Omega \to \mathbb{R}$ is a measurable function.

Remark 7 *Since $\varphi(x,\cdot)$ satisfies condition (φ), then $\Phi(x,\cdot)$ is a convex and increasing function from $\mathbb{R}^+$ to $\mathbb{R}^+$.*

We associate to the function Φ introduced above the *generalized Orlicz class* defined by

$$K_\Phi(\Omega) = \{u : \Omega \to \mathbb{R},\ \text{measurable;}\ \int_\Omega \Phi(x, |u(x)|)\, dx < \infty\}$$

and the *generalized Orlicz space*,

$$L^\Phi(\Omega) = \{u : \Omega \to \mathbb{R},\ \text{measurable;}\ \lim_{\lambda\to 0^+} \int_\Omega \Phi(x, \lambda|u(x)|)\, dx = 0\}\,.$$

The space $L^\Phi(\Omega)$ is a Banach space endowed with the *Luxemburg norm*

$$|u|_\Phi = \inf\left\{\mu > 0;\ \int_\Omega \Phi\left(x, \frac{|u(x)|}{\mu}\right)\, dx \leq 1\right\}$$

or the equivalent norm (the *Orlicz norm*)

$$|u|_{(\Phi)} = \sup\left\{\left|\int_\Omega uv\, dx\right|;\ v \in L^{\overline{\Phi}}(\Omega),\ \int_\Omega \overline{\Phi}(x, |v(x)|)\, dx \leq 1\right\},$$

where $\overline{\Phi}$ denotes the *conjugate Young* function of Φ, that is,

$$\overline{\Phi}(x,t) = \sup_{s>0}\{ts - \Phi(x,s);\ s \in \mathbb{R}\}, \quad \forall\, x \in \overline{\Omega},\ t \geq 0\,.$$

In this section, we assume that there exist positive constants φ_0 and φ^0 such that

$$1 < \varphi_0 \leq \frac{t\varphi(x,t)}{\Phi(x,t)} \leq \varphi^0 < \infty, \quad \forall\, x \in \overline{\Omega},\ t \geq 0\,. \tag{4.113}$$

The above relation implies that Φ satisfies the Δ_2-*condition*, namely

$$\Phi(x, 2t) \leq K \cdot \Phi(x,t), \quad \forall\, x \in \overline{\Omega},\ t \geq 0\,, \tag{4.114}$$

where K is a positive constant. Relation (4.114) implies that $L^\Phi(\Omega) = K_\Phi(\Omega)$.

Furthermore, we assume that Φ satisfies the following condition

$$\text{for each } x \in \overline{\Omega}, \text{ the function } [0,\infty) \ni t \to \Phi(x, \sqrt{t}) \text{ is convex}. \tag{4.115}$$

Relation (4.115) assures that $L^{\Phi}(\Omega)$ is a uniformly convex space and thus, a reflexive space.

On the other hand, we point out that assuming that Φ and Ψ belong to the class Φ and

$$\Psi(x,t) \leq K_1 \cdot \Phi(x, K_2 \cdot t) + h(x), \quad \forall\, x \in \overline{\Omega},\ t \geq 0\,, \tag{4.116}$$

where $h \in L^1(\Omega)$, $h(x) \geq 0$ a.e. $x \in \Omega$ and K_1, K_2 are positive constants, then by Theorem 8.5 in [74], there exists a continuous embedding $L^{\Phi}(\Omega) \subset L^{\Psi}(\Omega)$.

We recall the notion of *modular* of the $L^{\Phi}(\Omega)$ space, which is the mapping $\rho_\Phi : L^{\Phi}(\Omega) \to \mathbb{R}$ defined by

$$\rho_\Phi(u) = \int_\Omega \Phi(x, |u(x)|)\, dx.$$

If (u_n), $u \in L^{\Phi}(\Omega)$, then the following relations hold:

$$|u|_\Phi > 1 \quad \Rightarrow \quad |u|_\Phi^{\varphi_0} \leq \rho_\Phi(u) \leq |u|_\Phi^{\varphi^0}\,, \tag{4.117}$$

$$|u|_\Phi < 1 \quad \Rightarrow \quad |u|_\Phi^{\varphi^0} \leq \rho_\Phi(u) \leq |u|_\Phi^{\varphi_0}\,, \tag{4.118}$$

$$|u_n - u|_\Phi \to 0 \quad \Leftrightarrow \quad \rho_\Phi(u_n - u) \to 0\,, \tag{4.119}$$

$$|u_n|_\Phi \to \infty \quad \Leftrightarrow \quad \rho_\Phi(u_n) \to \infty\,. \tag{4.120}$$

Next, we define the *generalized Orlicz-Sobolev space*

$$W^{1,\Phi}(\Omega) = \left\{ u \in L^{\Phi}(\Omega);\ \frac{\partial u}{\partial x_i} \in L^{\Phi}(\Omega),\ i = 1, ..., N \right\}.$$

On $W^{1,\Phi}(\Omega)$ we define equivalent norms

$$\begin{aligned}
\|u\|_{1,\Phi} &= |\,|\nabla u|\,|_\Phi + |u|_\Phi \\
\|u\|_{2,\Phi} &= \max\{|\,|\nabla u|\,|_\Phi, |u|_\Phi\} \\
\|u\| &= \inf\left\{ \mu > 0;\ \int_\Omega \left[\Phi\left(x, \frac{|u(x)|}{\mu}\right) + \Phi\left(x, \frac{|\nabla u(x)|}{\mu}\right) \right] dx \leq 1 \right\},
\end{aligned}$$

(see Proposition 9).

Finally, we point out that assuming that Φ and Ψ belong to the class Φ, satisfying relation (4.116) and $\inf_{x\in\Omega} \Phi(x,1) > 0$, $\inf_{x\in\Omega} \Psi(x,1) > 0$ then there exists a continuous embedding $W^{1,\Phi}(\Omega) \subset W^{1,\Psi}(\Omega)$.

Remark 8 *a) Assuming that $\Phi(x,t) = \Phi(t)$, that is, Φ is independent of the variable x, we say that L^Φ and $W^{1,\Phi}$ are Orlicz spaces, respectively Orlicz-Sobolev spaces.*
b) Assuming that $\Phi(x,t) = |t|^{p(x)}$ with $p(x) \in C(\overline{\Omega})$, $p(x) > 1$ for all $x \in \overline{\Omega}$ we denote L^Φ by $L^{p(x)}$ and $W^{1,\Phi}$ by $W^{1,p(x)}$ and we refer to them as variable exponents Lebesgue spaces, respectively variable exponents Sobolev spaces.

c) Our framework enables us to work with spaces which are more general than those described in a) and b) (see the examples at the end of this section).

4.5.1 More about generalized Orlicz-Sobolev spaces

In what follows, we establish some useful properties of the generalized Orlicz-Sobolev spaces.

Proposition 6 *Assume that condition (4.113) is satisfied. Then the following relations hold:*

$$|u|_\Phi^{\varphi_0} \le \rho_\Phi(u) \le |u|_\Phi^{\varphi^0}, \quad \forall\, u \in L^\Phi(\Omega) \text{ with } |u|_\Phi > 1\,, \tag{4.121}$$

$$|u|_\Phi^{\varphi^0} \le \rho_\Phi(u) \le |u|_\Phi^{\varphi_0}, \quad \forall\, u \in L^\Phi(\Omega) \text{ with } |u|_\Phi < 1\,. \tag{4.122}$$

Proof. We first show that $\rho_\Phi(u) \le |u|_\Phi^{\varphi^0}$ for all $u \in L^\Phi(\Omega)$ with $|u|_\Phi > 1$.
Indeed, since $\varphi^0 \ge (t\varphi(x,t))/\Phi(x,t)$ for all $x \in \overline{\Omega}$ and all $t \ge 0$ it follows that letting $\sigma > 1$ we have

$$\begin{aligned}\log(\Phi(x,\sigma\cdot t)) - \log(\Phi(x,t)) &= \int_t^{\sigma\cdot t} \frac{\varphi(x,s)}{\Phi(x,s)}\, ds \\ &\le \int_t^{\sigma\cdot t} \frac{\varphi^0}{s}\, ds = \log(\sigma^{\varphi^0})\,.\end{aligned}$$

Therefore

$$\Phi(x,\sigma\cdot t) \le \sigma^{\varphi^0}\cdot\Phi(x,t), \quad \forall\, x \in \overline{\Omega},\ t > 0,\ \sigma > 1\,. \tag{4.123}$$

Let now $u \in L^\Phi(\Omega)$ with $|u|_\Phi > 1$. Using the definition of the Luxemburg norm and relation (4.123) we deduce

$$\begin{aligned}\int_\Omega \Phi(x,|u(x)|)\, dx &= \int_\Omega \Phi\left(x, |u|_\Phi\cdot\frac{|u(x)|}{|u|_\Phi}\right) dx \\ &\le |u|_\Phi^{\varphi^0}\cdot\int_\Omega \Phi\left(x,\frac{|u(x)|}{|u|_\Phi}\right) dx \\ &\le |u|_\Phi^{\varphi^0}\,.\end{aligned}$$

Now, we show that $\rho_\Phi(u) \ge |u|_\Phi^{\varphi_0}$ for all $u \in L^\Phi(\Omega)$ with $|u|_\Phi > 1$.

Since $\varphi_0 \leq (t\varphi(x,t))/\Phi(x,t)$ for all $x \in \overline{\Omega}$ and all $t \geq 0$, similar techniques as those used in the proof of relation (4.123) imply

$$\Phi(x, \sigma \cdot t) \geq \sigma^{\varphi_0} \cdot \Phi(x,t), \quad \forall\, x \in \overline{\Omega},\ t > 0,\ \sigma > 1\,. \tag{4.124}$$

Let $u \in L^\Phi(\Omega)$ with $|u|_\Phi > 1$. We consider $\beta \in (1, |u|_\Phi)$. Since $\beta < |u|_\Phi$ it follows that $\int_\Omega \Phi\left(x, \frac{|u(x)|}{\beta}\right)\, dx > 1$ otherwise we obtain a contradiction to the definition of the Luxemburg norm. We deduce that

$$\int_\Omega \Phi(x, |u(x)|)\, dx = \int_\Omega \Phi\left(x, \beta \cdot \frac{|u(x)|}{\beta}\right) dx \geq \beta^{\varphi_0} \cdot \int_\Omega \Phi\left(x, \frac{|u(x)|}{\beta}\right) dx \geq \beta^{\varphi_0}\,.$$

Letting $\beta \nearrow |u|_\Phi$ we conclude that relation (4.121) holds.

Next, we show that $\rho_\Phi(u) \leq |u|_\Phi^{\varphi_0}$ for all $u \in L^\Phi(\Omega)$ with $|u|_\Phi < 1$. As in the proof of relations (4.123) and (4.124), we deduce that

$$\Phi(x,t) \leq \tau^{\varphi_0} \cdot \Phi(x, t/\tau), \quad \forall\, x \in \overline{\Omega},\ t > 0,\ \tau \in (0,1)\,. \tag{4.125}$$

Let $u \in L^\Phi(\Omega)$ with $|u|_\Phi < 1$. The definition of the Luxemburg norm in combination with relation (4.125) yields

$$\begin{aligned}
\int_\Omega \Phi(x, |u(x)|)\, dx &= \int_\Omega \Phi\left(x, |u|_\Phi \cdot \frac{|u(x)|}{|u|_\Phi}\right)\, dx \\
&\leq |u|_\Phi^{\varphi_0} \cdot \int_\Omega \Phi\left(x, \frac{|u(x)|}{|u|_\Phi}\right)\, dx \\
&\leq |u|_\Phi^{\varphi_0}\,.
\end{aligned}$$

Finally, we show that $\rho_\Phi(u) \geq |u|_\Phi^{\varphi^0}$ for all $u \in L^\Phi(\Omega)$ with $|u|_\Phi < 1$. As in the proof of (4.123) we deduce that

$$\Phi(x,t) \geq \tau^{\varphi^0} \cdot \Phi(x, t/\tau), \quad \forall\, x \in \overline{\Omega},\ t > 0,\ \tau \in (0,1)\,. \tag{4.126}$$

Let $u \in L^\Phi(\Omega)$ with $|u|_\Phi < 1$ and $\xi \in (0, |u|_\Phi)$. By (4.126) we find

$$\int_\Omega \Phi(x, |u(x)|)\, dx \geq \xi^{\varphi^0} \cdot \int_\Omega \Phi\left(x, \frac{|u(x)|}{\xi}\right)\, dx\,. \tag{4.127}$$

Define $v(x) = u(x)/\xi$, for all $x \in \Omega$. We have $|v|_\Phi = |u|_\Phi/\xi > 1$. Using relation (4.121) we find

$$\int_\Omega \Phi(x, |v(x)|)\, dx \geq |v|_\Phi^{\varphi_0} > 1\,. \tag{4.128}$$

Combining (4.127) and (4.128) we obtain

$$\int_\Omega \Phi(x, |u(x)|)\, dx \geq \xi^{\varphi^0}, \quad \forall\, \xi \in (0, |u|_\Phi)\,.$$

Letting $\xi \nearrow |u|_\Phi$ we deduce that relation (4.122) holds. The proof of Proposition 6 is complete. □

Proposition 7 *Assume that Φ satisfies conditions (4.114) and (4.115). Then the space $L^{\Phi}(\Omega)$ is uniformly convex.*

Proof. From our hypotheses we deduce that for all $t, s \in \mathbb{R}$

$$\frac{1}{2}[\Phi(x,|t|)+\Phi(x,|s|)] \geq \Phi\left(x, \frac{|t+s|}{2}\right)+\Phi\left(x, \frac{|t-s|}{2}\right).$$

Thus, for all $u, v \in L^{\Phi}(\Omega)$

$$\frac{1}{2}[\rho_\Phi(u)+\rho_\Phi(v)] \geq \rho_\Phi\left(\frac{u+v}{2}\right)+\rho_\Phi\left(\frac{u-v}{2}\right). \tag{4.129}$$

Assume that $|u|_\Phi < 1$ and $|v|_\Phi < 1$ and $|u-v|_\Phi > \epsilon$ (with $\epsilon \in (0, 1/K)$). Then we have

$$\begin{aligned} \rho_\Phi(u-v) &\geq |u-v|_\Phi^{\varphi_0} \ \text{ if } |u-v|_\Phi > 1 \\ \rho_\Phi(u-v) &\geq |u-v|_\Phi^{\varphi^0} \ \text{ if } |u-v|_\Phi < 1, \end{aligned}$$

and

$$\rho_\Phi(u) < 1, \quad \rho_\Phi(v) < 1.$$

The above information and relation (4.113) yield

$$\rho_\Phi\left(\frac{u-v}{2}\right) \geq \frac{1}{K}\cdot\rho_\Phi(u-v) \geq \begin{cases} \dfrac{1}{K}\cdot\epsilon^{\varphi_0}, & \text{if } \ |u-v|_\Phi > 1 \\ \dfrac{1}{K}\cdot\epsilon^{\varphi^0}, & \text{if } \ |u-v|_\Phi < 1. \end{cases}$$

By (4.129) and the above inequality we have

$$\rho_\Phi\left(\frac{u+v}{2}\right) < \begin{cases} 1-\dfrac{1}{K}\cdot\epsilon^{\varphi_0}, & \text{if } \ |u-v|_\Phi > 1 \\ 1-\dfrac{1}{K}\cdot\epsilon^{\varphi^0}, & \text{if } \ |u-v|_\Phi < 1. \end{cases} \tag{4.130}$$

On the other hand, we have

$$\rho_\Phi\left(\frac{u+v}{2}\right) \geq \begin{cases} \left|\dfrac{u+v}{2}\right|_\Phi^{\varphi_0}, & \text{if } \ \left|\dfrac{u+v}{2}\right|_\Phi > 1 \\ \left|\dfrac{u+v}{2}\right|_\Phi^{\varphi^0}, & \text{if } \ \left|\dfrac{u+v}{2}\right|_\Phi < 1. \end{cases} \tag{4.131}$$

Relations (4.130) and (4.131) show that there exists $\delta > 0$ such that

$$\left|\frac{u+v}{2}\right|_\Phi < 1-\delta.$$

Thus $L^{\Phi}(\Omega)$ is a uniformly convex space. □

Remark 9 *Condition (4.115) in combination with relation (4.129) implies that for every $x \in \Omega$ fixed, the function $\Phi(x, \cdot)$ is convex from $\mathbb{R}^+$ to $\mathbb{R}^+$.*

Proposition 8 *Condition (4.113) implies condition (4.114).*

Proof. Since relation (4.113) holds, it follows by Proposition 6 that condition (4.123) is satisfied. We deduce that

$$\Phi(x, 2 \cdot t) \leq 2^{\varphi^0} \cdot \Phi(x, t), \quad \forall\, x \in \Omega,\ t > 0\,.$$

Thus relation (4.114) holds with $K = 2^{\varphi^0}$. The proof of Proposition 8 is complete. □

Proposition 9 *On $W^{1,\Phi}(\Omega)$, the following norms*

$$\|u\|_{1,\Phi} = |\,|\nabla u|\,|_\Phi + |u|_\Phi\,,$$

$$\|u\|_{2,\Phi} = \max\{|\,|\nabla u|\,|_\Phi, |u|_\Phi\}\,,$$

$$\|u\| = \inf\left\{\mu > 0;\ \int_\Omega \left[\Phi\left(x, \frac{|u(x)|}{\mu}\right) + \Phi\left(x, \frac{|\nabla u(x)|}{\mu}\right)\right]\, dx \leq 1\right\},$$

are equivalent.

Proof. We remark that $\|\ \|_{1,\Phi}$ and $\|\ \|_{2,\Phi}$ are equivalent, since

$$2 \cdot \|u\|_{2,\Phi} \geq \|u\|_{1,\Phi} \geq \|u\|_{2,\Phi}, \quad \forall\, u \in W^{1,\Phi}(\Omega)\,. \tag{4.132}$$

Next, we point out that

$$\int_\Omega \Phi\left(x, \frac{|u(x)|}{|u|_\Phi}\right)\, dx \leq 1 \quad \text{and} \quad \int_\Omega \Phi\left(x, \frac{|\nabla u(x)|}{|\,|\nabla u|\,|_\Phi}\right)\, dx \leq 1\,,$$

and

$$\int_\Omega \left[\Phi\left(x, \frac{|u(x)|}{\|u\|}\right) + \Phi\left(x, \frac{|\nabla u(x)|}{\|u\|}\right)\right]\, dx \leq 1\,.$$

Using these relations we obtain

$$\int_\Omega \Phi\left(x, \frac{|u(x)|}{\|u\|}\right)\, dx \leq 1 \quad \text{and} \quad \int_\Omega \Phi\left(x, \frac{|\nabla u(x)|}{\|u\|}\right)\, dx \leq 1.$$

Taking into account the definition of $|\ |_\Phi$, we have

$$2\|u\| \geq (|u|_\Phi + |\,|\nabla u|\,|_\Phi) = \|u\|_{1,\Phi}, \quad \forall\, u \in W^{1,\Phi}(\Omega)\,. \tag{4.133}$$

On the other hand, from relation (4.124) we deduce that

$$\Phi(x, 2 \cdot t) \geq 2 \cdot \Phi(x, t), \quad \forall\, x \in \Omega,\ t > 0\,.$$

Therefore

$$2 \cdot \Phi\left(x, \frac{|u(x)|}{2 \cdot \|u\|_{2,\Phi}}\right) \leq \Phi\left(x, \frac{|u(x)|}{\|u\|_{2,\Phi}}\right), \quad \forall\, u \in W^{1,\Phi}(\Omega),\ x \in \Omega$$

and

$$2 \cdot \Phi\left(x, \frac{|\nabla u(x)|}{2 \cdot \|u\|_{2,\Phi}}\right) \leq \Phi\left(x, \frac{|\nabla u(x)|}{\|u\|_{2,\Phi}}\right), \quad \forall\, u \in W^{1,\Phi}(\Omega),\ x \in \Omega\,.$$

It follows that

$$\begin{aligned}&\int_\Omega \left[\Phi\left(x, \frac{|u(x)|}{2\|u\|_{2,\Phi}}\right) + \Phi\left(x, \frac{|\nabla u(x)|}{2\|u\|_{2,\Phi}}\right)\right] dx \leq \\ &\frac{1}{2}\left\{\int_\Omega \left[\Phi\left(x, \frac{|u(x)|}{\|u\|_{2,\Phi}}\right) + \Phi\left(x, \frac{|\nabla u(x)|}{\|u\|_{2,\Phi}}\right)\right] dx\right\}.\end{aligned} \tag{4.134}$$

However, since

$$\|u\|_{2,\Phi} \geq |u|_\Phi \quad \text{and} \quad \|u\|_{2,\Phi} \geq |\,|\nabla u|\,|_\Phi, \quad \forall\, u \in W^{1,\Phi}(\Omega)\,,$$

we obtain for all $u \in W^{1,\Phi}(\Omega)$ and $x \in \Omega$

$$\frac{|u(x)|}{|u|_\Phi} \geq \frac{|u(x)|}{\|u\|_{2,\Phi}} \quad \text{and} \quad \frac{|\nabla u(x)|}{|\,|\nabla u|\,|_\Phi} \geq \frac{|\nabla u(x)|}{\|u\|_{2,\Phi}}\,. \tag{4.135}$$

Taking into account that Φ is increasing, we deduce from relations (4.134) and (4.135) that

$$\begin{aligned}&\int_\Omega \left[\Phi\left(x, \frac{|u(x)|}{2\|u\|_{2,\Phi}}\right) + \Phi\left(x, \frac{|\nabla u(x)|}{2\|u\|_{2,\Phi}}\right)\right] dx \leq \\ &\frac{1}{2}\left\{\int_\Omega \left[\Phi\left(x, \frac{|u(x)|}{|u|_\Phi}\right) + \Phi\left(x, \frac{|\nabla u(x)|}{|\,|\nabla u\,|_\Phi}\right)\right] dx\right\} \leq 1\,,\end{aligned}$$

for all $u \in W^{1,\Phi}(\Omega)$.

We conclude that

$$2 \cdot \|u\|_{1,\Phi} \geq 2 \cdot \|u\|_{2,\Phi} \geq \|u\|, \quad \forall\, u \in W^{1,\Phi}(\Omega)\,. \tag{4.136}$$

We deduce the conclusion of Proposition 9 from relations (4.132), (4.133) and (4.136). □

Proposition 10 *The following relations hold:*

$$\int_\Omega [\Phi(x, |u(x)|) + \Phi(x, |\nabla u(x)|)]\, dx \geq \|u\|^{\varphi_0}, \quad \forall\, u \in W^{1,\Phi}(\Omega),\ \|u\| > 1\,; \tag{4.137}$$

$$\int_\Omega [\Phi(x, |u(x)|) + \Phi(x, |\nabla u(x)|)]\, dx \geq \|u\|^{\varphi^0}, \quad \forall\, u \in W^{1,\Phi}(\Omega),\ \|u\| < 1\,. \tag{4.138}$$

Proof. First, we assume that $\|u\| > 1$. Let $\beta \in (1, \|u\|)$. By relation (4.124) we have

$$\int_\Omega [\Phi(x,|u(x)|) + \Phi(x,|\nabla u(x)|)]\, dx \geq$$
$$\beta^{\varphi_0} \cdot \int_\Omega \left[\Phi\left(x, \frac{|u(x)|}{\beta}\right) + \Phi\left(x, \frac{|\nabla u(x)|}{\beta}\right)\right].$$

Since $\beta < \|u\|$ we find

$$\int_\Omega \left[\Phi\left(x, \frac{|u(x)|}{\beta}\right) + \Phi\left(x, \frac{|\nabla u(x)|}{\beta}\right)\right] > 1.$$

Therefore

$$\int_\Omega [\Phi(x,|u(x)|) + \Phi(x,|\nabla u(x)|)]\, dx \geq \beta^{\varphi_0}.$$

Letting $\beta \nearrow \|u\|$ we deduce that relation (4.137) holds.

Next, we assume that $\|u\| < 1$. Let $\xi \in (0, \|u\|)$. By relation (4.126) we obtain

$$\int_\Omega [\Phi(x,|u(x)|) + \Phi(x,|\nabla u(x)|)]\, dx \geq$$
$$\xi^{\varphi^0} \cdot \int_\Omega \left[\Phi\left(x, \frac{|u(x)|}{\xi}\right) + \Phi\left(x, \frac{|\nabla u(x)|}{\xi}\right)\right] dx. \tag{4.139}$$

Defining $v(x) = u(x)/\xi$, for all $x \in \Omega$, we have $\|v\| = \|u\|/\xi > 1$. Using relation (4.137) we find

$$\int_\Omega [\Phi(x,|v(x)|) + \Phi(x,|\nabla v(x)|)]\, dx \geq \|v\|^{\varphi_0} > 1. \tag{4.140}$$

Relations (4.139) and (4.140) yield

$$\int_\Omega [\Phi(|u(x)|) + \Phi(|\nabla u(x)|)]\, dx \geq \xi^{p^0}.$$

Letting $\xi \nearrow \|u\|$ in the above inequality we obtain relation (4.138). □

4.5.2 Conditions for concentration of the spectrum

We study problem (4.112) in the special case when Φ satisfies

$$M \cdot |t|^{p(x)} \leq \Phi(x,t), \quad \forall\, x \in \overline{\Omega},\ t \geq 0, \tag{4.141}$$

where $p(x) \in C(\overline{\Omega})$ with $p(x) > 1$ for all $x \in \overline{\Omega}$ and $M > 0$ is a constant.

Remark 10 *By relation (4.141), the function space $W^{1,\Phi}(\Omega)$ is continuously embedded in $W^{1,p(x)}(\Omega)$ (see relation (4.116) with $\Psi(x,t) = |t|^{p(x)}$). On the other hand, $W^{1,p(x)}(\Omega)$ is compactly embedded in $L^{r(x)}(\Omega)$ for any $r(x) \in C(\overline{\Omega})$ with $1 < r^- \leq r^+ < \frac{Np^-}{N-p^-}$. Thus $W^{1,\Phi}(\Omega)$ is compactly embedded in $L^{r(x)}(\Omega)$ for any $r(x) \in C(\overline{\Omega})$ with $1 < r(x) < \frac{Np^-}{N-p^-}$ for all $x \in \overline{\Omega}$.*

We assume that the nonlinearity g arising in problem (4.112) satisfies the hypotheses

$$|g(x,t)| \leq C_0 \cdot |t|^{q(x)-1}, \quad \forall\, x \in \Omega,\ t \in \mathbb{R} \tag{4.142}$$

and

$$C_1 \cdot |t|^{q(x)} \leq G(x,t) := \int_0^t g(x,s)\, ds \leq C_2 \cdot |t|^{q(x)}, \quad \forall\, x \in \Omega,\ t \in \mathbb{R}, \tag{4.143}$$

where C_0, C_1 and C_2 are positive constants and $q(x) \in C(\overline{\Omega})$ satisfies $1 < q(x) < \frac{Np^-}{N-p^-}$ for all $x \in \overline{\Omega}$.

Examples. We list some functions g and G which satisfy hypotheses (4.142) and (4.143).

(1) $g(x,t) = q(x) \cdot |t|^{q(x)-2}t$ and $G(x,t) = |t|^{q(x)}$, where $q(x) \in C(\overline{\Omega})$ satisfies $2 \leq q(x) < \frac{Np^-}{N-p^-}$ for all $x \in \overline{\Omega}$;

(2) $g(x,t) = q(x)\cdot|t|^{q(x)-2}t+(q(x)-2)\cdot[\log(1+t^2)]\cdot|t|^{q(x)-4}t+\frac{t}{1+t^2}|t|^{q(x)-2}$ and $G(x,t) = |t|^{q(x)} + \log(1+t^2)\cdot|t|^{q(x)-2}$, where $q(x) \in C(\overline{\Omega})$ satisfies $4 \leq q(x) < \frac{Np^-}{N-p^-}$ for all $x \in \overline{\Omega}$;

(3) $g(x,t) = q(x)\cdot|t|^{q(x)-2}t+(q(x)-1)\cdot\sin(\sin(t))\cdot|t|^{q(x)-3}t+\cos(\sin(t))\cdot\cos(t)\cdot|t|^{q(x)-1}$ and $G(x,t) = |t|^{q(x)}+\sin(\sin(t))\cdot|t|^{q(x)-1}$, where $q(x) \in C(\overline{\Omega})$ satisfies $3 \leq q(x) < \frac{Np^-}{N-p^-}$ for all $x \in \overline{\Omega}$.

Fig. 4.1 gives the shape of the function g in Example (3), provided that $q(x) \equiv 3$.

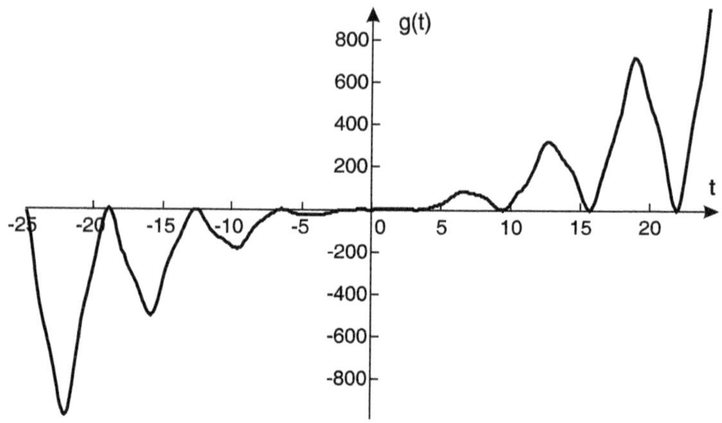

FIGURE 4.1: Graph of the function $g(t) = 3t|t| + 2t\sin(\sin(t)) + t^2\cos(t)\cos(\sin(t))$, $t \in [-25, 25]$

We say that $u \in W^{1,\Phi}(\Omega)$ is a *weak solution* of problem (4.112) if

$$\int_\Omega a(x,|\nabla u|)\nabla u \nabla v\, dx + \int_\Omega a(x,|u|)uv\, dx - \lambda \int_\Omega g(x,u)v\, dx = 0,$$

for all $v \in W^{1,\Phi}(\Omega)$.

The main results of this section establish sufficient conditions for the existence of continuous spectra, concentrating in neighborhoods of the origin or of infinity. The first property asserts that any small enough $\lambda > 0$ is an eigenvalue of problem (4.112).

Theorem 33 *Assume that φ and Φ satisfy conditions (φ), (Φ_1), (Φ_2), (4.113), (4.115) and (4.141), and the functions g and G satisfy conditions (4.142) and (4.143). Furthermore, we assume that $q^- < \varphi_0$. Then there exists $\lambda_\star > 0$ such that for any $\lambda \in (0, \lambda_\star)$ problem (4.112) has a nontrivial weak solution.*

The next result establishes that the spectrum can concentrate on both sides of the positive semi-axis.

Theorem 34 *Assume that φ and Φ satisfy conditions (φ), (Φ_1), (Φ_2), (4.113), (4.115) and (4.141), and the functions g and G satisfy conditions (4.142) and (4.143). Furthermore, we assume that $q^+ < \varphi_0$. Then there exists $\lambda_\star > 0$ and $\lambda^\star > 0$ such that for any $\lambda \in (0, \lambda_\star) \cup (\lambda^\star, \infty)$ problem (4.112) has a nontrivial weak solution.*

Let E denote the generalized Orlicz-Sobolev space $W^{1,\Phi}(\Omega)$.

For each $\lambda > 0$ we define the energy functional $J_\lambda : E \to \mathbb{R}$ by

$$J_\lambda(u) = \int_\Omega [\Phi(x, |\nabla u|) + \Phi(x, |u|)]\, dx - \lambda \int_\Omega G(x, u)\, dx, \quad \forall\, u \in E\,.$$

We first establish some basic properties of J_λ.

Proposition 11 *For each $\lambda > 0$ the functional J_λ is well-defined on E and $J_\lambda \in C^1(E, \mathbb{R})$ with the derivative given by*

$$\begin{aligned} \langle J_\lambda'(u), v\rangle &= \int_\Omega a(x, |\nabla u|)\nabla u \cdot \nabla v\, dx \\ &\quad + \int_\Omega a(x, |u|)uv\, dx - \lambda \int_\Omega g(x, u)v\, dx\,, \end{aligned}$$

for all u, $v \in E$.

We define the functional $\Lambda : E \to \mathbb{R}$ by

$$\Lambda(u) = \int_\Omega [\Phi(x, |\nabla u|) + \Phi(x, |u|)]\, dx, \quad \forall\, u \in E\,.$$

Our hypotheses imply that Λ is well defined on E.

Lemma 39 *We have $\Lambda \in C^1(E, \mathbb{R})$ and*

$$\langle \Lambda'(u), v\rangle = \int_\Omega a(x, |\nabla u|)\nabla u \cdot \nabla v\, dx + \int_\Omega a(x, |u|)uv\, dx\,,$$

for all u, $v \in E$.

Proof. We first establish the existence of the Gâteaux derivative. Let u, $v \in E$. Fix $x \in \Omega$ and $0 < |r| < 1$. Then, by the mean value theorem, there exist $\nu, \theta \in [0,1]$ such that

$$\begin{aligned}&|\Phi(x, |\nabla u(x) + r\nabla v(x)|) - \Phi(x, |\nabla u(x)|)|/|r| = \\ &|\varphi(x, |(1-\nu)|\nabla u(x) + r\nabla v(x)| + \nu|\nabla u(x)||)| \cdot \\ &||\nabla u(x) + r\nabla v(x)| - |\nabla u(x)||\end{aligned} \tag{4.144}$$

and

$$\begin{aligned}&|\Phi(x, |u(x) + rv(x)|) - \Phi(x, |u(x)|)|/|r| = \\ &|\varphi(x, |(1-\theta)|u(x) + rv(x)| + \theta|u(x)||)| \cdot \\ &||u(x) + rv(x)| - |u(x)||\,.\end{aligned} \tag{4.145}$$

Next, we claim that $\varphi(x, |u(x)|) \in L^{\overline{\Phi}}(\Omega)$ provided that $u \in L^{\Phi}(\Omega)$, where $\overline{\Phi}$ is the conjugate Young function of Φ.

Indeed, we know that

$$\overline{\Phi}(x,t) = \sup_{s>0}\{ts - \Phi(x,s);\ s \in \mathbb{R}\}, \quad \forall\, x \in \overline{\Omega},\ t \geq 0$$

or

$$\overline{\Phi}(x,t) = \int_0^t \overline{\varphi}(x,s)\, ds, \quad \forall\, x \in \overline{\Omega},\ t \geq 0\,,$$

where $\overline{\varphi}(x,t) = \sup\limits_{\varphi(x,s) \leq t} s$, for all $x \in \overline{\Omega}$ and $t \geq 0$.

On the other hand, by relation (φ) we know that for all $x \in \Omega$, $\varphi(x,\cdot) : \mathbb{R} \to \mathbb{R}$ is an odd, increasing homeomorphism from $\mathbb{R}$ onto $\mathbb{R}$ and thus, an increasing homeomorphism from $\mathbb{R}^+$ onto $\mathbb{R}^+$. It follows that for each $x \in \overline{\Omega}$ we can denote by $\varphi^{-1}(x,t)$ the inverse function of $\varphi(x,t)$ relative to variable t. Thus, we deduce that $\varphi(x,s) \leq t$ if and only if $s \leq \varphi^{-1}(x,t)$. Taking into account the above pieces of information we deduce that $\overline{\varphi}(x,t) = \varphi^{-1}(x,t)$. Consequently we find

$$\overline{\Phi}(x,t) = \int_0^t \varphi^{-1}(x,s)\, ds, \quad \forall\, x \in \overline{\Omega},\ t \geq 0\,.$$

Next, since

$$\overline{\Phi}(x, \varphi^{-1}(x,s)) = \int_0^{\varphi^{-1}(x,s)} \varphi(x,\theta)\, d\theta, \quad \forall\, x \in \overline{\Omega},\ s \geq 0\,,$$

taking $\varphi(x,\theta) = r$ we find

$$\overline{\Phi}(x, \varphi^{-1}(x,s)) = \int_0^s r \cdot (\varphi^{-1}(x,r))'_r\, dr = s \cdot \varphi^{-1}(x,s) - \overline{\Phi}(x,s),\ \forall\, x \in \overline{\Omega},\ s \geq 0\,.$$

The above relation implies

$$\overline{\Phi}(x,s) \leq s \cdot \varphi^{-1}(x,s), \quad \forall\, x \in \overline{\Omega},\ s \geq 0\,.$$

Taking here $s = \varphi(x,t)$ we find

$$\overline{\Phi}(x, \varphi(x,t)) \leq t \cdot \varphi(x,t), \quad \forall\, x \in \overline{\Omega},\ t \geq 0\,.$$

The last inequality and relation (4.113) yield

$$\overline{\Phi}(x, \varphi(x,t)) \leq \varphi^0 \cdot \Phi(x,t), \quad \forall\, x \in \overline{\Omega},\ t \geq 0\,.$$

Thus, for any $u \in L^{\Phi}(\Omega)$ we have $\varphi(x, |u(x)|) \in L^{\overline{\Phi}}(\Omega)$ and the claim is verified. By applying relations (4.144), (4.145), the above claim and the Hölder inequality, we infer that

$$\begin{aligned}
&|\Phi(x, |\nabla u(x) + r\nabla v(x)|) + \Phi(x, |u(x) + rv(x)|) - \Phi(x, |\nabla u(x)|) - \Phi(x, |u(x)|)||/|r| \leq \\
&\quad |\varphi(x, |(1-\nu)|\nabla u(x) + r\nabla v(x)| + \nu|\nabla u(x)|)| \cdot ||\nabla u(x) + r\nabla v(x)| - |\nabla u(x)|| + \\
&\quad\quad |\varphi(x, |(1-\theta)|u(x) + rv(x)| + \theta|\nabla u(x)|)| \cdot ||u(x) + rv(x)| - |u(x)|| \in L^1(\Omega)\,,
\end{aligned}$$

for all $u, v \in E$, $x \in \overline{\Omega}$ and $|r| \in (0,1)$. It follows from the Lebesgue dominated convergence theorem that

$$\langle \Lambda'(u), v\rangle = \int_\Omega a(x, |\nabla u|)\nabla u \cdot \nabla v\, dx + \int_\Omega a(x, |u|)uv\, dx\,.$$

Next, we establish the continuity of the Gâteaux derivative. Assume that $u_n \to u$ in E. The above claim and the Lebesgue dominated convergence theorem yield

$$a(x, |\nabla u_n|)\nabla u_n \to a(x, |\nabla u|)\nabla u, \quad \text{in} \quad (L^{\overline{\Phi}}(\Omega))^N$$

and

$$a(x, |u_n|)u_n \to a(x, |u|)u, \quad \text{in} \quad L^{\overline{\Phi}}(\Omega)\,.$$

These facts in combination with the Hölder inequality imply that

$$\begin{aligned}
|\langle \Lambda'(u_n) - \Lambda'(u), v\rangle| \ &\leq |a(x, |\nabla u_n|)\nabla u_n - a(x, |\nabla u|)\nabla u|_{\overline{\Phi}} \cdot |\ |\nabla v|\ |_{\Phi} \\
&+ |a(x, |u_n|)u_n - a(x, |u|)u|_{\overline{\Phi}} \cdot |v|_{\Phi}\,,
\end{aligned}$$

for all $v \in E$, and so

$$\begin{aligned}
\|\Lambda'(u_n) - \Lambda'(u)\| \ &\leq |a(x, |\nabla u_n|)\nabla u_n - a(x, |\nabla u|)\nabla u|_{\overline{\Phi}} \\
&+ |a(x, |u_n|)u_n - a(x, |u|)u|_{\overline{\Phi}} \to 0, \quad \text{as } n \to \infty\,.
\end{aligned}$$

The proof of Lemma 39 is complete. □

Combining Lemma 39 and Remark 10 we infer that Proposition 11 is true.

Lemma 40 *The functional Λ is weakly lower semi-continuous.*

Proof. It is enough to show that Λ is lower semi-continuous. For this purpose, we fix $u \in E$ and $\epsilon > 0$. Since Λ is convex (because Φ is convex) we deduce that for any $v \in E$ the following inequality holds

$$\Lambda(v) \geq \Lambda(u) + \langle \Lambda'(u), v - u \rangle ,$$

or

$$\begin{aligned} \Lambda(v) &\geq \Lambda(u) - \int_\Omega [a(x, |\nabla u|)|\nabla u| \cdot |\nabla v - \nabla u| + a(x, |u|)|u| \cdot |v - u|] \, dx \\ &= \Lambda(u) - \int_\Omega [\varphi(x, |\nabla u|)|\nabla v - \nabla u| + \varphi(x, |u|)|v - u|] \, dx . \end{aligned}$$

However, by the claim proved in Proposition 11, we know that for any $u \in L^\Phi(\Omega)$ we have $\varphi(x, |u|)$, $\varphi(x, |\nabla u|) \in L^{\overline{\Phi}}(\Omega)$. Thus, by the Hölder inequality we find

$$\begin{aligned} \Lambda(v) &\geq \Lambda(u) - C \cdot [|\varphi(x, |\nabla u|)|_{\overline{\Phi}} \cdot | \, |\nabla v - \nabla u| \, |_\Phi + |\varphi(x, |u|)|_{\overline{\Phi}} \cdot |v - u|_\Phi] \\ &\geq \Lambda(u) - C' \cdot \|u - v\| \\ &\geq \Lambda(u) - \epsilon , \end{aligned}$$

for all $v \in E$ with $\|v - u\| < \delta = \epsilon / C'$, where C and C' are positive constants. The proof of Lemma 40 is complete. □

Proposition 12 *The functional J_λ is weakly lower semi-continuous.*

Proof. Using Lemma 40, it is enough to show that J_λ is weakly lower semi-continuous. Let $\{u_n\} \subset E$ be a sequence which weakly converges to u in E. By Lemma 40, we deduce that

$$\Lambda(u) \leq \liminf_{n \to \infty} \Lambda(u_n) .$$

On the other hand, Remark 10 and conditions (4.142) and (4.143) yield

$$\lim_{n \to \infty} \int_\Omega G(x, u_n) \, dx = \int_\Omega G(x, u) \, dx .$$

Thus we find

$$J_\lambda(u) \leq \liminf_{n \to \infty} J_\lambda(u_n) ,$$

which shows that J_λ is weakly lower semi-continuous. □

Proposition 13 *Assume that the sequence $\{u_n\}$ weakly converges to u in E and*

$$\limsup_{n \to \infty} \langle \Lambda'(u_n), u_n - u \rangle \leq 0.$$

Then $\{u_n\}$ strongly converges to u in E.

Proof. Since $\{u_n\}$ weakly converges to u in E, it follows that $\{\|u_n\|\}$ is a bounded sequence of real numbers. That fact and Proposition 9 imply that $\{|u_n|_\Phi\}$ and $\{|\,|\nabla u_n|\,|_\Phi\}$ are bounded sequences of real numbers. Thus relations (4.117) and (4.118) yield that the sequence $\{\Lambda(u_n)\}$ is bounded. Then, up to a subsequence, we deduce that $\Lambda(u_n) \to c$.

By Lemma 40 we obtain

$$\Lambda(u) \leq \liminf_{n\to\infty} \Lambda(u_n) = c\,.$$

On the other hand, since Λ is convex, we have

$$\Lambda(u) \geq \Lambda(u_n) + \langle \Lambda'(u_n), u - u_n\rangle\,.$$

Using the above hypothesis we conclude that $\Lambda(u) = c$. Taking into account that $\{(u_n + u)/2\}$ weakly converges to u in E and using Lemma 40 we get

$$c = \Lambda(u) \leq \Lambda\left(\frac{u_n + u}{2}\right). \tag{4.146}$$

We assume by contradiction that $\{u_n\}$ does not converge to u in E or $\{(u_n - u)/2\}$ does not converge to 0 in E. It follows that there exist $\epsilon > 0$ and a subsequence $\{u_{n_m}\}$ of $\{u_n\}$ such that

$$\left\|\frac{u_{n_m} - u}{2}\right\| \geq \epsilon, \quad \forall m\,. \tag{4.147}$$

Furthermore, relations (4.137), (4.138) and (4.147) imply that there exists $\epsilon_1 > 0$ such that

$$\Lambda\left(\frac{u_{n_m} - u}{2}\right) \geq \epsilon_1, \quad \forall m\,. \tag{4.148}$$

On the other hand, relations (4.129) and (4.148) yield

$$\frac{1}{2}\Lambda(u) + \frac{1}{2}\Lambda(u_{n_m}) - \Lambda\left(\frac{u_{n_m} + u}{2}\right) \geq \Lambda\left(\frac{u_{n_m} - u}{2}\right) \geq \epsilon_1, \quad \forall m\,.$$

Letting $m \to \infty$ in the above inequality we obtain

$$c - \epsilon_1 \geq \limsup_{m\to\infty} \Lambda\left(\frac{u_{n_m} + u}{2}\right),$$

and this is a contradiction to (4.146). We conclude that $\{u_n\}$ strongly converges to u in E and Proposition 13 is proved. □

Lemma 41 *Assume that hypotheses of Theorem 33 are fulfilled. Then there exists $\lambda_\star > 0$ such that for any $\lambda \in (0, \lambda_\star)$ there exist ρ, $\alpha > 0$ such that $J_\lambda(u) \geq \alpha > 0$ for any $u \in E$ with $\|u\| = \rho$.*

Proof. By Remark 10 and conditions (4.142) and (4.143), it follows that E is continuously embedded in $L^{q(x)}(\Omega)$. So there exists a positive constant c_1 such that

$$|u|_{q(x)} \leq c_1 \cdot \|u\|, \quad \forall\, u \in E\,. \tag{4.149}$$

where by $|\cdot|_{q(x)}$ we denoted the norm on $L^{q(x)}(\Omega)$.

We fix $\rho \in (0,1)$ such that $\rho < 1/c_1$. Then relation (4.149) implies

$$|u|_{q(x)} < 1, \quad \forall\, u \in E, \text{ with } \|u\| = \rho\,.$$

Furthermore, relation (4.118) applied to $\Phi(x,t) = |t|^{q(x)}$ yields

$$\int_\Omega |u|^{q(x)}\,dx \leq |u|^{q^-}_{q(x)}, \quad \forall\, u \in E, \text{ with } \|u\| = \rho\,. \tag{4.150}$$

Relations (4.149) and (4.150) imply

$$\int_\Omega |u|^{q(x)}\,dx \leq c_1^{q^-}\,\|u\|^{q^-}, \quad \forall\, u \in E, \text{ with } \|u\| = \rho\,. \tag{4.151}$$

Taking into account relations (4.138), (4.151) and (4.143), we deduce that for any $u \in E$ with $\|u\| = \rho$

$$J_\lambda(u) \geq \|u\|^{\varphi^0} - \lambda \cdot C_2 \cdot c_1^{q^-} \cdot \|u\|^{q^-} = \rho^{q^-}\left(\rho^{\varphi^0 - q^-} - \lambda \cdot C_2 \cdot c_1^{q^-}\right).$$

Thus, if we define

$$\lambda_\star = \frac{\rho^{\varphi^0 - q^-}}{2 \cdot C_2 \cdot c_1^{q^-}}\,, \tag{4.152}$$

then for any $\lambda \in (0, \lambda_\star)$ and any $u \in E$ with $\|u\| = \rho$ there exists $\alpha = \frac{\rho^{\varphi^0}}{2} > 0$ such that

$$J_\lambda(u) \geq \alpha > 0\,.$$

The proof of Lemma 41 is complete. □

Lemma 42 *Assume that hypotheses of Theorem 33 are fulfilled. Then there exists $\theta \in E$ such that $\theta \geq 0$, $\theta \neq 0$ and $J_\lambda(t\theta) < 0$, for small enough $t > 0$.*

Proof. By the hypotheses of Theorem 33 we have $q^- < \varphi_0$. Let $\epsilon_0 > 0$ be such that $q^- + \epsilon_0 < \varphi_0$. Since $q \in C(\overline{\Omega})$ it follows that there exists an open set $\Omega_0 \subset\subset \Omega$ such that $|q(x) - q^-| < \epsilon_0$ for all $x \in \Omega_0$. Thus, we conclude that $q(x) \leq q^- + \epsilon_0 < \varphi_0$ for all $x \in \Omega_0$.

Let $\theta \in C_0^\infty(\Omega) \subset E$ be such that $\operatorname{supp}(\theta) \supset \overline{\Omega}_0$, $\theta(x) = 1$ for all $x \in \overline{\Omega}_0$ and $0 \leq \theta \leq 1$ in Ω.

Taking into account all the above pieces of information and relations (4.125) and (4.143) we have

$$\begin{aligned} J_\lambda(t\theta) &= \int_\Omega [\Phi(x,t|\nabla\theta(x)|)+\Phi(x,t|\theta(x)|)]\,dx - \lambda\int_\Omega G(x,t\cdot\theta(x))\,dx \\ &\le t^{\varphi_0}\cdot\int_\Omega [\Phi(x,|\nabla\theta(x)|)+\Phi(x,|\theta(x)|)]\,dx - \lambda\cdot C_1\cdot\int_\Omega t^{q(x)}|\theta|^{q(x)}\,dx \\ &\le t^{\varphi_0}\cdot\Lambda(\theta)-\lambda\cdot C_1\cdot\int_{\Omega_0} t^{q(x)}|\theta|^{q(x)}\,dx \\ &\le t^{\varphi_0}\cdot\Lambda(\theta)-\lambda\cdot C_1\cdot t^{q^-+\epsilon_0}\cdot\int_{\Omega_0}|\theta|^{q(x)}\,dx\,, \end{aligned}$$

for any $t\in(0,1)$. Therefore

$$J_\lambda(t\theta)<0$$

for $t<\delta^{1/(\varphi_0-q^--\epsilon_0)}$ with

$$0<\delta<\min\left\{1,\ \frac{\lambda C_1\int_{\Omega_0}|\theta|^{q(x)}\,dx}{\Lambda(\theta)}\right\}.$$

Finally, we point out that $\Lambda(\theta)>0$. Indeed, we have

$$0<\int_{\Omega_0}|\theta|^{q(x)}\,dx\le\int_\Omega|\theta|^{q(x)}\,dx\int_\Omega|\theta|^{q^-}\,dx\le c_1^{q^-}\,\|u\|^{q^-}.$$

Thus, we infer that $\|\theta\|>0$. This fact and relations (4.137) and (4.138) imply that $\Lambda(\theta)>0$. The proof of Lemma 42 is complete. □

In what follows, we give a proof of Theorem 33. Let $\lambda_\star>0$ be defined as in (4.152) and $\lambda\in(0,\lambda_\star)$. It follows by Lemma 41 that on the boundary of the ball centered at the origin and of radius ρ in E, denoted by $B_\rho(0)$, we have

$$\inf_{\partial B_\rho(0)} J_\lambda>0.$$

On the other hand, by Lemma 42, there exists $\theta\in E$ such that $J_\lambda(t\theta)<0$ for all small enough $t>0$. Moreover, relations (4.138), (4.151) and (4.143) imply that for any $u\in B_\rho(0)$ we have

$$J_\lambda(u)\ge\|u\|^{\varphi^0}-\lambda\cdot C_2\cdot c_1^{q^-}\,\|u\|^{q^-}.$$

It follows that

$$-\infty<\underline{c}:=\inf_{B_\rho(0)} J_\lambda<0\,.$$

Fix $0<\epsilon<\inf_{\partial B_\rho(0)} J_\lambda-\inf_{B_\rho(0)} J_\lambda$. Applying the Ekeland variational principle to the functional $J_\lambda:\overline{B_\rho(0)}\to\mathbb{R}$, we find $u_\epsilon\in\overline{B_\rho(0)}$ such that

$$\begin{aligned} J_\lambda(u_\epsilon) &< \inf_{B_\rho(0)} J_\lambda+\epsilon \\ J_\lambda(u_\epsilon) &< J_\lambda(u)+\epsilon\cdot\|u-u_\epsilon\|,\quad u\neq u_\epsilon. \end{aligned}$$

Since

$$J_\lambda(u_\epsilon) \leq \inf_{B_\rho(0)} J_\lambda + \epsilon \leq \inf_{B_\rho(0)} J_\lambda + \epsilon < \inf_{\partial B_\rho(0)} J_\lambda \,,$$

we deduce that $u_\epsilon \in B_\rho(0)$.

Now we define $I_\lambda : \overline{B_\rho(0)} \to \mathbb{R}$ by $I_\lambda(u) = J_\lambda(u) + \epsilon \cdot \|u - u_\epsilon\|$. Then u_ϵ is a minimum point of I_λ and hence

$$\frac{I_\lambda(u_\epsilon + t \cdot v) - I_\lambda(u_\epsilon)}{t} \geq 0$$

for small $t > 0$ and any $v \in B_1(0)$. The above relation yields

$$\frac{J_\lambda(u_\epsilon + t \cdot v) - J_\lambda(u_\epsilon)}{t} + \epsilon \cdot \|v\| \geq 0.$$

Letting $t \to 0$ it follows that $\langle J'_\lambda(u_\epsilon), v\rangle + \epsilon \cdot \|v\| > 0$ and we infer that $\|J'_\lambda(u_\epsilon)\| \leq \epsilon$.

We deduce that there exists a sequence $\{w_n\} \subset B_\rho(0)$ such that

$$J_\lambda(w_n) \to \underline{c} \quad \text{and} \quad J'_\lambda(w_n) \to 0. \tag{4.153}$$

Then the sequence $\{w_n\}$ is bounded in E. Thus, there exists $w \in E$ such that, up to a subsequence, $\{w_n\}$ weakly converges to w in E. Since, by Remark 10, E is compactly embedded in $L^{q(x)}(\Omega)$ it follows that $\{w_n\}$ strongly converges to w in $L^{q(x)}(\Omega)$. The above information combined with relation (4.142) and the Hölder inequality implies

$$\begin{aligned} &\left|\int_\Omega g(x, w_n) \cdot (w_n - w)\, dx\right| \leq C_0 \cdot \int_\Omega |w_n|^{q(x)-1} |w_n - w|\, dx \leq \\ &C_0 \cdot \big|\, |w_n|^{q(x)-1} \,\big|_{\frac{q(x)}{q(x)-1}} \cdot |w_n - w|_{q(x)} \to 0, \quad \text{as } n \to \infty\,. \end{aligned} \tag{4.154}$$

On the other hand, by (4.153) we have

$$\lim_{n\to\infty} \langle J'_\lambda(w_n), w_n - w\rangle = 0\,. \tag{4.155}$$

Relations (4.154) and (4.155) imply

$$\lim_{n\to\infty} \langle \Lambda'(w_n), w_n - w\rangle = 0\,.$$

Thus, by Proposition 13, we deduce that $\{w_n\}$ strongly converges to w in E. So by (4.153),

$$J_\lambda(w) = \underline{c} < 0 \quad \text{and} \quad J'_\lambda(w) = 0\,.$$

We conclude that w is a nontrivial weak solution of problem (4.112) for any $\lambda \in (0, \lambda_\star)$. The proof of Theorem 33 is complete. □

Lemma 43 *Assume the hypotheses of Theorem 34 are fulfilled. Then for any $\lambda > 0$ the functional J_λ is coercive.*

Proof. For each $u \in E$ with $\|u\| > 1$ and $\lambda > 0$ relations (4.137), (4.142) and Remark 10 imply

$$\begin{aligned} J_\lambda(u) &\geq \|u\|^{\varphi_0} - \lambda \cdot C_2 \cdot \int_\Omega |u|^{q(x)}\, dx \\ &\geq \|u\|^{\varphi_0} - \lambda \cdot C_2 \cdot \left[\int_\Omega |u|^{q^-}\, dx + \int_\Omega |u|^{q^+}\, dx\right] \\ &\geq \|u\|^{\varphi_0} - \lambda \cdot C_3 \cdot [\|u\|^{q^-} + \|u\|^{q^+}], \end{aligned}$$

where C_3 is a positive constant. Since $q^+ < \varphi_0$ the above inequality implies that $J_\lambda(u) \to \infty$ as $\|u\| \to \infty$, that is, J_λ is coercive. The proof of Lemma 43 is complete. □

Proof of Theorem 34. Since $q^+ < \varphi_0$ it follows that $q^- < \varphi_0$ and thus, by Theorem 33 there exists $\lambda_\star > 0$ such that for any $\lambda \in (0, \lambda_\star)$ problem (4.112) has a nontrivial weak solution.

Next, by Lemma 43 and Proposition 12, we infer that J_λ is coercive and weakly lower semi-continuous in E, for all $\lambda > 0$. Thus, there exists a global minimizer $u_\lambda \in E$ of I_λ and hence a weak solution of problem (4.112).

We show that u_λ is not trivial for large enough λ. Indeed, letting $t_0 > 1$ be a fixed real and $u_0(x) = t_0$, for all $x \in \Omega$ we have $u_0 \in E$ and

$$\begin{aligned} J_\lambda(u_0) &= \Lambda(u_0) - \lambda \int_\Omega G(x, u_0)\, dx \\ &\leq \int_\Omega \Phi(x, t_0)\, dx - \lambda \cdot C_1 \cdot \int_\Omega |t_0|^{q(x)}\, dx \\ &\leq L - \lambda \cdot C_1 \cdot t_0^{q^+} \cdot |\Omega_1|, \end{aligned}$$

where L is a positive constant. Thus there exists $\lambda^\star > 0$ such that $J_\lambda(u_0) < 0$ for any $\lambda \in [\lambda^\star, \infty)$. It follows that $J_\lambda(u_\lambda) < 0$ for any $\lambda \geq \lambda^\star$ and thus u_λ is a nontrivial weak solution of problem (4.112) for large enough λ. The proof of Theorem 34 is complete. □

We conclude this section with some examples of functions $\varphi(x,t)$ and $\Phi(x,t)$ satisfying our hypotheses.

(i) We take

$$\varphi(x,t) = p(x)|t|^{p(x)-2}t \quad \text{and} \quad \Phi(x,t) = |t|^{p(x)},$$

with $p(x) \in C(\overline{\Omega})$ satisfying $2 \leq p(x) < N$, for all $x \in \overline{\Omega}$. Then φ and Φ satisfy conditions (φ), (Φ_1), (Φ_2), (4.113), (4.115) and (4.141), since in this case we can take $\varphi_0 = p^-$ and $\varphi^0 = p^+$.

(ii) Consider

$$\varphi(x,t) = p(x)\frac{|t|^{p(x)-2}t}{\log(1+|t|)}$$

and

$$\Phi(x,t) = \frac{|t|^{p(x)}}{\log(1+|t|)} + \int_0^{|t|} \frac{s^{p(x)}}{(1+s)(\log(1+s))^2}\, ds\,,$$

with $p(x) \in C(\overline{\Omega})$ satisfying $3 \le p(x) < N$, for all $x \in \overline{\Omega}$.

Then relations (φ), (Φ_1) and (Φ_2) are verified.

For each $x \in \overline{\Omega}$ fixed by Example 3 on p. 243 of [19] we have

$$p(x) - 1 \le \frac{t \cdot \varphi(x,t)}{\Phi(x,t)} \le p(x), \quad \forall\, t \ge 0\,.$$

Thus relation (4.113) holds with $\varphi_0 = p^- - 1$ and $\varphi^0 = p^+$.

Next, Φ satisfies condition (4.141) since

$$\Phi(x,t) \ge t^{p(x)-1}, \quad \forall\, x \in \overline{\Omega},\ t \ge 0\,.$$

Finally, we point out that $\frac{d^2(\Phi(x,\sqrt{t}))}{dt^2} \ge 0$ for all $x \in \overline{\Omega}$ and $t \ge 0$. Thus relation (4.115) is satisfied.

Figure 4.2 illustrates the graph of the function Φ defined in (ii).

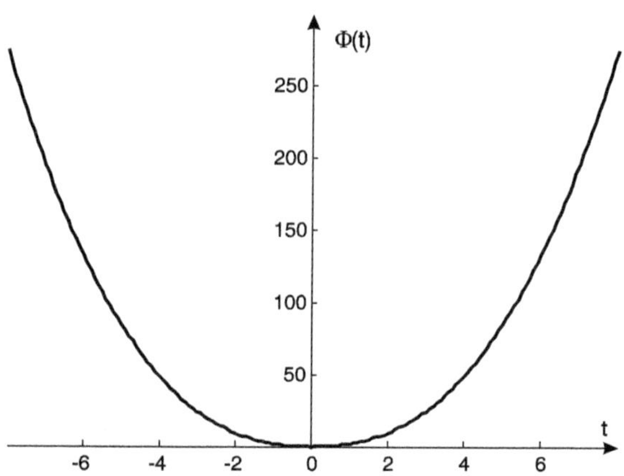

FIGURE 4.2: Graph of the function $\Phi(t)$, $t \in [-8, 8]$

(iii) Set

$$\varphi(x,t) = p(x) \cdot \log(1+\alpha+|t|) \cdot |t|^{p(x)-1} t\,,$$

and

$$\Phi(x,t) = \log(1+\alpha+|t|) \cdot |t|^{p(x)} - \int_0^{|t|} \frac{s^{p(x)}}{1+\alpha+s}\, dx\,,$$

where $\alpha > 0$ is a constant and $p(x) \in C(\overline{\Omega})$ satisfying $2 \le p(x) < N$, for all $x \in \overline{\Omega}$.

Then relations (φ), (Φ_1) and (Φ_2) are verified.

Next, we remark that for each $x \in \overline{\Omega}$ fixed we have

$$p(x) \leq \frac{t \cdot \varphi(x,t)}{\Phi(x,t)}, \quad \forall\, t \geq 0\,.$$

This information shows that taking $\varphi_0 = p^-$ we have

$$1 < p^- \leq \frac{t \cdot \varphi(x,t)}{\Phi(x,t)}, \quad \forall\, x \in \overline{\Omega},\ t \geq 0\,.$$

On the other hand, we have

$$\lim_{t\to\infty} \frac{t \cdot \varphi(x,t)}{\Phi(x,t)} = p(x), \quad \forall\, x \in \overline{\Omega}$$

and

$$\lim_{t\to 0} \frac{t \cdot \varphi(x,t)}{\Phi(x,t)} = p(x), \quad \forall\, x \in \overline{\Omega}\,.$$

Thus, defining $H(x,t) = \frac{t\cdot\varphi(x,t)}{\Phi(x,t)}$ we observe that $H(x,t)$ is continuous on $\overline{\Omega} \times [0,\infty)$ and $1 < p^- \leq \lim_{t\to 0} H(x,t) \leq p^+ < \infty$ and $1 < p^- \leq \lim_{t\to\infty} H(x,t) \leq p^+ < \infty$. It follows that

$$\varphi^0 = \sup_{t>0,\ x\in\overline{\Omega}} \frac{t \cdot \varphi(x,t)}{\Phi(x,t)} < \infty\,.$$

We conclude that relation (4.113) is satisfied.

On the other hand, since

$$\varphi(x,t) \geq p^- \cdot \log(1+\alpha) \cdot t^{p(x)-1}, \quad \forall\, x \in \overline{\Omega},\ t \geq 0\,,$$

it follows that

$$\Phi(x,t) \geq \frac{p^-}{p^+} \cdot (1+\alpha) \cdot t^{p(x)}, \quad \forall\, x \in \overline{\Omega},\ t \geq 0\,.$$

This relation shows that hypothesis (4.141) is verified.

Finally, we point out that $\frac{d^2(\Phi(x,\sqrt{t}))}{dt^2} \geq 0$ for all $x \in \overline{\Omega}$ and $t \geq 0$ and thus, relation (4.115) is satisfied.

4.6 Glossary

Indefinite Potential: This notion refers to sign-changing potentials and appears frequently in the qualitative analysis of partial differential equations.

Mountain Pass Theorem: This is an existence theorem from the calculus of variations. It establishes sufficient conditions for the existence of a saddle point of a real-valued functional.

Orlicz Space: It is a Banach space of measurable functions, which generalizes the L^p space. The notion is named after the Polish mathematician Wladyslaw Orlicz (1903–1990).

Palais-Smale Condition: This is a compactness condition, which is essential in the development of critical point theory on Banach spaces or Banach manifolds.

Three Critical Points Theorem: It is an abstract property that establishes a sufficient condition for the existence of at least three critical points of a functional defined on a Banach space. The result is originally due to Patrizia Pucci (b. 1952) and James Serrin (1926–2012) and it has been extended by Biaggio Ricceri (b. 1955).

Part III

Anisotropic Problems: Continuous and Discrete

Chapter 5

Anisotropic Problems

5.1 Eigenvalue problems for anisotropic elliptic equations 196
5.1.1 Existence of weak solutions 199
5.1.2 Concentration properties of the spectrum 203
5.2 Combined effects in anisotropic elliptic equations 207
5.2.1 Existence of infinitely many high-energy solutions 212
5.2.2 Case of small positive parameters 216
5.3 Anisotropic problems with no-flux boundary condition 220
5.3.1 The first domination case 223
5.3.2 The second domination case 229
5.4 Bifurcation for a singular problem modelling the equilibrium of anisotropic continuous media 233
5.4.1 Existence of low-energy solutions 238
5.4.2 Additional properties and examples 243
5.5 Glossary 247

> Truth is ever to be found in the simplicity, and not in the multiplicity and confusion of things.
>
> Sir Isaac Newton (1643–1727)

In this chapter, we consider a new class of quasilinear elliptic problems, namely elliptic equations that involve a differential operator having partial derivatives with different variable exponents. In this case, the underlying functional-analytic framework relies upon anisotropic Sobolev and Lebesgue spaces. We are concerned with the qualitative analysis of solutions and we establish several existence and nonexistence results. These properties are influenced by the competition between the growth rates of the anisotropic coefficients. Some of the results in this chapter extend original contributions in [85, 87, 88, 62, 63, 91, 9, 8]. We also refer to the book by Repovš and Semenov [92] for some material related to this subject.

5.1 Eigenvalue problems for anisotropic elliptic equations

In this section, we analyze the existence of solutions of the nonhomogeneous anisotropic eigenvalue problem

$$\begin{cases} -\sum_{i=1}^{N} \partial_{x_i} \left(|\partial_{x_i} u|^{p_i(x)-2} \partial_{x_i} u \right) = \lambda |u|^{q(x)-2} u & \text{in } \Omega, \\ u = 0 & \text{on } \partial\Omega, \end{cases} \tag{5.1}$$

where $\Omega \subset \mathbb{R}^N$ $(N \geq 3)$ is a bounded domain with smooth boundary, λ is a positive number, and p_i, q are continuous functions on $\overline{\Omega}$ such that $2 \leq p_i(x) < N$ and $q(x) > 1$ for any $x \in \overline{\Omega}$ and $i \in \{1, ..., N\}$.

In the special case when $p_i = p$ for any $i \in \{1, ..., N\}$ the operator involved in problem (5.1) is the $p(\cdot)$–Laplace operator, namely $\Delta_{p(\cdot)} u := \operatorname{div}(|\nabla u|^{p(x)-2} \nabla u)$. This differential operator is a natural generalization of the isotropic p–Laplace operator $\Delta_p u := \operatorname{div}(|\nabla u|^{p-2} \nabla u)$, where $p > 1$ is a real constant. However, the $p(\cdot)$–Laplace operator possesses more complicated nonlinearities than the p–Laplace operator, due to the fact that $\Delta_{p(\cdot)}$ is *not homogeneous*.

We first recall a natural generalization of the variable exponent Sobolev space $W_0^{1,p(\cdot)}(\Omega)$ that will enable us to study problem (5.1) with sufficient accuracy. For this purpose, let $\vec{p} : \overline{\Omega} \to \mathbb{R}^N$ denote the vector function $\vec{p} = (p_1, ..., p_N)$. We define the *anisotropic variable exponent Sobolev space* $W_0^{1,\vec{p}(\cdot)}(\Omega)$ as the closure of $C_0^\infty(\Omega)$ with respect to the norm

$$\|u\|_{\vec{p}(\cdot)} = \sum_{i=1}^{N} |\partial_{x_i} u|_{p_i(\cdot)} .$$

In the case when $p_i \in C_+(\overline{\Omega})$ are constant functions for any $i \in \{1, .., N\}$ the resulting anisotropic Sobolev space is denoted by $W_0^{1,\vec{p}}(\Omega)$, where $\vec{p}$ is the constant vector $(p_1, ..., p_N)$. Then $W_0^{1,\vec{p}}(\Omega)$ is a reflexive Banach space for any $\vec{p} \in \mathbb{R}^N$ with $p_i > 1$ for all $i \in \{1, .., N\}$. This result can be easily extended to $W_0^{1,\vec{p}(\cdot)}(\Omega)$. Indeed, denoting by $X = L^{p_1(\cdot)}(\Omega) \times ... \times L^{p_N(\cdot)}(\Omega)$ and considering the operator $T : W_0^{1,\vec{p}(\cdot)}(\Omega) \to X$, defined by $T(u) = \nabla u$, it is clear that $W_0^{1,\vec{p}(\cdot)}(\Omega)$ and X are isometric by T, since $\|Tu\|_X = \sum_{i=1}^N |\partial_{x_i} u|_{p_i(\cdot)} = \|u\|_{\vec{p}(\cdot)}$. Thus $T(W_0^{1,\vec{p}(\cdot)}(\Omega))$ is a closed subspace of X, which is a reflexive Banach space. It follows that $T(W_0^{1,\vec{p}(\cdot)}(\Omega))$ is reflexive and consequently also $W_0^{1,\vec{p}(\cdot)}(\Omega)$ is a reflexive Banach space.

On the other hand, in order to facilitate the manipulation of the space $W_0^{1,\vec{p}(\cdot)}(\Omega)$ we introduce $\vec{P}_+$, $\vec{P}_- \in \mathbb{R}^N$ as

$$\vec{P}_+ = (p_1^+, ..., p_N^+), \quad \vec{P}_- = (p_1^-, ..., p_N^-),$$

and P_+^+, P_-^+, $P_-^- \in \mathbb{R}^+$ as

$$P_+^+ = \max\{p_1^+, ..., p_N^+\}, \quad P_-^+ = \max\{p_1^-, ..., p_N^-\}, \quad P_-^- = \min\{p_1^-, ..., p_N^-\}.$$

Throughout this section we assume that

$$\sum_{i=1}^{N} \frac{1}{p_i^-} > 1, \tag{5.2}$$

and define $P_-^\star \in \mathbb{R}^+$ and $P_{-,\infty} \in \mathbb{R}^+$ by

$$P_-^\star = \frac{N}{\sum_{i=1}^N 1/p_i^- - 1}, \qquad P_{-,\infty} = \max\{P_-^+, P_-^\star\}.$$

Note that $P_-^\star$ is well-defined thanks to (5.2) and that it coincides with the usual critical exponent $\overline{P}^\star = N\overline{p}/(N - \overline{p})$ for the harmonic mean $\overline{p}$ of p_i^-. We also notice that it may happen that $P_-^+ > P_-^\star$, thus it is meaningful to define the maximal exponent $P_{-,\infty}$. Actually, $P_{-,\infty}$ turns out to be the "true" critical exponent. We refer to Fragalà, Gazzola and Kawohl [28] for details about anisotropic function spaces.

We start with a compact embedding property for anisotropic spaces.

Theorem 35 *Assume that $\Omega \subset \mathbb{R}^N$ ($N \geq 3$) is a bounded domain with smooth boundary. Suppose that relation (5.2) is fulfilled. Then for any $q \in C(\overline{\Omega})$ satisfying*

$$1 < q(x) < P_{-,\infty} \quad \textit{for all } x \in \overline{\Omega}, \tag{5.3}$$

the embedding

$$W_0^{1,\vec{p}(\cdot)}(\Omega) \hookrightarrow L^{q(\cdot)}(\Omega)$$

is compact.

Proof. The function space $L^{p_i(\cdot)}(\Omega)$ is continuously embedded in $L^{p_i^-}(\Omega)$ for any $i \in \{1, ..., N\}$, since $p_i^- \leq p_i(x)$ for all $x \in \overline{\Omega}$. Thus for each $i \in \{1, ..., N\}$ there exists a positive constant $C_i > 0$ such that

$$|\phi|_{p_i^-} \leq C_i |\phi|_{p_i(\cdot)} \quad \text{for all } \phi \in L^{p_i(\cdot)}(\Omega).$$

If $u \in W_0^{1,\vec{p}(\cdot)}(\Omega)$ then $\partial_{x_i} u \in L^{p_i(\cdot)}(\Omega)$ for each $i \in \{1, ..., N\}$. The above inequalities imply

$$\|u\|_{\vec{P}_-} = \sum_{i=1}^{N} |\partial_{x_i} u|_{p_i^-} \leq C \sum_{i=1}^{N} |\partial_{x_i} u|_{p_i(\cdot)} = C\|u\|_{\vec{p}(\cdot)},$$

where $C = \max\{C_1, ..., C_N\}$. Thus we deduce that $W_0^{1,\vec{p}(\cdot)}(\Omega)$ is continuously embedded in $W_0^{1,\vec{P}_-}(\Omega)$. On the other hand, using relation (5.3) we obtain that $q^+ < P_{-,\infty}$. This fact combined with the result of Theorem 1 from [28] implies that $W_0^{1,\vec{P}_-}(\Omega)$ is compactly embedded in $L^{q^+}(\Omega)$. Finally, since $q(x) \leq q^+$ for each $x \in \overline{\Omega}$ we deduce that $L^{q^+}(\Omega)$ is continuously embedded in $L^{q(\cdot)}(\Omega)$. This shows that $W_0^{1,\vec{p}(\cdot)}(\Omega)$ is compactly embedded in $L^{q(\cdot)}(\Omega)$. □

By a *weak solution* of problem (5.1) we understand a function $u \in W_0^{1,\vec{p}(\cdot)}(\Omega)$ such that

$$\int_\Omega \left\{ \sum_{i=1}^N |\partial_{x_i} u|^{p_i(x)-2} \partial_{x_i} u \partial_{x_i} \phi - \lambda |u|^{q(x)-2} u\phi \right\} dx = 0$$

for all $\phi \in W_0^{1,\vec{p}(\cdot)}(\Omega)$.

In this section we provide several existence results for problem (5.1). The first property establishes a sufficient condition for the existence of solutions for all $\lambda > 0$.

Theorem 36 *Assume that the function $q \in C(\overline{\Omega})$ satisfies the hypothesis*

$$P_+^+ < \min_{x\in\overline{\Omega}} q(x) \leq \max_{x\in\overline{\Omega}} q(x) < P_-^\star. \tag{5.4}$$

Then for any $\lambda > 0$ problem (5.1) possesses a nontrivial weak solution.

Next, we consider a case when the spectrum concentrates at infinity.

Theorem 37 *Assume that $q \in C(\overline{\Omega})$ satisfies*

$$1 < \min_{x\in\overline{\Omega}} q(x) \leq \max_{x\in\overline{\Omega}} q(x) < P_-^-. \tag{5.5}$$

Then there exists $\lambda^{**} > 0$ such that for any $\lambda > \lambda^{**}$ problem (5.1) possesses a nontrivial weak solution.

The following property establishes a case when all values in a neighborhood of the origin are eigenvalues of problem (5.1).

Theorem 38 *Assume that $q \in C(\overline{\Omega})$, with*

$$1 < \min_{x\in\overline{\Omega}} q(x) < P_-^- \quad \text{and} \quad \max_{x\in\overline{\Omega}} q(x) < P_{-,\infty}. \tag{5.6}$$

Then there exists $\lambda^\star > 0$ such that for any $\lambda \in (0, \lambda^\star)$, problem (5.1) has a nontrivial weak solution.

If $q \in C(\overline{\Omega})$ satisfies (5.5) then it satisfies condition (5.6), too. Consequently, the assertion of Theorem 37 can be completed with the conclusion of Theorem 38. More precisely, we deduce the following property, which asserts that the spectrum can concentrate both near the origin and in a neighborhood of infinity.

Corollary 1 *Assume that $q \in C(\overline{\Omega})$ satisfies the following inequalities*

$$1 < \min_{x\in\overline{\Omega}} q(x) \leq \max_{x\in\overline{\Omega}} q(x) < P_-^- .$$

Then there exist $\lambda^\star > 0$ and $\lambda^{\star\star} > 0$ such that for any $\lambda \in (0, \lambda^\star)$ and $\lambda > \lambda^{\star\star}$, problem (5.1) *has a nontrivial weak solution.*

Remark 11 *The result of Theorem 38 remains true in situations that extend relation* (5.5), *since in relation* (5.6) *we can have*

$$1 < \min_{x\in\overline{\Omega}} q(x) < P_-^- < \max_{x\in\overline{\Omega}} q(x) < P_{-,\infty}.$$

5.1.1 Existence of weak solutions

In this section, we present the proof of Theorem 36.

Let E denote the anisotropic variable exponent Orlicz–Sobolev space $W_0^{1,\vec{p}(\cdot)}(\Omega)$. For any $\lambda > 0$ the energy functional corresponding to problem (5.1) is defined by $J_\lambda : E \to \mathbb{R}$,

$$J_\lambda(u) = \int_\Omega \left\{ \sum_{i=1}^N \frac{|\partial_{x_i} u|^{p_i(x)}}{p_i(x)} - \frac{\lambda}{q(x)} |u|^{q(x)} \right\} dx. \tag{5.7}$$

Theorem 35 implies that $J_\lambda \in C^1(E, \mathbb{R})$ and

$$\langle J_\lambda'(u), v\rangle = \int_\Omega \left\{ \sum_{i=1}^N |\partial_{x_i} u|^{p_i(x)-2} \partial_{x_i} u \partial_{x_i} v - \lambda |u|^{q(x)-2} uv \right\} dx$$

for all u, $v \in E$. Thus the weak solutions of (5.1) coincide with the critical points of J_λ.

In order to prove that the energy functional J_λ has a nontrivial critical point, we show that J_λ actually has a mountain pass geometry. It turns out that also the application of this by now standard tool is not straightforward, due to the fact that the kinetic functional J_λ is no longer homogeneous, as in the isotropic case.

We start with two auxiliary results.

Lemma 44 *There exist $\eta > 0$ and $\alpha > 0$ such that $J_\lambda(u) \geq \alpha > 0$ for any $u \in E$ with $\|u\|_{\vec{p}(\cdot)} = \eta$.*

Proof. First, we point out that

$$|u(x)|^{q^-}+|u(x)|^{q^+}\geq |u(x)|^{q(x)} \quad \text{for all } x\in\overline{\Omega}. \tag{5.8}$$

Using this inequality and (5.7), we find

$$J_\lambda(u)\geq \frac{1}{P_+^+}\sum_{i=1}^N\int_\Omega |\partial_{x_i}u|^{p_i(x)}\,dx-\frac{\lambda}{q^-}\left(|u|_{q^-}^{q^-}+|u|_{q^+}^{q^+}\right) \tag{5.9}$$

for any $u\in E$.

Since (5.4) holds, the function space E is continuously embedded in both $L^{q^-}(\Omega)$ and $L^{q^+}(\Omega)$ by Theorem 35. It follows that there exist positive constants B_1 and B_2 such that

$$B_1\|u\|_{\vec{p}(\cdot)}\geq |u|_{q^+},\quad B_2\|u\|_{\vec{p}(\cdot)}\geq |u|_{q^-} \quad \text{for all } u\in E. \tag{5.10}$$

Next, we focus on the case when $u\in E$ and $\|u\|_{\vec{p}(\cdot)}<1$. For such an element u we have $|\partial_{x_i}u|_{p_i(\cdot)}<1$ and, by relation (1.5), we obtain

$$\begin{aligned}\frac{\|u\|_{\vec{p}(\cdot)}^{P_+^+}}{N^{P_+^+-1}} &= N\left(\frac{\sum_{i=1}^N|\partial_{x_i}u|_{p_i(\cdot)}}{N}\right)^{P_+^+}\leq\sum_{i=1}^N|\partial_{x_i}u|_{p_i(\cdot)}^{P_+^+}\\ &\leq\sum_{i=1}^N|\partial_{x_i}u|_{p_i(\cdot)}^{p_i^+}\leq\sum_{i=1}^N\int_\Omega|\partial_{x_i}u|^{p_i(x)}\,dx.\end{aligned} \tag{5.11}$$

Relations (5.9), (5.10) and (5.11) imply

$$\begin{aligned}J_\lambda(u) &\geq \frac{1}{P_+^+N^{P_+^+-1}}\|u\|_{\vec{p}(\cdot)}^{P_+^+}-\frac{\lambda}{q^-}\left[\left(B_1\|u\|_{\vec{p}(\cdot)}\right)^{q^+}+\left(B_2\|u\|_{\vec{p}(\cdot)}\right)^{q^-}\right]\\ &= (B_3-B_4\|u\|_{\vec{p}(\cdot)}^{q^+-P_+^+}-B_5\|u\|_{\vec{p}(\cdot)}^{q^--P_+^+})\|u\|_{\vec{p}(\cdot)}^{P_+^+}\end{aligned}$$

for any $u\in E$ with $\|u\|_{\vec{p}(\cdot)}<1$, where B_3, B_4 and B_5 are positive constants.

Since the function $g:[0,1]\to\mathbb{R}$ defined by

$$g(t)=B_3-B_4t^{q^+-P_+^+}-B_5t^{q^--P_+^+},$$

is positive in a neighborhood of the origin, the conclusion of the lemma follows at once. □

Lemma 45 *There exists* $e\in E$ *by* $\|e\|_{\vec{p}(\cdot)}>\eta$ *(where* η *is given in Lemma* 44*) such that* $J_\lambda(e)<0$.

Proof. Let $\psi \in C_0^\infty(\Omega)$, $\psi \geq 0$ and $\psi \not\equiv 0$, be fixed and let $t > 1$. By (5.7) we have

$$\begin{aligned} J_\lambda(t\psi) &= \int_\Omega \left\{ \sum_{i=1}^N \frac{t^{p_i(x)}}{p_i(x)} |\partial_{x_i}\psi|^{p_i(x)} - \lambda \frac{t^{q(x)}}{q(x)} |\psi|^{q(x)} \right\} dx \\ &\leq \frac{t^{P_+^+}}{P_-^-} \sum_{i=1}^N \int_\Omega |\partial_{x_i}\psi|^{p_i(x)}\, dx - \frac{\lambda t^{q^-}}{q^+} \int_\Omega |\psi|^{q(x)} dx. \end{aligned}$$

Since $q^- > P_+^+$, it is clear by (5.4) that $\lim_{t\to\infty} J_\lambda(t\psi) = -\infty$. Therefore, for large enough $t > 1$, we can take $e = t\psi$ such that $\|e\|_{\vec{p}(\cdot)} > \eta$ and $J_\lambda(e) < 0$. This completes the proof. □

Proof of Theorem 36. By Lemmas 44 and 45 and the mountain pass theorem of Ambrosetti and Rabinowitz [4] (see also Pucci and Serrin [84] for the case of mountains of zero altitude), we deduce the existence of a sequence $(u_n) \subset E$ such that

$$J_\lambda(u_n) \to \overline{c} > 0 \quad \text{and} \quad J'_\lambda(u_n) \to 0 \ \text{(in } E^\star) \quad \text{as } n \to \infty. \tag{5.12}$$

We prove that (u_n) is bounded in E. To this end, assume to the contrary that passing eventually to a subsequence, still denoted by (u_n), we have $\|u_n\|_{\vec{p}(\cdot)} \to \infty$ and that $\|u_n\|_{\vec{p}(\cdot)} > 1$ for all n.

Relation (5.12) and the above considerations imply that for large enough n we have

$$\begin{aligned} 1 + \overline{c} + \|u_n\|_{\vec{p}(\cdot)} &\geq J_\lambda(u) - \frac{1}{q^-}\langle J'_\lambda(u_n), u_n\rangle \\ &\geq \left(\frac{1}{P_+^+} - \frac{1}{q^-}\right) \sum_{i=1}^N \int_\Omega |\partial_{x_i} u_n|^{p_i(x)}\ dx. \end{aligned}$$

For each $i \in \{1, ..., N\}$ and n we define

$$\alpha_{i,n} = \begin{cases} P_+^+, & \text{if } \ |\partial_{x_i} u_n|_{p_i(\cdot)} < 1 \\ P_-^-, & \text{if } \ |\partial_{x_i} u_n|_{p_i(\cdot)} > 1\,. \end{cases}$$

Using relations (1.4) and (1.5) we infer that for large enough n we have

$$\begin{aligned} 1+\bar{c}+\|u_n\|_{\vec{p}(\cdot)} &\geq \left(\frac{1}{P_+^+}-\frac{1}{q^-}\right)\sum_{i=1}^N\int_\Omega |\partial_{x_i}u_n|^{p_i(x)}\,dx \\ &\geq \left(\frac{1}{P_+^+}-\frac{1}{q^-}\right)\sum_{i=1}^N |\partial_{x_i}u_n|_{p_i(\cdot)}^{\alpha_{i,n}} \\ &\geq \left(\frac{1}{P_+^+}-\frac{1}{q^-}\right)\sum_{i=1}^N |\partial_{x_i}u_n|_{p_i(\cdot)}^{P_-^-} \\ &- \left(\frac{1}{P_+^+}-\frac{1}{q^-}\right)\sum_{\{i;\ \alpha_{i,n}=P_+^+\}}\left(|\partial_{x_i}u_n|_{p_i(\cdot)}^{P_-^-}-|\partial_{x_i}u_n|_{p_i(\cdot)}^{P_+^+}\right) \\ &\geq \left(\frac{1}{P_+^+}-\frac{1}{q^-}\right)\frac{1}{N^{P_-^-}}\|u_n\|_{\vec{p}(\cdot)}^{P_-^-}-N\left(\frac{1}{P_+^+}-\frac{1}{q^-}\right). \end{aligned} \tag{5.13}$$

Dividing by $\|u_n\|_{\vec{p}(\cdot)}^{P_-^-}$ in the above inequality and passing to the limit as $n\to\infty$ we obtain a contradiction. It follows that (u_n) is bounded in E. This information combined with the fact that E is reflexive implies that there exist a subsequence, still denoted by (u_n), and $u_0\in E$ such that (u_n) weakly converges to u_0 in E. Since by Theorem 35, the space E is compactly embedded in $L^{q(\cdot)}(\Omega)$, it follows that (u_n) strongly converges to u_0 in $L^{q(\cdot)}(\Omega)$. Then by the Hölder inequality we deduce that

$$\lim_{n\to\infty}\int_\Omega |u_n|^{q(x)-2}u_n(u_n-u_0)\,dx=0.$$

This fact and relation (5.12) yield

$$\lim_{n\to\infty}\langle J_\lambda'(u_n),u_n-u_0\rangle=0.$$

Therefore

$$\lim_{n\to\infty}\sum_{i=1}^N\int_\Omega |\partial_{x_i}u_n|^{p_i(x)-2}\,\partial_{x_i}u_n\,(\partial_{x_i}u_n-\partial_{x_i}u_0)\,dx=0. \tag{5.14}$$

Since (u_n) weakly converges to u_0 in E, we get by relation (5.14)

$$\lim_{n\to\infty}\sum_{i=1}^N\int_\Omega\left(|\partial_{x_i}u_n|^{p_i(x)-2}\,\partial_{x_i}u_n-|\partial_{x_i}u_0|^{p_i(x)-2}\,\partial_{x_i}u_0\right)(\partial_{x_i}u_n-\partial_{x_i}u_0)\,dx=0. \tag{5.15}$$

Next, we apply the following inequality (see Simons [98, formula (2.2)]):

$$(|\xi|^{r-2}\xi-|\psi|^{r-2}\psi)\cdot(\xi-\psi)\geq 2^{-r}|\xi-\psi|^r,\quad \xi,\psi\in\mathbb{R}^N, \tag{5.16}$$

valid for all $r \geq 2$. Relations (5.15) and (5.16) show that (u_n) actually strongly converges to u_0 in E. Hence, by relation (5.12) we have

$$J_\lambda(u_0) = \overline{c} > 0 \quad \text{and} \quad J'_\lambda(u_0) = 0,$$

that is, u_0 is a nontrivial weak solution of equation (5.1). □

5.1.2 Concentration properties of the spectrum

In this section, we give the proof of Theorem 37, which establishes a sufficient condition for the existence of eigenvalues for all sufficiently large λ. This time, the main idea is to show that J_λ has a nontrivial global minimum point in E.

Lemma 46 *The functional J_λ is coercive on E.*

Proof. By relations (5.9) and (5.10) we deduce that for all $u \in E$,

$$\begin{aligned} J_\lambda(u) \ &\geq \frac{1}{P_+^+}\sum_{i=1}^N \int_\Omega |\partial_{x_i} u|^{p_i(x)}\,dx \\ &-\frac{\lambda}{q^-}\left[\left(B_1\|u\|_{\vec{p}(\cdot)}\right)^{q^+} + \left(B_2\|u\|_{\vec{p}(\cdot)}\right)^{q^-}\right]. \end{aligned} \tag{5.17}$$

Assume that $u \in E$ and $\|u\|_{\vec{p}(\cdot)} > 1$. Using the same ideas as in the proof of relation (5.13) combined with estimate (5.17), we find

$$J_\lambda(u) \geq \frac{1}{P_+^+ N^{P_-^-}}\|u\|_{\vec{p}(\cdot)}^{P_-^-} - \frac{N}{P_+^+} - \frac{\lambda}{q^-}\left[\left(B_1\|u\|_{\vec{p}(\cdot)}\right)^{q^+} + \left(B_2\|u\|_{\vec{p}(\cdot)}\right)^{q^-}\right]$$

for any $u \in E$ with $\|u\|_{\vec{p}(\cdot)} > 1$. Since by relation (5.5) we have $P_-^- > q^+ \geq q^-$, we infer that $J_\lambda(u) \to \infty$ as $\|u\|_{\vec{p}(\cdot)} \to \infty$. This shows that J_λ is coercive in E, which completes the proof. □

Proof of Theorem 37. The functional J_λ is weakly lower semicontinuous on E. By Lemma 46, J_λ is also coercive on E. Thus there exists $u_\lambda \in E$ a global minimizer of J_λ and hence a weak solution of problem (5.1).

We show that u_λ is nontrivial for large enough λ. Indeed, letting $t_0 > 1$ be a fixed real and Ω_1 be an open subset of Ω with $|\Omega_1| > 0$ we deduce that there exists $v_0 \in C_0^\infty(\Omega) \subset E$ such that $v_0(x) = t_0$ for any $x \in \overline{\Omega}_1$ and $0 \leq v_0(x) \leq t_0$ in $\Omega \setminus \Omega_1$. We have

$$\begin{aligned} J_\lambda(v_0) \ &= \ \int_\Omega \left\{\sum_{i=1}^N \frac{1}{p_i(x)}\,|\partial_{x_i} v_0|^{p_i(x)} - \frac{\lambda}{q(x)}|v_0|^{q(x)}\right\} dx \\ &\leq \ L - \frac{\lambda}{q^+}\int_{\Omega_1} |v_0|^{q(x)}\,dx \leq L - \frac{\lambda}{q^+} t_0^{q^-}\,|\Omega_1|, \end{aligned}$$

where L is a positive constant. Thus there exists $\lambda^{**} > 0$ such that $J_\lambda(u_0) < 0$ for any $\lambda \in [\lambda^{**}, \infty)$. It follows that $J_\lambda(u_\lambda) < 0$ for any $\lambda \geq \lambda^{**}$ and thus u_λ is a nontrivial weak solution of problem (5.1) for large enough λ. The proof of Theorem 37 is complete. □

Next, we give the proof of Theorem 38. The idea is to use the Ekeland variational principle in order to show that assumption (5.6) implies that the energy functional J_λ has a nontrivial critical point. We start with two auxiliary results.

Lemma 47 *There exists $\lambda^\star > 0$ such that for any $\lambda \in (0, \lambda^\star)$ there are ρ, $a > 0$ such that $J_\lambda(u) \geq a > 0$ for any $u \in E$ with $\|u\|_{\vec{p}(\cdot)} = \rho$.*

Proof. By hypothesis (5.6), it follows by Theorem 35 that E is continuously embedded in $L^{q(\cdot)}(\Omega)$. Thus there exists a positive constant c_1 such that

$$|u|_{q(\cdot)} \leq c_1 \|u\|_{\vec{p}(\cdot)} \quad \text{for all} \quad u \in E. \tag{5.18}$$

We fix $\rho \in (0, 1)$ such that $\rho < 1/c_1$. Then relation (5.18) implies

$$|u|_{q(\cdot)} < 1 \quad \text{for all} \quad u \in E, \quad \text{with} \quad \|u\|_{\vec{p}(\cdot)} = \rho.$$

Furthermore, relation (1.5) yields

$$\int_\Omega |u|^{q(x)}\, dx \leq |u|_{q(\cdot)}^{q^-} \quad \text{for all} \quad u \in E, \quad \text{with} \quad \|u\|_{\vec{p}(\cdot)} = \rho. \tag{5.19}$$

Relations (5.18) and (5.19) imply

$$\int_\Omega |u|^{q(x)}\, dx \leq c_1^{q^-} \|u\|_{\vec{p}(\cdot)}^{q^-} \quad \text{for all} \quad u \in E, \quad \text{with} \quad \|u\|_{\vec{p}(\cdot)} = \rho. \tag{5.20}$$

Taking into account relations (5.11) and (5.20), we deduce that for any $u \in E$ with $\|u\|_{\vec{p}(\cdot)} = \rho$ the following inequalities hold:

$$\begin{aligned} J_\lambda(u) &\geq \frac{1}{P_+^+ N^{P_+^+ - 1}} \|u\|_{\vec{p}(\cdot)}^{P_+^+} - \frac{\lambda}{q^-} \int_\Omega |u|^{q(x)}\, dx \\ &\geq \frac{1}{P_+^+ N^{P_+^+ - 1}} \|u\|_{\vec{p}(\cdot)}^{P_+^+} - \frac{\lambda c_1^{q^-}}{q^-} \|u\|_{\vec{p}(\cdot)}^{q^-} \\ &= \rho^{q^-} \left(\frac{1}{P_+^+ N^{P_+^+ - 1}} \rho^{P_+^+ - q^-} - \frac{\lambda c_1^{q^-}}{q^-} \right). \end{aligned}$$

Set

$$\lambda^\star = \frac{q^-}{2 c_1^{q^-} P_+^+ N^{P_+^+ - 1}} \rho^{P_+^+ - q^-}. \tag{5.21}$$

Thus for any $\lambda \in (0, \lambda^\star)$ and $u \in E$ with $\|u\|_{\vec{p}(\cdot)} = \rho$ the number $a = \rho^{P_+^+}/2P_+^+ N^{P_+^+ - 1}$ satisfies

$$J_\lambda(u) \geq a > 0.$$

This completes the proof. □

Lemma 48 *There exists $\phi \in E$ such that $\phi \geq 0$, $\phi \not\equiv 0$ and $J_\lambda(t\phi) < 0$ for small enough $t > 0$.*

Proof. Assumption (5.6) implies that $q^- < P_-^-$. Let $\epsilon_0 > 0$ be such that $q^- + \epsilon_0 < P_-^-$. On the other hand, since $q \in C(\overline{\Omega})$ it follows that there exists an open set $\Omega_2 \subset \Omega$ such that $|q(x) - q^-| < \epsilon_0$ for all $x \in \Omega_2$. Thus, we conclude that $q(x) \leq q^- + \epsilon_0 < P_-^-$ for all $x \in \Omega_2$.

Let $\phi \in C_0^\infty(\Omega)$ be such that $\text{supp}(\phi) \supset \overline{\Omega}_2$, $\phi(x) = 1$ for all $x \in \overline{\Omega}_2$ and $0 \leq \phi \leq 1$ in Ω. Then by (5.7), we have for any $t \in (0, 1)$

$$\begin{aligned} J_\lambda(t\phi) &= \int_\Omega \left\{ \sum_{i=1}^N \frac{t^{p_i(x)}}{p_i(x)} |\partial_{x_i}\phi|^{p_i(x)} - \lambda \frac{t^{q(x)}}{q(x)} |\phi|^{q(x)} \right\} dx \\ &\leq \frac{t^{P_-^-}}{P_-^-} \sum_{i=1}^N \int_\Omega |\partial_{x_i}\phi|^{p_i(x)}\, dx - \frac{\lambda}{q^+} \int_\Omega t^{q(x)} |\phi|^{q(x)} dx \\ &\leq \frac{t^{P_-^-}}{P_-^-} \sum_{i=1}^N \int_\Omega |\partial_{x_i}\phi|^{p_i(x)}\, dx - \frac{\lambda}{q^+} \int_{\Omega_2} t^{q(x)} |\phi|^{q(x)} dx \\ &\leq \frac{t^{P_-^-}}{P_-^-} \sum_{i=1}^N \int_\Omega |\partial_{x_i}\phi|^{p_i(x)}\, dx - \frac{\lambda\, t^{q^- + \epsilon_0}}{q^+} \int_{\Omega_2} |\phi|^{q(x)} dx. \end{aligned}$$

Therefore

$$J_\lambda(t\phi) < 0$$

for $t < \delta^{1/(P_-^- - q^- - \epsilon_0)}$ with

$$0 < \delta < \min \left\{ 1, \frac{\lambda\, P_-^-}{q^+} \int_{\Omega_2} |\phi|^{q(x)} dx \Big/ \sum_{i=1}^N \int_\Omega |\partial_{x_i}\phi|^{p_i(x)}\, dx \right\}.$$

This is possible since we claim that $\sum_{i=1}^N \int_\Omega |\partial_{x_i}\phi|^{p_i(x)}\, dx > 0$. Indeed, it is clear that

$$\int_{\Omega_2} |\phi|^{q(x)}\, dx \leq \int_\Omega |\phi|^{q(x)}\, dx \leq \int_\Omega |\phi|^{q^-}\, dx.$$

On the other hand, $E = W_0^{1,\vec{p}(x)}(\Omega)$ is continuously embedded in $L^{q^-}(\Omega)$ and thus there exists a positive constant c_2 such that

$$|\phi|_{q^-} \leq c_2 \|\phi\|_{\vec{p}(\cdot)}.$$

The last two inequalities imply that

$$\|\phi\|_{\vec{p}(\cdot)} > 0$$

and combining this fact with relations (1.4) or (1.5) the claim follows at once. The proof of the lemma is now complete. □

Proof of Theorem 38. Let $\lambda^\star > 0$ be defined as in (5.21) and let $\lambda \in (0, \lambda^\star)$. By Lemma 47 it follows that on the boundary of the ball centered at the origin and of radius ρ in E, denoted by $B_\rho(0)$, we have

$$\inf_{\partial B_\rho(0)} J_\lambda > 0. \tag{5.22}$$

On the other hand, by Lemma 48, there exists $\phi \in E$ such that $J_\lambda(t\phi) < 0$ for all small enough $t > 0$. Moreover, relations (5.11), (5.20) and (1.5) imply that for any $u \in B_\rho(0)$ we have

$$J_\lambda(u) \geq \frac{1}{P_+^+ N^{P_+^+ - 1}} \|u\|_{\vec{p}(\cdot)}^{P_+^+} - \frac{\lambda c_1^{q^-}}{q^-} \|u\|_{\vec{p}(\cdot)}^{q^-}.$$

It follows that

$$-\infty < \underline{c} := \inf_{\overline{B_\rho(0)}} J_\lambda < 0.$$

Fix $0 < \epsilon < \inf_{\partial B_\rho(0)} J_\lambda - \inf_{\overline{B_\rho(0)}} J_\lambda$. Applying the Ekeland variational principle to the functional $J_\lambda : \overline{B_\rho(0)} \to \mathbb{R}$, we find $u_\epsilon \in \overline{B_\rho(0)}$ such that

$$\begin{aligned} J_\lambda(u_\epsilon) &< \inf_{\overline{B_\rho(0)}} J_\lambda + \epsilon \\ J_\lambda(u_\epsilon) &< J_\lambda(u) + \epsilon \|u - u_\epsilon\|_{\vec{p}(\cdot)}, \quad u \neq u_\epsilon. \end{aligned}$$

Since

$$J_\lambda(u_\epsilon) \leq \inf_{\overline{B_\rho(0)}} J_\lambda + \epsilon \leq \inf_{B_\rho(0)} J_\lambda + \epsilon < \inf_{\partial B_\rho(0)} J_\lambda\,,$$

we deduce that $u_\epsilon \in B_\rho(0)$. Now, we define $I_\lambda : \overline{B_\rho(0)} \to \mathbb{R}$ by $I_\lambda(u) = J_\lambda(u) + \epsilon \|u - u_\epsilon\|_{\vec{p}(\cdot)}$. Since u_ϵ is a minimum point of I_λ we obtain

$$\frac{I_\lambda(u_\epsilon + t\,v) - I_\lambda(u_\epsilon)}{t} \geq 0$$

for small $t > 0$ and any $v \in B_1(0)$. This relation yields

$$\frac{J_\lambda(u_\epsilon + t\,v) - J_\lambda(u_\epsilon)}{t} + \epsilon \|v\|_{\vec{p}(\cdot)} \geq 0.$$

Letting $t \to 0$ it follows that $\langle J_\lambda'(u_\epsilon), v\rangle + \epsilon \|v\|_{\vec{p}(\cdot)} > 0$ and we infer that $\|J_\lambda'(u_\epsilon)\| \leq \epsilon$.

We deduce that there exists a sequence $(w_n) \subset B_\rho(0)$ such that

$$J_\lambda(w_n) \to \underline{c} \quad \text{and} \quad J'_\lambda(w_n) \to 0. \tag{5.23}$$

It is clear that (w_n) is bounded in E. Thus there exists $w \in E$ such that, up to a subsequence, (w_n) weakly converges to w in E. Actually, with similar arguments as those used in the proof of Theorem 36 we can show that (w_n) strongly converges to w in E. Thus by (5.23)

$$J_\lambda(w) = \underline{c} < 0 \quad \text{and} \quad J'_\lambda(w) = 0,$$

hence w is a nontrivial weak solution of problem (5.1). □

5.2 Combined effects in anisotropic elliptic equations

Let $\Omega \subset \mathbb{R}^N$ $(N \geq 3)$ be a bounded domain with smooth boundary. The purpose of this section is to analyze the existence of one or several weak solutions of the anisotropic problem

$$\begin{cases} -\displaystyle\sum_{i=1}^{N} \partial_{x_i} a_i(x, \partial_{x_i} u) + b(x)|u|^{P_+^+ - 2} u = f(x, u) & \text{in } \Omega \\ u = 0 & \text{on } \partial\Omega, \end{cases} \tag{5.24}$$

where $b \in L^\infty(\overline{\Omega})$, $f : \overline{\Omega} \times \mathbb{R} \to \mathbb{R}$ and $a_i : \overline{\Omega} \times \mathbb{R} \to \mathbb{R}$ are Carathéodory functions fulfilling some natural hypotheses.

The anisotropic differential operator $\sum_{i=1}^{N} \partial_{x_i} a_i(x, \partial_{x_i} u)$ is a $\overrightarrow{p}(\cdot)$–Laplace type operator, where $\overrightarrow{p}(x) = (p_1(x), p_2(x), \ldots, p_N(x))$, and $P_+^+ = \max_{i \in \{1,\ldots,N\}}\{\sup_{x \in \Omega} p_i(x)\}$. For $i = 1, \ldots, N$, we assume that p_i is a continuous function on $\overline{\Omega}$.

Let $a_i(x, \eta)$ denote the continuous derivative with respect to η of the mapping $A_i : \overline{\Omega} \times \mathbb{R}^N \to \mathbb{R}^N$, $A_i = A_i(x, \eta)$, that is, $a_i(x, \eta) = \frac{\partial}{\partial \eta} A_i(x, \eta)$. Throughout this section we assume that the following hypotheses are fulfilled:

(A₀) $A_i(x, 0) = 0$ for a.e. $x \in \Omega$.

(A₁) There exists a positive constant $\overline{c}_i$ such that a_i satisfies the growth condition

$$|a_i(x, \eta)| \leq \overline{c}_i (1 + |\eta|^{p_i(x) - 1}),$$

for all $x \in \overline{\Omega}$ and $\eta \in \mathbb{R}^N$.

(A₂) The inequalities

$$|\eta|^{p_i(x)} \leq a_i(x, \eta)\eta \leq p_i(x) A_i(x, \eta),$$

hold for all $x \in \overline{\Omega}$ and $\eta \in \mathbb{R}^N$.

($\mathbf{A_3}$) There exists $k_i > 0$ such that

$$A_i\left(x, \frac{\eta+\xi}{2}\right) \le \frac{1}{2}A_i(x,\eta) + \frac{1}{2}A_i(x,\xi) - k_i|\eta-\xi|^{p_i(x)},$$

for all $x \in \overline{\Omega}$ and $\eta, \xi \in \mathbb{R}^N$, with equality if and only if $\eta = \xi$.

($\mathbf{A_4}$) The mapping A_i is even with respect to its second variable, that is,

$$A_i(x,-\eta) = A_i(x,\eta),$$

for all $x \in \overline{\Omega}$ and $\eta \in \mathbb{R}^N$.

The differential operator $\sum_{i=1}^N \partial_{x_i} a_i(x, \partial_{x_i} u)$ is the anisotropic $\overrightarrow{p}(x)$–Laplace type operator (where $\overrightarrow{p}(x) = (p_1(x), \ldots, p_N(x))$ because if we take

$$a_i(x,\eta) = |\eta|^{p_i(x)-2}\eta,$$

for all $i \in \{1, \ldots, N\}$, we have $A_i(x,\eta) = \frac{1}{p_i(x)}|\eta|^{p_i(x)}$ for all $i \in \{1, \ldots, N\}$, that is,

$$\Delta_{\overrightarrow{p}(x)}(u) = \sum_{i=1}^N \partial_{x_i}\left(|\partial_{x_i} u|^{p_i(x)-2}\partial_{x_i} u\right).$$

There are many other operators which can be derived from $\sum_{i=1}^N \partial_{x_i} a_i(x, \partial_{x_i} u)$. Indeed, if we take

$$a_i(x,\eta) = (1+|\eta|^2)^{\frac{(p_i(x)-2)}{2}}\eta,$$

for all $i \in \{1, \ldots, N\}$, we have

$$A_i(x,\eta) = \frac{1}{p_i(x)}[(1+|\eta|^2)^{\frac{p_i(x)}{2}} - 1] \quad \text{for all } i \in \{1, \ldots, N\}.$$

Thus we obtain the anisotropic variable mean curvature operator

$$\sum_{i=1}^N \partial_{x_i}\left[(1+|\partial_{x_i} u|^2)^{\frac{(p_i(x)-2)}{2}}\partial_{x_i} u\right].$$

Set $C_+(\overline{\Omega}) = \{h \in C(\overline{\Omega}) : \min_{x\in\overline{\Omega}} h(x) > 1\}$. For any $h \in C_+(\overline{\Omega})$ we define

$$h^+ = \sup_{x\in\Omega} h(x) \qquad \text{and} \qquad h^- = \inf_{x\in\Omega} h(x).$$

If $p \in C_+(\overline{\Omega})$, we define the variable exponent Lebesgue space

$$L^{p(x)}(\Omega) = \{u : u \text{ is measurable and } \int_\Omega |u(x)|^{p(x)}\,dx < \infty\},$$

endowed with the *Luxemburg norm* defined by

$$|u|_{p(x)} = \inf\left\{\mu > 0;\ \int_\Omega \left|\frac{u(x)}{\mu}\right|^{p(x)} dx \le 1\right\}.$$

If $p \in C_+(\overline{\Omega})$, the variable exponent Sobolev space $W^{1,p(x)}(\Omega)$ contains all functions $u \in L^{p(x)}(\Omega)$ such that gradient ∇u exists almost everywhere and belongs to $[L^{p(x)}(\Omega)]^N$. Then $W^{1,p(x)}(\Omega)$ is a separable and reflexive Banach space with respect to the norm

$$\|u\| = |u|_{p(x)} + |\nabla u|_{p(x)},$$

If $p \in C_+(\overline{\Omega})$ is *logarithmic Hölder continuous*, that is,

$$|p(x) - p(y)| \le -\frac{M}{\log(|x-y|)} \quad \text{for all } x,\, y \in \Omega \text{ with } |x-y| \le 1/2, \tag{5.25}$$

then the smooth functions are dense in $W^{1,p(x)}(\Omega)$. Let $W_0^{1,p(x)}(\Omega)$ denote the Sobolev space of functions with zero boundary values under the norm $\|\cdot\|$. Furthermore, if $p \in C_+(\overline{\Omega})$ satisfies (5.25), then $C_0^\infty(\Omega)$ is dense in $W_0^{1,p(x)}(\Omega)$. Due to the Poincaré inequality,

$$\|u\|_{1,p(x)} = |\nabla u|_{p(x)}$$

is an equivalent norm in $W_0^{1,p(x)}(\Omega)$. Of course, also the norm

$$\|u\|_{p(x)} = \sum_{i=1}^N |\partial_{x_i} u|_{p(x)}$$

is an equivalent norm in $W_0^{1,p(x)}(\Omega)$.

Let $\overrightarrow{p} : \overline{\Omega} \to \mathbb{R}^N$ denote the vectorial function $\overrightarrow{p} = (p_1, \ldots, p_N)$. We define $W_0^{1,\overrightarrow{p}(x)}(\Omega)$, the *anisotropic variable exponent Sobolev space*, as the closure of $C_0^\infty(\Omega)$ with respect to the norm

$$\|u\|_{\overrightarrow{p}(x)} = \sum_{i=1}^N |\partial_{x_i} u|_{p_i(\cdot)}\,.$$

In the case when $p_i \in C_+(\overline{\Omega})$ are constant functions for any $i \in \{1,..,N\}$ the resulting anisotropic Sobolev space is denoted by $W_0^{1,\overrightarrow{p}}(\Omega)$, where $\overrightarrow{p}$ is the constant vector $(p_1,...,p_N)$. The function space $W_0^{1,\overrightarrow{p}}(\Omega)$ is a reflexive Banach space for any $\overrightarrow{p} \in \mathbb{R}^N$ with $p_i > 1$ for all $i \in \{1,..,N\}$. This result can be extended to $W_0^{1,\overrightarrow{p}(x)}(\Omega)$. Indeed, denoting $Z = L^{p_1(\cdot)}(\Omega) \times \ldots \times L^{p_N(\cdot)}(\Omega)$ and considering the operator $T : W_0^{1,\vec{p}(x)}(\Omega) \to Z$, defined by $T(u) = \nabla u$, then $W_0^{1,\vec{p}(x)}(\Omega)$ and Z are isometric by T, since

$$\|Tu\|_Z = \sum_{i=1}^{N} |\partial_{x_i} u|_{p_i(\cdot)} = \|u\|_{\vec{p}(x)}.$$

Thus $T(W_0^{1,\vec{p}(x)}(\Omega))$ is a closed subspace of Z, which is a reflexive Banach space, hence $T(W_0^{1,\vec{p}(x)}(\Omega))$ is reflexive and consequently also $W_0^{1,\vec{p}(x)}(\Omega)$ is a reflexive Banach space.

We define $X := W_0^{1,\vec{p}(x)}(\Omega)$. Since X is reflexive there exist $\{e_j\} \subset X$ and $\{e_j^*\} \subset X^*$ such that

$$X = \overline{\operatorname{span}\{e_j \ : j = 1, 2, ...\}}, \qquad X^* = \overline{\operatorname{span}\{e_j^* \ : j = 1, 2, ...\}},$$

and

$$\langle e_i, e_j^* \rangle = \begin{cases} 1 & \text{if} \quad i = j, \\ 0 & \text{if} \quad i \neq j, \end{cases}$$

where $\langle .,. \rangle$ denotes the duality product between X and X^*. We define

$$X_j = \operatorname{span}\{e_j\}, \qquad Y_k = \bigoplus_{j=1}^{k} X_j, \qquad Z_k = \overline{\bigoplus_{j=k}^{\infty} X_j}.$$

Define $\vec{P}_+, \vec{P}_- \in \mathbb{R}^N$ and the real numbers $P_+^+,\ P_-^+,\ P_+^-,\ P_-^- \in \mathbb{R}^+$ as follows:

$$\begin{aligned} \vec{P}_+ &= (p_1^+, p_2^+, \ldots, p_N^+), & \vec{P}_- &= (p_1^-, p_2^-, \ldots, p_N^-), \\ P_+^+ &= \max\{p_1^+, p_2^+, \ldots, p_N^+\}, & P_-^+ &= \max\{p_1^-, p_2^-, \ldots, p_N^-\}, \\ P_+^- &= \min\{p_1^+, p_2^+, \ldots, p_N^+\}, & P_-^- &= \min\{p_1^-, p_2^-, \ldots, p_N^-\}. \end{aligned}$$

Throughout this section, we assume that

$$\sum_{i=1}^{N} \frac{1}{p_i^-} > 1. \tag{5.26}$$

Define $P_-^* \in \mathbb{R}^+$ and $P_{-,\infty} \in \mathbb{R}^+$ by

$$P_-^* = \frac{N}{\sum_{i=1}^{N} \frac{1}{p_i^-} - 1}, \qquad P_{-,\infty} = \max\{P_-^+, P_-^*\}.$$

For the Carathéodory function $f : \Omega \times \mathbb{R} \to \mathbb{R}$, we consider the antiderivative $F : \Omega \times \mathbb{R} \to \mathbb{R}$,

$$F(x, s) = \int_0^s f(x, t)dt.$$

We assume that b and f satisfy the following hypotheses:

(B) $b \in L^\infty(\Omega)$ and there exists $b_0 > 0$ such that $b(x) \geq b_0$ for all $x \in \Omega$.

(F$_1$) There exist a positive constant c_1 and $\alpha(x) \in C_+(\overline{\Omega})$ with $\alpha(x) < P_{-,\infty}$ such that

$$|f(x,t)| \leq c_1(1 + |t|^{\alpha(x)-1}) \qquad \text{for all} \quad (x,t) \in \Omega \times \mathbb{R}\,. \tag{5.27}$$

(F$_2$) There exist $M > 0,\ \theta > P_+^+$ such that for all $x \in \Omega$ and all $t \in \mathbb{R}$ with $|t| \geq M$,

$$0 < \theta F(x,t) \leq t f(x,t).$$

(F$_3$) The function f is odd with respect to its second variable, that is,

$$f(x,-t) = -f(x,t),$$

for all $x \in \Omega$ and $t \in \mathbb{R}$.

A function $u \in W_0^{1,\vec{p}(x)}(\Omega)$ that satisfies

$$\int_\Omega \Big\{ \sum_{i=1}^N a_i(x, \partial_{x_i})\partial_{x_i}\varphi + b(x)|u|^{P_+^+-2}u\varphi - f(x,u)\varphi \Big\} dx = 0,$$

for all $\varphi \in W_0^{1,\vec{p}(x)}(\Omega)$ is called a weak solution of problem (5.24).

The following result establishes the existence of infinitely many high-energy solutions of problem (5.24), provided that the right-hand side is odd.

Theorem 39 *Problem* (5.24) *admits a sequence* $(\pm u_n)$ *of weak solutions with high energies.*

In addition, we consider the case when $f(x,u) = \lambda |u|^{q(x)-2}u$, where the parameter λ is positive and $q(x)$ is continuous function on $\overline{\Omega}$. Problem (5.24) becomes

$$\begin{cases} -\sum_{i=1}^N \partial_{x_i} a_i(x, \partial_{x_i} u) + b(x)|u|^{P_+^+-2}u = \lambda |u|^{q(x)-2}u & \text{in } \Omega, \\ u = 0 \quad \text{on } \partial\Omega. \end{cases} \tag{5.28}$$

A function $u \in W_0^{1,\vec{p}(x)}(\Omega)$ is said to be a weak solution of problem (5.28) if

$$\int_\Omega \Big\{ \sum_{i=1}^N a_i(x, \partial_{x_i})\partial_{x_i}\varphi + b(x)|u|^{P_+^+-2}u\varphi - \lambda |u|^{q(x)-2}u\varphi \Big\} dx = 0,$$

for all $\varphi \in W_0^{1,\vec{p}(x)}(\Omega)$.

The following property establishes the existence of weak solutions of problem (5.28), provided that the positive parameter λ is small enough.

Theorem 40 *Assume that* $q^- < P_-^- \leq P_+^+ < P_{-,\infty}$ *for all* $x \in \overline{\Omega}$. *Then there exists a positive constant* λ^* *such that for any* $\lambda \in (0, \lambda^*)$, *problem* (5.28) *has at least one nontrivial weak solution.*

5.2.1 Existence of infinitely many high-energy solutions

In what follows, we are concerned with the existence of multiple weak solutions of problem (5.24). We associate to problem (5.24) the energy functional $I : X \to \mathbb{R}$ defined by

$$I(u) = \int_\Omega \Big\{ \sum_{i=1}^N A_i(x, \partial_{x_i} u) + \frac{b(x)}{P_+^+} |u|^{P_+^+} - F(x,u) \Big\} dx.$$

Then I is well-defined and $I \in C^1(X, \mathbb{R})$ with

$$\langle I'(u), \varphi \rangle = \int_\Omega \Big\{ \sum_{i=1}^N a_i(x, \partial_{x_i} u) \partial_{x_i} \varphi + b(x)|u|^{P_+^+ - 2} u\varphi - f(x,u)\varphi \Big\} dx,$$

for all $u, \varphi \in X$. Hence any critical point $u \in X$ of I is a weak solution of problem (5.24).

We prove three auxiliary properties.

Lemma 49 *For every $k \in \mathbb{N}$, there exists $r_k > 0$ such that $\inf_{u \in Z_k,\ ||u|| = r_k} I(u) \to \infty$, as $k \to \infty$.*

Proof. By (**A2**) and ($\mathbf{F_1}$) for any $u \in Z_k$, $||u|| = r_k > 1$ (r_k will be specified below), we have

$$\begin{aligned} I(u) &= \int_\Omega \Big\{ \sum_{i=1}^N A_i(x, \partial_{x_i} u) + \frac{b(x)}{P_+^+} |u|^{P_+^+} - F(x,u) \Big\} dx \\ &\geq \frac{1}{P_+^+} \sum_{i=1}^N \int_\Omega |\partial_{x_i} u|^{p_i(x)} dx + \frac{b_0}{P_+^+} \int_\Omega |u|^{P_+^+} dx - c_1 \int_\Omega (1 + |u|^{\alpha(x)}) dx \\ &\geq \frac{1}{P_+^+} \sum_{i=1}^N \int_\Omega |\partial_{x_i} u|^{p_i(x)} dx + \frac{b_0}{P_+^+} |u|_{L^{P_+^+}(\Omega)}^{P_+^+} \qquad (5.29) \\ &- c_2 \max\{ |u|_{L^{\alpha(x)}(\Omega)}^{\alpha^+}, |u|_{L^{\alpha(x)}(\Omega)}^{\alpha^-} \} - c_3. \qquad (5.30) \end{aligned}$$

Using (**B**), we can write

$$\frac{1}{P_+^+} \int_\Omega b(x)|u|^{P_+^+} dx \geq \frac{b_0}{P_+^+} |u|_{L^{P_+^+}(\Omega)}^{P_+^+} \geq 0. \qquad (5.31)$$

For each $i \in \{1, 2, \ldots, N\}$, we define

$$\alpha_i = \begin{cases} P_+^+ & \text{if } |\partial_{x_i} u|_{p_i(x)} < 1, \\ P_-^- & \text{if } |\partial_{x_i} u|_{p_i(x)} > 1. \end{cases}$$

Using the Jensen inequality applied to the convex function $g : \mathbb{R}^+ \to \mathbb{R}^+$,

$g(t) = t^{P_-^-}$, $P_-^- > 1$, we have

$$\begin{aligned}\sum_{i=1}^{N}\int_\Omega |\partial_{x_i}u|^{p_i(x)}dx &\geq \sum_{i=1}^{N}|\partial_{x_i}u|_{p_i(x)}^{\alpha_i} \\ &\geq \sum_{i=1}^{N}|\partial_{x_i}u|_{p_i(x)}^{P_-^-} - \sum_{\{i;\alpha_i=P_+^+\}}\left(|\partial_{x_i}u|_{p_i(x)}^{P_-^-} - |\partial_{x_i}u|_{p_i(x)}^{P_+^+}\right) \\ &\geq N\Big(\frac{\sum_{i=1}^{N}|\partial_{x_i}u|^{p_i(x)}}{N}\Big)^{P_-^-} - N \\ &= \frac{||u||^{P_-^-}}{N^{P_-^- -1}} - N.\end{aligned} \tag{5.32}$$

Taking into account relations (5.31) and (5.32), the inequality (5.29) reduces to

$$I(u) \geq \frac{||u||^{P_-^-}}{P_+^+ N^{P_-^- -1}} - c_2\max\{|u|_{L^{\alpha(x)}(\Omega)}^{\alpha^+}, |u|_{L^{\alpha(x)}(\Omega)}^{\alpha^-}\} - c_4.$$

Denote

$$\alpha_k = \sup\{|u|_{L^{\alpha(x)}(\Omega)};\ ||u|| = 1,\ u\in Z_k\}.$$

By Proposition 3.5 in [29], we know that $\lim_{k\to\infty}\alpha_k = 0$.
If $\max\{|u|_{L^{\alpha(x)}(\Omega)}^{\alpha^+}, |u|_{L^{\alpha(x)}(\Omega)}^{\alpha^-}\} = |u|_{L^{\alpha(x)}(\Omega)}^{\alpha^+}$, we have

$$I(u) \geq \frac{||u||^{P_-^-}}{P_+^+ N^{P_-^- -1}} - c_2\alpha_k^{\alpha^+}||u||^{\alpha^+} - c_4. \tag{5.33}$$

Depending on the relation between P_-^- and α^+ we distinguish between the following cases:
Case 1 : $P_-^- > \alpha^+$. For sufficiently large k, we have $\alpha_k < \frac{1}{2c_2P_+^+N^{P_-^- -1}}$, so relation (5.33) yields

$$I(u) \geq \frac{1}{2c_2P_+^+N^{P_-^- -1}}||u||^{P_-^-}.$$

By choosing r_k such that $r_k\to\infty$ as $k\to\infty$, for $u\in Z_k$ with $||u|| = r_k$, we have that $I(u)\to\infty$.
Case 2 : $\alpha^- > P_+^+$. Choose $r_k = \left(c_2N^{P_-^- -1}\alpha^+\alpha_k^{\alpha^+}\right)^{\frac{1}{P_-^- -\alpha^+}}$, we deduce that

$$I(u) \geq \frac{1}{N^{P_-^- -1}}\Big(\frac{1}{P_+^+} - \frac{1}{\alpha^+}\Big)r_k^{P_-^-} - c_4.$$

Since $\alpha_k\to 0$, $r_k\to\infty$ as $k\to\infty$, we obtain $I(u)\to\infty$.

Similarly, if $\max\{|u|_{L^{\alpha(x)}(\Omega)}^{\alpha^+}, |u|_{L^{\alpha(x)}(\Omega)}^{\alpha^-}\} = |u|_{L^{\alpha(x)}(\Omega)}^{\alpha^-}$, we can deduce for $u\in Z_k$ with $||u|| = r_k > 1$, $I(u)\to\infty$ as $k\to+\infty$ and the proof is complete.
□

Lemma 50 *For every $k \in \mathbb{N}$ there exists $\rho_k > r_k$ (r_k given by Lemma 49) such that*

$$\max_{u \in Y_k,\ ||u||=\rho_k} I(u) \leq 0.$$

Proof. From $(\mathbf{A_0})$ and $(\mathbf{A_1})$ we have

$$A_i(x,\eta) = \int_0^1 a_i(x,t\eta)\eta dt \leq c_5(|\eta| + \frac{1}{p_i(x)}|\eta|^{p_i(x)}),$$

for all $x \in \overline{\Omega}$ and $\eta \in \mathbb{R}^N$, where $c_5 = \max_{i\in\{1,\ldots,N\}} \overline{c}_i$. Therefore

$$\int_\Omega \sum_{i=1}^N A_i(x,\partial_{x_i} u)dx \leq c_5\Big(\sum_{i=1}^N \int_\Omega |\partial_{x_i} u|dx + \sum_{i=1}^N \int_\Omega \frac{|\partial_{x_i} u|^{p_i(x)}}{p_i(x)}dx\Big).$$

Moreover, by rewriting condition $(\mathbf{F_2})$, we can obtain the existence of a positive constant c_6 such that

$$F(x,s) \geq c_6|s|^\theta,$$

for all $x \in \Omega$ and $s \in \mathbb{R}$. Then for any $u \in Y_k \backslash \{0\}$ with $||u|| = 1$ and $1 < \rho_k = t_k$ with $t_k \to \infty$, we have

$$\begin{aligned}
I(t_k u) &= \int_\Omega \sum_{i=1}^N A_i(x,\partial_{x_i}(t_k u))dx + \frac{1}{P_+^+}\int_\Omega b(x)|t_k u|^{P_+^+}dx - \int_\Omega F(x,t_k u)dx \\
&\leq c_5\Big(\sum_{i=1}^N \int_\Omega |\partial_{x_i}(t_k u)|dx + \sum_{i=1}^N \int_\Omega \frac{|\partial_{x_i}(t_k u)|^{p_i(x)}}{p_i(x)}dx\Big) + \frac{1}{P_+^+}\int_\Omega b(x)|t_k u|^{P_+^+}dx \\
&\quad - c_6\int_\Omega |t_k u|^\theta dx + c_7 \\
&\leq c_5 t_k^{P_+^+} \sum_{i=1}^N \int_\Omega \Big(|\partial_{x_i} u| + \frac{|\partial_{x_i} u|^{p_i(x)}}{P_-^-}\Big)dx + \frac{t_k^{P_+^+}}{P_+^+}\int_\Omega b(x)|u|^{P_+^+}dx \\
&\quad - c_6 t_k^\theta \int_\Omega |u|^\theta dx + c_7.
\end{aligned}$$

Since $\dim Y_k < \infty$ and all norms are equivalent in the finite dimensional space, it is easy to see that $I(t_k u) \to -\infty$ as $k \to \infty$ for $u \in Y_k$, due to $\theta > P_+^+$. Therefore, we deduce that for ρ_k large enough ($\rho_k > r_k$),

$$\max_{u \in Y_k,\ ||u||=\rho_k} I(u) \leq 0.$$

This completes the proof. □

Lemma 51 *The energy functional I satisfies the Palais-Smale condition.*

Proof. Let $(u_n) \subset X$ be a sequence such that

$$|I(u_n)| < c_8 \qquad \text{and} \qquad I'(u_n) \to 0 \text{ as} \qquad n \to \infty. \tag{5.34}$$

We claim that (u_n) is bounded. Arguing by contradiction, we assume that, passing eventually to a subsequence still denoted by (u_n), $||u_n|| \to \infty$ as $n \to \infty$.
Using relation (5.34), for large enough n we have

$$\begin{aligned}
1 + c_8 + ||u_n|| &\geq I(u_n) - \frac{1}{\theta}\langle I'(u_n), u_n\rangle \\
&\geq \sum_{i=1}^{N} \int_\Omega \Big[A_i(x, \partial_{x_i} u_n) - \frac{1}{\theta} a_i(x, \partial_{x_i} u_n)\partial_{x_i} u_n\Big] dx \\
&\quad + \Big(\frac{1}{P_+^+} - \frac{1}{\theta}\Big) \int_\Omega b(x)|u|^{P_+^+} dx \\
&\quad - \Big[\int_\Omega F(x, u_n) - \frac{1}{\theta} u_n f(x, u_n)\Big] dx.
\end{aligned} \tag{5.35}$$

From $(\mathbf{A_2})$, for all $x \in \Omega$ and $i \in \{1, \ldots, N\}$ we have

$$a_i(x, \partial_{x_i} u_n)\partial_{x_i} u_n \leq p_i(x) A_i(x, \partial_{x_i} u_n) \leq P_+^+ A_i(x, \partial_{x_i} u_n),$$

which implies

$$-\frac{1}{\theta} a_i(x, \partial_{x_i} u_n)\partial_{x_i} u_n \geq -\frac{P_+^+}{\theta} A_i(x, \partial_{x_i} u_n).$$

Joining the previous inequality in relation (5.35) and using $(\mathbf{F_2})$ we obtain

$$1 + c_8 + ||u_n|| \geq \Big(1 - \frac{P_+^+}{\theta}\Big) \sum_{i=1}^{N} \int_\Omega A_i(x, \partial_{x_i} u_n) dx.$$

Again by $(\mathbf{A_2})$ we have

$$A_i(x, \partial_{x_i} u_n) \geq \frac{1}{p_i(x)} |\partial_{x_i} u_n|^{p_i(x)} \geq \frac{1}{P_+^+} |\partial_{x_i} u_n|^{p_i(x)},$$

for all $x \in \Omega$ and $i \in \{1, \ldots, N\}$.
Taking into consideration relation (5.32), we obtain

$$1 + c_8 + ||u_n|| \geq \Big(\frac{1}{P_+^+} - \frac{1}{\theta}\Big)\Big(\frac{||u_n||^{P_-^-}}{N^{P_-^- - 1}} - N\Big).$$

Dividing the above inequality by $||u_n||^{P_-^-}$ and passing to the limit as $n \to \infty$ we obtain a contradiction.

It follows that (u_n) is bounded in X. This information combined with the fact that X is reflexive implies that there exist a subsequence, still denoted by (u_n), and $u_0 \in X$ such that (u_n) weakly converges to u_0 in X.

Since the embeddings $X \hookrightarrow L^{\alpha(x)}(\Omega)$ and $X \hookrightarrow L^{P_+^+}(\Omega)$ are compact, (u_n) strongly converges to u_0 in $L^{\alpha(x)}(\Omega)$ and also in $L^{P_+^+}(\Omega)$.

Using the Hölder inequality and $(\mathbf{F_1})$, we deduce that

$$\begin{aligned} &\lim_{n\to\infty}\int_\Omega f(x,u_n)(u_n-u_0)dx=0,\\ &\lim_{n\to\infty}\int_\Omega b(x)|u_n|^{P_+^+-2}u_n(u_n-u_0)dx=0. \end{aligned} \tag{5.36}$$

Using (5.34) we infer that

$$\lim_{n\to\infty}\langle I'(u_n),u_n-u_0\rangle=0,$$

more precisely,

$$\lim_{n\to\infty}\int_\Omega\Big[\sum_{i=1}^N a_i(x,\partial_{x_i}u_n)(\partial_{x_i}u_n-\partial_{x_i}u_0)+b(x)|u_n|^{P_+^+-2}u_n(u_n-u_0) \\ -f(x,u_n)(u_n-u_0)\Big]dx=0.$$

Combining the above relation with (5.36) we have

$$\lim_{n\to\infty}\int_\Omega\sum_{i=1}^N a_i(x,\partial_{x_i}u_n)(\partial_{x_i}u_n-\partial_{x_i}u_0)dx=0.$$

We deduce that (u_n) strongly converges to u_0 in X, in other words I satisfies the Palais-Smale condition. □

Completion of the proof of Theorem 39. Since the mapping A_i is even and f is odd with respect to their second variable, we deduce that I is even. The proof follows immediately from Lemma 49, Lemma 50, Lemma 51 and the fountain theorem.

5.2.2 Case of small positive parameters

Define the functional $I_\lambda : X \to \mathbb{R}$ by

$$I_\lambda(u)=\int_\Omega\Big\{\sum_{i=1}^N A_i(x,\partial_{x_i}u)+\frac{b(x)}{P_+^+}|u|^{P_+^+}-\frac{\lambda}{q(x)}|u|^{q(x)}\Big\}dx.$$

Then I_λ is well defined and of C^1 class on X. Moreover, we have

$$\langle I_\lambda'(u),\varphi\rangle=\int_\Omega\Big\{\sum_{i=1}^N a_i(x,\partial_{x_i}u)\partial_{x_i}\varphi+b(x)|u|^{P_+^+-2}u\varphi-\lambda|u|^{q(x)-2}u\varphi\Big\}dx,$$

for all $u, \varphi \in X$. Thus weak solutions of problem (5.28) are exactly the critical points of the functional I_λ. Next, I_λ is weakly lower semi-continuous in X.

We establish the following two auxiliary properties.

Lemma 52 *There exists $\lambda^* > 0$ such that for any $\lambda \in (0, \lambda^*)$ there exist $\rho, a > 0$ such that $I_\lambda(u) \geq a > 0$ for any $u \in X$ with $||u|| = \rho$.*

Proof. Under the hypotheses of Theorem 40, X is continuously embedded in $L^{q(x)}(\Omega)$. Thus there exists a positive constant c_9 such that

$$|u|_{q(x)} \leq c_9||u|| \qquad \text{for all} \qquad u \in X. \tag{5.37}$$

Now, let us assume that $||u|| \leq \min\{1, \frac{1}{c_9}\}$, where c_9 is the positive constant from above. Then we have $|u|_{q(x)} < 1$. Therefore

$$\int_\Omega |u|^{q(x)} dx \leq |u|^{q^-}_{q(x)} \qquad \text{for all} \quad u \in X \quad \text{with} \quad ||u|| = \rho \in (0,1). \tag{5.38}$$

Relations (5.37) and (5.38) yield

$$\int_\Omega |u|^{q(x)} dx \leq c_9^{q^-} ||u||^{q^-} \qquad \text{for all} \quad u \in X \quad \text{with} \quad ||u|| = \rho. \tag{5.39}$$

Using the hypothesis $(\mathbf{A_2})$, $(\mathbf{B})$ and (5.38), we deduce that for any $u \in X$ with $||u|| = \rho$, the following holds:

$$\begin{aligned} I_\lambda(u) &= \int_\Omega \Big\{ \sum_{i=1}^N A_i(x, \partial_{x_i} u) + \frac{b(x)}{P_+^+} |u|^{P_+^+} - \frac{\lambda}{q(x)} |u|^{q(x)} \Big\} dx \\ &\geq \frac{1}{P_+^+} \sum_{i=1}^N \int_\Omega |\partial_{x_i} u|^{p_i(x)} dx + \frac{b_0}{P_+^+} |u|^{P_+^+}_{L^{P_+^+}(\Omega)} - \frac{\lambda}{q^-} c_9^{q^-} ||u||^{q^-}. \end{aligned} \tag{5.40}$$

Here, we let $||u| < 1$, so $|\partial_{x_i} u|_{p_i(x)} < 1$, $i \in \{1, \dots, N\}$. Therefore

$$\begin{aligned} \sum_{i=1}^N \int_\Omega |\partial_{x_i} u|^{p_i(x)} dx &\geq \sum_{i=1}^N |\partial_{x_i} u|^{p_i^+}_{p_i(x)} \geq \sum_{i=1}^N |\partial_{x_i} u|^{P_+^+}_{p_i(x)} \\ &\geq N \Big(\frac{\sum_{i=1}^N |\partial_{x_i} u|_{p_i(x)}}{N} \Big)^{P_+^+} = \frac{||u||^{P_+^+}}{N^{P_+^+ - 1}}. \end{aligned} \tag{5.41}$$

Taking into account relations (5.31) and (5.41), the inequality (5.40) reduces to

$$\begin{aligned} I_\lambda(u) &\geq \frac{||u||^{P_+^+}}{P_+^+ N^{P_+^+ - 1}} - \frac{\lambda}{q^-} c_9^{q^-} ||u||^{q^-} \\ &= \frac{\rho^{P_+^+}}{P_+^+ N^{P_+^+ - 1}} - \frac{\lambda}{q^-} c_9^{q^-} \rho^{q^-} \\ &= \rho^{q^-} \Big(\frac{1}{P_+^+ N^{P_+^+ - 1}} \rho^{P_+^+ - q^-} - \frac{\lambda}{q^-} c_9^{q^-} \Big). \end{aligned}$$

Set

$$\lambda^* = \frac{q^-}{2P_+^+ N^{P_+^+-1} c_9^{q^-}} \rho^{P_+^+ - q^-}. \tag{5.42}$$

Then for any $\lambda \in (0, \lambda^*)$ and $u \in X$ with $||u|| = \rho$, there exists $a = \frac{\rho^{P_+^+}}{2P_+^+ N^{P_+^+-1}}$ such that $I_\lambda(u) \geq a > 0$. □

Lemma 53 *For any $\lambda \in (0, \lambda^*)$ where λ^* is given by (5.42), there exist $\psi \in X$ such that $\psi \geq 0$, $\psi \neq 0$ and $I_\lambda(t\psi) < 0$ for all small enough $t > 0$.*

Proof. From $(\mathbf{A_0})$ and $(\mathbf{A_1})$, we have

$$A_i(x, \eta) = \int_0^1 a_i(x, t\eta)\eta dt \leq c_{10}(|\eta| + \frac{1}{p_i(x)}|\eta|^{p_i(x)}),$$

for all $x \in \overline{\Omega}$ and $\eta \in \mathbb{R}^N$, where $c_{10} = \max_{i\in\{1,\dots,N\}} \bar{c}_i$. Therefore

$$\int_\Omega \sum_{i=1}^N A_i(x, \partial_{x_i} u)dx \leq c_{10} \sum_{i=1}^N \int_\Omega \Big(|\partial_{x_i} u| + \frac{|\partial_{x_i} u|^{p_i(x)}}{p_i(x)}\Big)dx.$$

By the hypotheses of Theorem 40, $q^- < P_-^-$. Let $\epsilon_0 > 0$ such that $q^- + \epsilon_0 < P_-^-$. Since $q \in C(\overline{\Omega})$, there exists an open set $\Omega_0 \subset \Omega$ such that $|q(x) - q^-| < \epsilon_0$ for all $x \in \Omega_0$. It follows that $q(x) < q^- + \epsilon_0 < P_-^-$ for all $x \in \Omega_0$.

Let $\psi \in C_0^\infty(\Omega)$ be such that $\operatorname{supp}(\psi) \supset \overline{\Omega_0}$, $\psi(x) = 1$ for all $x \in \overline{\Omega_0}$ and $0 \leq \psi \leq 1$ in Ω. Then for any $t \in (0, 1)$, we have

$$\begin{aligned}
I_\lambda(t\psi) &= \int_\Omega \Big\{\sum_{i=1}^N A_i(x, \partial_{x_i}(t\psi)) + \frac{b(x)}{P_+^+}|t\psi|^{P_+^+} - \frac{\lambda}{q(x)}|t\psi|^{q(x)}\Big\}dx \\
&\leq c_{10} \sum_{i=1}^N \int_\Omega \Big(|\partial_{x_i}(t\psi)| + \frac{|\partial_{x_i}(t\psi)|^{p_i(x)}}{p_i(x)}\Big)dx + \frac{1}{P_+^+}\int_\Omega b(x)|t\psi|^{P_+^+}dx \\
&\quad -\lambda \int_\Omega \frac{1}{q(x)}|t\psi|^{q(x)}dx \\
&\leq c_{10} t^{P_-^-} \sum_{i=1}^N \int_\Omega \Big(|\partial_{x_i}\psi| + \frac{1}{P_-^-}|\partial_{x_i}\psi|^{p_i(x)}\Big)dx + \frac{t^{P_+^+}}{P_+^+}\int_\Omega b(x)|\psi|^{P_+^+}dx \\
&\quad -\frac{\lambda}{q^+}\int_{\Omega_0} t^{q(x)}|\psi|^{q(x)}dx \\
&\leq c_{10} t^{P_-^-} \sum_{i=1}^N \int_\Omega \Big(|\partial_{x_i}\psi| + \frac{1}{P_-^-}|\partial_{x_i}\psi|^{p_i(x)}\Big)dx + \frac{t^{P_+^+}}{P_+^+}\int_\Omega b(x)|\psi|^{P_+^+}dx \\
&\quad -\frac{\lambda t^{q^-+\epsilon_0}}{q^+}\int_{\Omega_0} |\psi|^{q(x)}dx.
\end{aligned}$$

So, $I_\lambda(t\psi) < 0$ for $t < \delta^{\frac{1}{P_-^- - q^- - \epsilon_0}}$, with

$$0 < \delta < \min\left\{1, \frac{\lambda}{q^+} \frac{\int_{\Omega_0} |\psi|^{q(x)} dx}{c_{10} \sum_{i=1}^N \int_\Omega \left(|\partial_{x_i}\psi| + \frac{1}{P_-^-} |\partial_{x_i}\psi|^{p_i(x)} \right) dx + \frac{1}{P_+^+} \int_\Omega b(x) |\psi|^{P_+^+} dx} \right\}.$$

This completes the proof. □

Completion of the proof of Theorem 40. Let λ^* be defined as in (5.42) and $\lambda \in (0, \lambda^*)$. By Lemma 52, it follows that on the boundary of the ball centered at the origin and of radius ρ in X, we have

$$\inf_{\partial B_\rho(0)} I_\lambda(u) > 0.$$

On the other hand, by Lemma 53, there exists $\psi \in X$ such that

$$I_\lambda(t\psi) < 0 \quad \text{for} \quad \text{small enough } t > 0.$$

Moreover, for $u \in B_\rho(0)$,

$$I_\lambda(u) \geq \frac{||u||^{P_+^+}}{P_+^+ N^{P_+^+ - 1}} - \frac{\lambda}{q^-} c_9^{q^-} ||u||^{q^-}.$$

It follows that

$$-\infty < c_{11} = \inf_{\overline{B_\rho(0)}} I_\lambda(u) < 0.$$

We let now $0 < \varepsilon < \inf_{\partial B_\rho(0)} I_\lambda - \inf_{B_\rho(0)} I_\lambda$. Applying the Ekeland variational principle to the functional $I_\lambda : \overline{B_\rho(0)} \to \mathbb{R}$, we find $u_\varepsilon \in \overline{B_\rho(0)}$ such that

$$\begin{aligned} I_\lambda(u_\varepsilon) &< \inf_{\overline{B_\rho(0)}} I_\lambda + \varepsilon \\ I_\lambda(u_\varepsilon) &< I_\lambda(u) + \varepsilon ||u - u_\varepsilon||, \quad u \neq u_\varepsilon. \end{aligned}$$

Since

$$I_\lambda(u_\varepsilon) \leq \inf_{\overline{B_\rho(0)}} I_\lambda + \varepsilon \leq \inf_{B_\rho(0)} I_\lambda + \varepsilon < \inf_{\partial B_\rho(0)} I_\lambda,$$

we deduce that $u_\varepsilon \in B_\rho(0)$. Now, we define $K_\lambda : \overline{B_\rho(0)} \to \mathbb{R}$ by $K_\lambda(u) = I_\lambda(u) + \varepsilon ||u - u_\varepsilon||$. Since u_ε is a minimum point of K_λ we have

$$\frac{K_\lambda(u_\varepsilon + tv) - K_\lambda(u_\varepsilon)}{t} \geq 0,$$

for small $t > 0$ and $v \in B_\rho(0)$. This relation yields

$$\frac{I_\lambda(u_\varepsilon + tv) - I_\lambda(u_\varepsilon)}{t} + \varepsilon ||v|| \geq 0.$$

Letting $t \to 0$ it follows that $\langle I'_\lambda(u_\varepsilon), v \rangle + \varepsilon ||v|| > 0$ and we infer that

$||I'_\lambda(u_\varepsilon)|| \leq \varepsilon$. We deduce that there exists a sequence $(v_n) \subset B_1(0)$ such that

$$I_\lambda(v_n) \to c_{11} \quad \text{and} \quad I'_\lambda(v_n) \to 0. \tag{5.43}$$

The sequence (v_n) is bounded in X. Thus, there exists $u_1 \in X$ such that, up to a subsequence, (v_n) weakly converges to u_1 in X. Since the embeddings $X \hookrightarrow L^{q(x)}(\Omega)$ and $X \hookrightarrow L^{P_+^+}(\Omega)$ are compact, it follows that (v_n) strongly converges to u_1 in $L^{q(x)}(\Omega)$ and in $L^{P_+^+}(\Omega)$.

Using the Hölder inequality, we obtain that

$$\lim_{n\to\infty} \int_\Omega |v_n|^{q(x)-2} v_n(v_n - u_1)dx = 0,$$

$$\lim_{n\to\infty} \int_\Omega b(x)|v_n|^{P_+^+-2} v_n(v_n - u_1)dx = 0.$$

On the other hand, relation (5.43) yields

$$\lim_{n\to\infty} \langle I'(v_n), v_n - u_1\rangle = 0.$$

Using the above information, we find

$$\lim_{n\to\infty} \int_\Omega \sum_{i=1}^N a_i(x, \partial_{x_i} v_n)(\partial_{x_i} v_n - \partial_{x_i} u_1)dx = 0.$$

Next, the sequence (v_n) strongly converges to u_1 in X. So by (5.43)

$$I_\lambda(u_1) = c_{11} < 0 \quad \text{and} \quad I'_\lambda(u_1) = 0,$$

that is, u_1 is a nontrivial weak solution of problem (5.28). This completes the proof. □

5.3 Anisotropic problems with no-flux boundary condition

> What I cannot create, I do not understand.
>
> Richard Feynman (1918–1988)

Let $\Omega \subset \mathbb{R}^N$ be a bounded domain with smooth boundary. In this section, we are concerned with the nonlinear degenerate anisotropic problem

$$\begin{cases} -\sum_{i=1}^{N} \partial_{x_i} a_i(x, \partial_{x_i} u) + b(x)|u|^{p_M(x)-2}u = \lambda |u|^{q(x)-2}u \quad \text{in } \Omega, \\ u(x) = \text{constant} \quad \text{on } \partial\Omega \\ \sum_{i=1}^{N} \int_{\partial\Omega} a_i(x, \partial_{x_i} u)\nu_i d\sigma = 0. \end{cases} \tag{5.44}$$

The constant value of the boundary data is not specified, whereas the zero integral term corresponds to a no-flux boundary condition. In the case when $|u|^{q(x)-2}u$ "dominates" the left-hand side, we show that a nontrivial solution exists for all positive values of λ. If the term $|u|^{q(x)-2}u$ is dominated by the left-hand side, we prove that a solution exists either for small or for large values of $\lambda > 0$.

No-flux problems were studied for the first time by Berestycki and Brezis [6], in relationship with models arising in plasma physics. As stated in [6], these problems stem "from a model describing the equilibrium of a plasma confined in a toroidal cavity (a Tokamak machine)".

We notice that the constant value of the boundary data in problem (5.44) is not specified and corresponds to the one-dimensional case $u(0) = u(1)$, whereas the requirement in one-dimension that $u'(0) = u'(1)$, corresponds to the *no-flux* boundary integral term, in the case that $a_i(x, \xi) = \xi$ for all $i = 1, \ldots, N$. We also point out that Zou, Li, Liu and Lv [107] studied the existence of solutions of the problem

$$\begin{cases} -\Delta u = f(x, u) \quad \text{in } \{u > 0\}, \\ -\Delta u = 0 \quad \text{in } \{u \leq 0\}, \\ u = c \text{ (a negative constant)} \quad \text{on } \partial\Omega \\ -\int_{\partial\Omega} \frac{\partial u}{\partial n} d\sigma = I \text{ (a given positive constant)}, \end{cases}$$

where Ω is bounded, open and connected subset of $\mathbb{R}^2$ with regular boundary with the outward unit normal n. Problems of this type are related to plasma fusion and plasma confinement in Tokamak devices. The set $\{u > 0\}$ represents the region filled by the plasma, the set $\{u < 0\}$ represents the vacuum region, and the set $\{u = 0\}$ corresponds to the free boundary that separates the plasma and the vacuum. The case studied in problem (5.44) corresponds to *nonresonant surfaces*, that is, no-flux surfaces on which the wave number of the perturbation parallel to the equilibrium magnetic field is zero. Unfortunately, we are not able to give some *a posteriori* information on the constant value of the solution on the boundary in problem (5.44).

The differential operator $\sum_{i=1}^{N} \partial_{x_i} a_i(x, \partial_{x_i} u)$ is a $\overrightarrow{p}(\cdot)$-Laplace type operator, $\overrightarrow{p}(x) = (p_1(x), p_2(x), \ldots, p_N(x))$. For $i = 1, \ldots, N$, $p_i(x)$ and $q(x)$ are continuous functions on $\overline{\Omega}$, while $a_i(x, \eta)$ is the continuous derivative with respect to η of the mapping $A_i : \overline{\Omega} \times \mathbb{R}^N \to \mathbb{R}^N$, $A_i = A_i(x, \eta)$, that is, $a_i(x, \eta) = \frac{\partial}{\partial \eta} A_i(x, \eta)$.

Throughout this section we assume that the following hypotheses are fulfilled:

(A_0) $A_i(x,0)=0$ for a.e. $x\in\Omega$.

(A_1) There exists a positive constant $\overline{c}_i$ such that a_i satisfies the growth condition

$$|a_i(x,\eta)|\le \overline{c}_i(1+|\eta|^{p_i(x)-1}),$$

for all $x\in\overline{\Omega}$ and $\eta\in\mathbb{R}^N$.

(A_2) The inequalities

$$|\eta|^{p_i(x)}\le a_i(x,\eta)\eta\le p_i(x)A_i(x,\eta),$$

hold for all $x\in\overline{\Omega}$ and $\eta\in\mathbb{R}^N$.

(A_3) There exists $k_i>0$ such that

$$A_i\left(x,\frac{\eta+\xi}{2}\right)\le\frac{1}{2}A_i(x,\eta)+\frac{1}{2}A_i(x,\xi)-k_i|\eta-\xi|^{p_i(x)},$$

for all $x\in\overline{\Omega}$ and $\eta,\xi\in\mathbb{R}^N$, with equality if and only if $\eta=\xi$.

(A_4) $a_i(x,0)=0$ for all $x\in\partial\Omega$.

(B) $b\in L^\infty(\Omega)$ and there exists $b_0>0$ such that $b(x)\ge b_0$ for all $x\in\Omega$.

The above operator is the *anisotropic* $\overrightarrow{p}(x)$*-Laplace operator* because when we take

$$a_i(x,\eta)=|\eta|^{p_i(x)-2}\eta,$$

for all $i\in\{1,\dots,N\}$, we get $A_i(x,\eta)=\frac{1}{p_i(x)}|\eta|^{p_i(x)}$ for all $i\in\{1,\dots,N\}$. Therefore

$$\Delta_{\overrightarrow{p}(x)}(u)=\sum_{i=1}^{N}\partial_{x_i}\left(|\partial_{x_i}u|^{p_i(x)-2}\partial_{x_i}u\right).$$

There are many other operators deriving from $\sum_{i=1}^{N}\partial_{x_i}a_i(x,\partial_{x_i}u)$. Indeed, if we take

$$a_i(x,\eta)=(1+|\eta|^2)^{\frac{(p_i(x)-2)}{2}}\eta,$$

for all $i\in\{1,\dots,N\}$, we have $A_i(x,\eta)=\frac{1}{p_i(x)}[(1+|\eta|^2)^{\frac{p_i(x)}{2}}-1]$ for all $i\in\{1,\dots,N\}$ and we obtain the *anisotropic variable mean curvature operator*

$$\sum_{i=1}^{N}\partial_{x_i}\left[(1+|\partial_{x_i}u|^2)^{\frac{(p_i(x)-2)}{2}}\partial_{x_i}u\right].$$

An important feature of the results in this section is that we do not assume a zero Dirichlet boundary condition, but we work in the anisotropic variable exponent space of functions that are constant on the boundary and satisfy a no-flux boundary integral condition.

Let $\overrightarrow{p} : \overline{\Omega} \to \mathbb{R}^N$ denote the vector function

$$\overrightarrow{p}(x) = (p_1(x), p_2(x), ..., p_N(x)) \quad \text{with} \quad p_i(x) \in C_+(\overline{\Omega}),\ i \in \{1, \ldots, N\}$$

and we set

$$p_M(x) = \max\{p_1(x), \ldots, p_N(x)\}, \quad p_m(x) = \min\{p_1(x), \ldots, p_N(x)\}.$$

The *anisotropic Sobolev space with variable exponent* is

$$W^{1,\overrightarrow{p}(x)}(\Omega) = \{u \in L^{p_M(x)}(\Omega) : \partial_{x_i} u \in L^{p_i(x)}(\Omega) \text{ for all } i \in \{1, \ldots, N\}\}.$$

This space is endowed with the norm

$$\|u\|_{W^{1,\overrightarrow{p}(x)}(\Omega)} = \|u\|_{L^{p_M(x)}(\Omega)} + \sum_{i=1}^{N} \|\partial_{x_i} u\|_{L^{p_i(x)}(\Omega)}.$$

Set

$$X = \left\{u \in W^{1,\overrightarrow{p}(x)}(\Omega) : u|_{\partial\Omega} \equiv \text{constant}\right\}.$$

Since X is a closed subset of $W^{1,\overrightarrow{p}(x)}(\Omega)$, it follows that X is a reflexive Banach space.

5.3.1 The first domination case

A function $u \in X$ that satisfies

$$\int_\Omega \left\{\sum_{i=1}^{N} a_i(x, \partial_{x_i} u)\partial_{x_i}\varphi + b(x)|u|^{p_M(x)-2}u\varphi - \lambda|u|^{q(x)-2}u\varphi\right\} dx = 0,$$

for all $\varphi \in X$ is called a weak solution of problem (5.44).

We associate to problem (5.44) the energy functional $I_\lambda : X \to \mathbb{R}$ defined by

$$I_\lambda(u) = \int_\Omega \left\{\sum_{i=1}^{N} A_i(x, \partial_{x_i} u) + \frac{b(x)}{p_M(x)}|u|^{p_M(x)} - \frac{\lambda}{q(x)}|u|^{q(x)}\right\} dx.$$

Then I_λ is well-defined and $I_\lambda \in C^1(X, \mathbb{R})$ has the following property

$$\langle I'_\lambda(u), \varphi\rangle = \int_\Omega \left\{\sum_{i=1}^{N} a_i(x, \partial_{x_i} u)\partial_{x_i}\varphi + b(x)|u|^{p_M(x)-2}u\varphi - \lambda|u|^{q(x)-2}u\varphi\right\} dx,$$

for all $u, \varphi \in X$. Hence any critical point $u \in X$ of I_λ is a weak solution of problem (5.44).

The first result of this section shows that if the variable exponent $q(\cdot)$ "dominates" $p(\cdot)$ then the solution exists for all positive values of the parameter λ. In other words, the right-hand side of problem (5.44) is "stronger" than the other side, even if $\lambda > 0$ is small.

Theorem 41 *Assume that the function $q \in C_+(\overline{\Omega})$ satisfies the hypothesis*

$$p_M^+ < q^- \leq q^+ < \frac{Np_m^-}{N - p_m^-}.$$

Then for any $\lambda > 0$ problem (5.44) has a nontrivial weak solution.

We first prove two auxiliary results.

Lemma 54 *There exist $\eta > 0$ and $\alpha > 0$ such that $I_\lambda(u) \geq \alpha > 0$ for any $u \in X$ with $\|u\|_{W^{1,\vec{p}(x)}(\Omega)} = \eta$.*

Proof. First, we point out that

$$|u(x)|^{q(x)} \leq |u(x)|^{q^-} + |u(x)|^{q^+}, \qquad \text{for all} \quad x \in \overline{\Omega}.$$

Using the above inequality and hypotheses $(\mathbf{A_2})$ and $(\mathbf{B})$, we obtain

$$\begin{aligned} I_\lambda(u) &= \int_\Omega \Big\{ \sum_{i=1}^N A_i(x, \partial_{x_i} u) + \frac{b(x)}{p_M(x)} |u|^{p_M(x)} - \frac{\lambda}{q(x)} |u|^{q(x)} \Big\}\, dx \\ &\geq \frac{1}{p_M^+} \sum_{i=1}^N \int_\Omega |\partial_{x_i} u|^{p_i(x)}\, dx + \frac{b_0}{p_M^+} \int_\Omega |u|^{p_M(x)}\, dx \\ &\quad - \frac{\lambda}{q^-} \Big(\|u\|^{q^-}_{L^{q^-}(\Omega)} + \|u\|^{q^+}_{L^{q^+}(\Omega)} \Big). \end{aligned} \tag{5.45}$$

By the hypotheses of Theorem 41, $W^{1,\vec{p}(x)}(\Omega)$ is continuously embedded in $L^{q^-}(\Omega)$ and $L^{q^+}(\Omega)$. Hence there exist positive constants c_1 and c_2 such that for all $u \in X$

$$|u(x)|_{L^{q^-}(\Omega)} \leq c_1 \|u\|_{W^{1,\vec{p}(x)}(\Omega)} \text{ and } |u(x)|_{L^{q^+}(\Omega)} \leq c_2 \|u\|_{W^{1,\vec{p}(x)}(\Omega)}. \tag{5.46}$$

Here, we let $\|u\|_{W^{1,\vec{p}(x)}(\Omega)} < 1$, so $\|u\|_{L^{p_M(x)}(\Omega)} < 1$ and $\|\partial_{x_i} u\|_{L^{p_i(x)}(\Omega)} < 1$, $i \in \{1, \ldots, N\}$.

Taking into account relation (5.46), the inequality (5.45) reduces to

$$\begin{aligned} I_\lambda(u) &\geq \frac{\min\{1, b_0\}}{p_M^+ (N+1)^{p_M^+ - 1}} \|u\|^{p_M^+}_{W^{1,\vec{p}(x)}(\Omega)} \\ &\quad - \frac{\lambda}{q^-} \Big[(c_1 \|u\|_{W^{1,\vec{p}(x)}(\Omega)})^{q^-} + (c_2 \|u\|_{W^{1,\vec{p}(x)}(\Omega)})^{q^+} \Big] \\ &= \Big(c_3 - c_4 \|u\|^{q^- - p_M^+}_{W^{1,\vec{p}(x)}(\Omega)} - c_5 \|u\|^{q^+ - p_M^+}_{W^{1,\vec{p}(x)}(\Omega)} \Big) \|u\|^{p_M^+}_{W^{1,\vec{p}(x)}(\Omega)}, \end{aligned}$$

for any $u \in X$ with $\|u\|_{W^{1,\vec{p}(x)}(\Omega)} < 1$. Since the function $g : [0,1] \to \mathbb{R}$, defined by

$$g(t) = c_3 - c_4 t^{q^- - p_M^+} - c_5 t^{q^+ - p_M^+}$$

is positive in a neighborhood of the origin, the conclusion follows. □

Lemma 55 *There exists $e \in X$ with $\|e\|_{W^{1,\vec{p}(x)}(\Omega)} > \eta$ (where η is given in Lemma 54) such that $I_\lambda(e) < 0$.*

Proof. From $(\mathbf{A_0})$ and $(\mathbf{A_1})$, we have

$$A_i(x,\eta) = \int_0^1 a_i(x, t\eta)\eta dt \leq c_6(|\eta| + \frac{1}{p_i(x)}|\eta|^{p_i(x)}),$$

for all $x \in \Omega$ and $\eta \in \mathbb{R}^N$, where $c_6 = \max_{i\in\{1,\ldots,N\}} \overline{c}_i$. Therefore

$$\int_\Omega \sum_{i=1}^N A_i(x, \partial_{x_i} u)dx \leq c_6 \sum_{i=1}^N \int_\Omega \left(|\partial_{x_i} u| + \frac{|\partial_{x_i} u|^{p_i(x)}}{p_i(x)}\right) dx. \tag{5.47}$$

Let $\varphi \in X$, $\varphi \neq 0$. For any $t > 1$, we find

$$\begin{aligned} I_\lambda(t\varphi) &= \int_\Omega \Big\{ \sum_{i=1}^N A_i(x, \partial_{x_i}(t\varphi)) + \frac{b(x)}{p_M(x)}|t\varphi|^{p_M(x)} - \frac{\lambda}{q(x)}|t\varphi|^{q(x)} \Big\} dx \\ &\leq c_6 \sum_{i=1}^N \int_\Omega \Big(|\partial_{x_i}(t\varphi)| + \frac{|\partial_{x_i}(t\varphi)|^{p_i(x)}}{p_i(x)}\Big)dx + \frac{1}{p_M^+}\int_\Omega b(x)|t\varphi|^{p_M(x)}\,dx \\ &\quad -\lambda \int_\Omega \frac{1}{q(x)}|t\varphi|^{q(x)}\,dx \\ &\leq c_6 t^{p_M^+} \sum_{i=1}^N \int_\Omega \Big(|\partial_{x_i}\varphi| + \frac{1}{p_m^-}|\partial_{x_i}\varphi|^{p_i(x)}\Big)\,dx + \frac{t^{p_M^+}}{p_M^+}\int_\Omega b(x)|\varphi|^{p_M(x)}\,dx \\ &\quad - \frac{\lambda t^{q^-}}{q^+}\int_\Omega |\varphi|^{q(x)}\,dx. \end{aligned}$$

Since $p_M^+ < q^-$, we infer that $\lim_{t\to\infty} I_\lambda(t\varphi) = -\infty$. Then for large enough $t > 1$, we can take $e = t\varphi$ such that $\|e\|_{W^{1,\vec{p}(x)}(\Omega)} > \eta$ and $I_\lambda(e) < 0$. □

The following property provides a sufficient condition for the strong convergence of a weakly convergent sequence.

Lemma 56 *Let $\Omega \subset \mathbb{R}^N$ $(N \geq 2)$ be a bounded domain with smooth boundary. Assume that the sequence (u_n) weakly converges to u in $W^{1,\vec{p}(x)}(\Omega)$ and*

$$\limsup_{n\to\infty} \int_\Omega \sum_{i=1}^N a_i(x, \partial_{x_i} u_n)(\partial_{x_i} u_n - \partial_{x_i} u)\,dx \leq 0.$$

Then (u_n) strongly converges to u in $W^{1,\vec{p}(x)}(\Omega)$.

Proof. We use the fact that $W^{1,\vec{p}(x)}(\Omega) \hookrightarrow L^{p_M(x)}(\Omega)$ compactly. Since $u_n \rightharpoonup u$ in $W^{1,\vec{p}(x)}(\Omega)$, we deduce that

$$u_n \to u \quad \text{in} \quad L^{p_M(x)}(\Omega). \tag{5.48}$$

Then, combining our hypotheses and (5.48), we conclude that $u_n \to u$ in $W^{1,\vec{p}(x)}(\Omega)$. □

Proof of Theorem 41. By Lemmas 54 and 55 and the mountain pass theorem of Ambrosetti and Rabinowitz, we deduce the existence of a sequence $(u_n) \subset X$ such that

$$I_\lambda(u_n) \to c_7 > 0 \qquad \text{and} \qquad I'_\lambda(u_n) \to 0 \text{ as} \qquad n \to \infty. \tag{5.49}$$

We claim that (u_n) is bounded. Arguing by contradiction, we assume that, up to a subsequence still denoted by (u_n), we have $\|u_n\|_{W^{1,\vec{p}(x)}(\Omega)} \to \infty$ as $n \to \infty$.

Using relation (5.49), for large enough n, we have

$$\begin{aligned}
&1 + c_7 + \|u_n\|_{W^{1,\vec{p}(x)}(\Omega)} \geq I_\lambda(u_n) - \frac{1}{q^-}\langle I'_\lambda(u_n), u_n\rangle \geq \\
&\sum_{i=1}^N \int_\Omega \Big[A_i(x, \partial_{x_i} u_n) + \frac{1}{p_M(x)} b(x)|u_n|^{p_M(x)} - \frac{\lambda}{q(x)}|u_n|^{q(x)}\Big]\, dx - \\
&\frac{1}{q^-}\sum_{i=1}^N \int_\Omega \Big[a_i(x, \partial_{x_i} u_n)\partial_{x_i} u_n + b(x)|u_n|^{p_M(x)} - \lambda|u_n|^{q(x)}\Big]\, dx \geq \\
&\sum_{i=1}^N \int_\Omega \Big[A_i(x, \partial_{x_i} u_n) - \frac{1}{q^-} a_i(x, \partial_{x_i} u_n)\partial_{x_i} u_n\Big]\, dx + \\
&\Big(\frac{1}{p_M^+} - \frac{1}{q^-}\Big)\int_\Omega b(x)|u_n|^{p_M(x)}\, dx + \lambda \int_\Omega \Big(\frac{1}{q^-} - \frac{1}{q(x)}\Big)|u_n|^{q(x)}\, dx.
\end{aligned} \tag{5.50}$$

From $(\mathbf{A_2})$, for all $x \in \Omega$ and $i \in \{1, \ldots, N\}$ we have

$$a_i(x, \partial_{x_i} u_n)\partial_{x_i} u_n \leq p_i(x) A_i(x, \partial_{x_i} u_n) \leq p_M^+ A_i(x, \partial_{x_i} u_n),$$

which implies

$$-\frac{1}{q^-} a_i(x, \partial_{x_i} u_n)\partial_{x_i} u_n \geq -\frac{p_M^+}{q^-} A_i(x, \partial_{x_i} u_n).$$

Inserting this inequality into relation (5.50) we obtain

$$\begin{aligned}
&1 + c_7 + \|u_n\|_{W^{1,\vec{p}(x)}(\Omega)} \geq \\
&\Big(1 - \frac{p_M^+}{q^-}\Big)\sum_{i=1}^N \int_\Omega A_i(x, \partial_{x_i} u_n)\, dx + b_0\Big(\frac{1}{p_M^+} - \frac{1}{q^-}\Big)\int_\Omega |u_n|^{p_M(x)}\, dx.
\end{aligned}$$

Again, by $(\mathbf{A_2})$ we get

$$A_i(x, \partial_{x_i} u_n) \geq \frac{1}{p_i(x)}|\partial_{x_i} u_n|^{p_i(x)} \geq \frac{1}{p_M^+}|\partial_{x_i} u_n|^{p_i(x)},$$

for all $x \in \Omega$ and $i \in \{1, \dots, N\}$. Therefore

$$1 + c_7 + \|u_n\|_{W^{1,\vec{p}(x)}(\Omega)} \geq \Big(\frac{1}{p_M^+} - \frac{1}{q^-}\Big) \sum_{i=1}^{N} \int_\Omega |\partial_{x_i} u_n|^{p_i(x)}\, dx + b_0 \Big(\frac{1}{p_M^+} - \frac{1}{q^-}\Big) \int_\Omega |u_n|^{p_M(x)}\, dx. \tag{5.51}$$

We denote

$$\ell_1 = \{i \in \{1, \dots, N\} : \|\partial_{x_i} u_n\|_{L^{p_i(x)}(\Omega)} \leq 1\}$$

and

$$\ell_2 = \{i \in \{1, \dots, N\} : \|\partial_{x_i} u_n\|_{L^{p_i(x)}(\Omega)} > 1\}.$$

By the Jensen inequality (applied to the convex function $h : \mathbb{R}^+ \to \mathbb{R}^+$, $h(t) = t^{p_m^-}$, $p_m^- > 1$), we have for large enough n

$$\begin{aligned} \sum_{i=1}^{N} \int_\Omega |\partial_{x_i} u_n|^{p_i(x)}\, dx &= \sum_{i \in \ell_1} \int_\Omega |\partial_{x_i} u_n|^{p_i(x)}\, dx + \sum_{i \in \ell_2} \int_\Omega |\partial_{x_i} u_n|^{p_i(x)}\, dx \\ &\geq \sum_{i \in \ell_1} \|\partial_{x_i} u_n\|^{p_M^+}_{L^{p_i(x)}(\Omega)} + \sum_{i \in \ell_2} \|\partial_{x_i} u_n\|^{p_m^-}_{L^{p_i(x)}(\Omega)} \\ &\geq \sum_{i=1}^{N} \|\partial_{x_i} u_n\|^{p_m^-}_{L^{p_i(x)}(\Omega)} - \sum_{i \in \ell_1} \|\partial_{x_i} u_n\|^{p_m^-}_{L^{p_i(x)}(\Omega)} \\ &\geq N \Big(\frac{\sum_{i=1}^{N} \|\partial_{x_i} u_n\|_{L^{p_i(x)}(\Omega)}}{N}\Big)^{p_m^-} - N. \end{aligned} \tag{5.52}$$

We now analyze the two cases depending on the value of $\|u\|_{L^{p_M(x)}(\Omega)}$.
Case 1: $\|u\|_{L^{p_M(x)}(\Omega)} \geq 1$. By (5.51) and (5.52) we have

$$\begin{aligned} 1 + c_7 + \|u_n\|_{W^{1,\vec{p}(x)}(\Omega)} &\geq \Big(\frac{1}{p_M^+} - \frac{1}{q^-}\Big)\Big[\frac{1}{N^{p_m^- - 1}}\Big(\sum_{i=1}^{N} \|\partial_{x_i} u_n\|_{L^{p_i(x)}(\Omega)}\Big)^{p_m^-} - N\Big] \\ &+ b_0 \Big(\frac{1}{p_M^+} - \frac{1}{q^-}\Big) \|u\|^{p_m^-}_{L^{p_M(x)}(\Omega)}, \end{aligned}$$

and thus

$$1 + c_7 + \|u_n\|_{W^{1,\vec{p}(x)}(\Omega)} \geq \frac{1}{2^{p_m^-}}\Big(\frac{1}{p_M^+} - \frac{1}{q^-}\Big) \min\Big\{\frac{1}{N^{p_m^- - 1}}, b_0\Big\} \|u_n\|^{p_m^-}_{W^{1,\vec{p}(x)}(\Omega)} - N\Big(\frac{1}{p_M^+} - \frac{1}{q^-}\Big). \tag{5.53}$$

Case 2: $\|u\|_{L^{p_M(x)}(\Omega)} < 1$. Then

$$\begin{aligned} &1 + c_7 + \|u_n\|_{W^{1,\vec{p}(x)}(\Omega)} \geq \\ &\Big(\frac{1}{p_M^+} - \frac{1}{q^-}\Big)\Big[\frac{1}{N^{p_m^- - 1}}\Big(\sum_{i=1}^{N} \|\partial_{x_i} u_n\|_{L^{p_i(x)}(\Omega)}\Big)^{p_m^-} - N\Big] \geq \\ &\Big(\frac{1}{p_M^+} - \frac{1}{q^-}\Big)\Big[\frac{1}{N^{p_m^- - 1}}\Big(\sum_{i=1}^{N} \|\partial_{x_i} u_n\|_{L^{p_i(x)}(\Omega)}\Big)^{p_m^-} + \|u_n\|^{p_m^-}_{L^{p_M(x)}(\Omega)} - N - 1\Big]. \end{aligned}$$

We obtain

$$1 + c_7 + \|u_n\|_{W^{1,\vec{p}(x)}(\Omega)} \geq \frac{1}{2^{p_m^-}}\left(\frac{1}{p_M^+} - \frac{1}{q^-}\right)\min\left\{\frac{1}{N^{p_m^- -1}}, 1\right\}\|u_n\|^{p_m^-}_{W^{1,\vec{p}(x)}(\Omega)} - \frac{N+1}{p_M^+}. \tag{5.54}$$

From relations (5.53) and (5.54), we deduce that there exist $\widetilde{c_1}, \widetilde{c_2} > 0$ such that

$$1 + c_7 + \|u_n\|_{W^{1,\vec{p}(x)}(\Omega)} \geq \widetilde{c_1}\|u_n\|^{p_m^-}_{W^{1,\vec{p}(x)}(\Omega)} - \widetilde{c_2}.$$

Dividing the above inequality by $\|u_n\|^{p_m^-}_{W^{1,\vec{p}(x)}(\Omega)}$ and passing to the limit as $n \to \infty$ we obtain a contradiction.

It follows that (u_n) is bounded in $W^{1,\vec{p}(x)}(\Omega)$. This information combined with the fact that $W^{1,\vec{p}(x)}(\Omega)$ is reflexive implies that there exists a subsequence, still denoted by (u_n), and $u_0 \in W^{1,\vec{p}(x)}(\Omega)$ such that (u_n) weakly converges to u_0 in $W^{1,\vec{p}(x)}(\Omega)$.

Using (5.49), we infer that

$$\lim_{n\to\infty} \langle I'_\lambda(u_n), u_n - u_0\rangle = 0,$$

more precisely,

$$\lim_{n\to\infty} \int_\Omega \Big[\sum_{i=1}^N a_i(x, \partial_{x_i} u_n)(\partial_{x_i} u_n - \partial_{x_i} u_0) \quad + b(x)|u_n|^{p_M(x)-2}u_n(u_n - u_0) - \lambda|u_n|^{q(x)-2}u_n(u_n - u_0)\Big]\, dx = 0. \tag{5.55}$$

Since the space $W^{1,\vec{p}(x)}(\Omega)$ is compactly embedded in $L^{p_M(x)}(\Omega)$ and $L^{q(x)}(\Omega)$, it follows that (u_n) strongly converges to u_0 in $L^{p_M(x)}(\Omega)$ and also in $L^{q(x)}(\Omega)$. Therefore

$$\left|\int_\Omega b(x)|u_n|^{p_M(x)-2}u_n(u_n - u_0)\, dx\right| \leq 2\|b\|_{L^\infty(\Omega)}\left\||u_n|^{p_M(x)-1}\right\|_{L^{\frac{p_M(x)}{p_M(x)-1}}(\Omega)}\|u_n - u_0\|_{L^{p_M(x)}(\Omega)}, \tag{5.56}$$

and

$$\left|\int_\Omega |u_n|^{q(x)-2}u_n(u_n - u_0)\, dx\right| \leq 2\left\||u_n|^{q(x)-1}\right\|_{L^{\frac{q(x)}{q(x)-1}}(\Omega)}\|u_n - u_0\|_{L^{q(x)}(\Omega)}. \tag{5.57}$$

By (5.56) and (5.57), using the strong convergence of (u_n) to u_0 in $L^{p_M(x)}(\Omega)$ and $L^{q(x)}(\Omega)$ we deduce

$$\lim_{n\to\infty} \int_\Omega b(x)|u_n|^{p_M(x)-2}u_n(u_n - u_0)\, dx = 0,$$

and

$$\lim_{n\to\infty} \int_\Omega |u_n|^{q(x)-2}u_n(u_n - u_0)\, dx = 0.$$

By the above relations, (5.55) reduces to

$$\lim_{n\to\infty}\int_\Omega \sum_{i=1}^N a_i(x, \partial_{x_i}u_n)(\partial_{x_i}u_n - \partial_{x_i}u_0)\,dx = 0.$$

Using Lemma 56, we deduce that (u_n) strongly converges to u_0 in $W^{1,\vec{p}(x)}(\Omega)$. Since X is a closed subspace of $W^{1,\vec{p}(x)}(\Omega)$ and $(u_n) \subset X$ we obtain that $u_0 \in X$. Then by relation (5.49)

$$I_\lambda(u_0) = c_7 > 0 \quad \text{and} \quad I'_\lambda(u_0) = 0,$$

that is, u_0 is a nontrivial weak solution of problem (5.44). □

5.3.2 The second domination case

The next result establishes an interesting concentration property in neighborhoods of the origin and of the infinity. More precisely, under an additional assumption, we show that there are positive numbers λ^* and λ^{**} such that problem (5.44) has a solution provided that either $\lambda \in (0, \lambda^*)$ or $\lambda \in (\lambda^{**}, \infty)$. The existence of a "gap" between λ^* and λ^{**} still remains an interesting open problem.

Theorem 42 *Assume that hypotheses* $(\mathbf{A_0})$–$(\mathbf{A_4})$, $(\mathbf{B})$ *are fulfilled and that* $q \in C_+(\overline{\Omega})$ *satisfies the condition*

$$1 < q^- \le q^+ < p_m^-.$$

Then the following properties hold.
(i) *There exists* $\lambda^* > 0$ *such that for any* $\lambda \in (0, \lambda^*)$ *problem* (5.44) *possesses a nontrivial weak solution.*
(ii) *There exists* $\lambda^{**} > 0$ *such that for any* $\lambda > \lambda^{**}$ *problem* (5.44) *possesses a nontrivial weak solution.*

The assumptions in Theorem 42 show that the weight $p(\cdot)$ is dominating with respect to the variable exponent $q(\cdot)$ that controls the right-hand side. The above results assert that in this case, a solution exists either if $\lambda > 0$ is sufficiently small or for large values of λ.

First, applying the Ekeland variational principle, we show that there exists $\lambda^* > 0$ such that for any $\lambda \in (0, \lambda^*)$ the functional I_λ has a nontrivial critical point. We start with two auxiliary results.

Lemma 57 *There exists* $\lambda^* > 0$ *such that for any* $\lambda \in (0, \lambda^*)$ *there are* $\rho, a > 0$ *such that* $I_\lambda(u) \ge a > 0$ *for any* $u \in X$ *with* $\|u\|_{W^{1,\vec{p}(x)}(\Omega)} = \rho$.

Proof. Under the conditions of Theorem 42, $W^{1,\vec{p}(x)}(\Omega)$ is continuously embedded in $L^{q(x)}(\Omega)$. Thus there exists a positive constant c_8 such that

$$\|u\|_{L^{q(x)}(\Omega)} \le c_8 \|u\|_{W^{1,\vec{p}(x)}(\Omega)} \quad \text{for all} \quad u \in X. \tag{5.58}$$

Now, let us assume that $\|u\|_{W^{1,\vec{p}(x)}(\Omega)} < \min\{1, \frac{1}{c_8}\}$, where c_8 is the positive constant above. Then we have $\|u\|_{L^{q(x)}(\Omega)} < 1$. Thus for all $u \in X$ with $\|u\|_{W^{1,\vec{p}(x)}(\Omega)} = \rho \in (0,1)$

$$\int_\Omega |u|^{q(x)}dx \leq \|u\|^{q^-}_{L^{q(x)}(\Omega)}. \tag{5.59}$$

Relations (5.58) and (5.59) imply that for all $u \in X$ with $\|u\|_{W^{1,\vec{p}(x)}(\Omega)} = \rho \in (0,1)$

$$\int_\Omega |u|^{q(x)}dx \leq c_8^{q^-} \|u\|^{q^-}_{W^{1,\vec{p}(x)}(\Omega)}. \tag{5.60}$$

Using the hypothesis $(\mathbf{A_2})$, $(\mathbf{B})$ and relation (5.60), we deduce that for any $u \in X$ with $\|u\|_{W^{1,\vec{p}(x)}(\Omega)} = \rho$, the following holds:

$$\begin{aligned}
I_\lambda(u) &= \int_\Omega \Big\{ \sum_{i=1}^N A_i(x, \partial_{x_i} u) + \frac{b(x)}{p_M(x)}|u|^{p_M(x)} - \frac{\lambda}{q(x)}|u|^{q(x)} \Big\}\, dx \\
&\geq \frac{1}{p_M^+} \sum_{i=1}^N \int_\Omega |\partial_{x_i} u|^{p_i(x)}dx + \frac{b_0}{p_M^+}\int_\Omega |u|^{p_M(x)}\, dx \\
&\quad - \frac{\lambda}{q^-} c_8^{q^-} \|u\|^{q^-}_{W^{1,\vec{p}(x)}(\Omega)} \\
&\geq \frac{\min\{1, b_0\}}{p_M^+ (N+1)^{p_M^+ - 1}} \|u\|^{p_M^+}_{W^{1,\vec{p}(x)}(\Omega)} - \frac{\lambda}{q^-} c_8^{q^-} \|u\|^{q^-}_{W^{1,\vec{p}(x)}(\Omega)} \\
&= \frac{\min\{1, b_0\}}{p_M^+ (N+1)^{p_M^+ - 1}} \rho^{p_M^+} - \frac{\lambda}{q^-} c_8^{q^-} \rho^{q^-} \\
&= \rho^{q^-} \Big(\frac{\min\{1, b_0\}}{p_M^+ (N+1)^{p_M^+ - 1}} \rho^{p_M^+ - q^-} - \frac{\lambda}{q^-} c_8^{q^-} \Big).
\end{aligned} \tag{5.61}$$

Set

$$\lambda^* = \frac{\min\{1, b_0\} q^-}{2p_M^+ (N+1)^{p_M^+ - 1} c_8^{q^-}} \rho^{p_M^+ - q^-}. \tag{5.62}$$

Then for any $\lambda \in (0, \lambda^*)$ and $u \in X$ with $\|u\|_{W^{1,\vec{p}(x)}(\Omega)} = \rho$, there exists $a = \frac{\min\{1, b_0\}\rho^{p_M^+}}{2p_M^+ (N+1)^{p_M^+ - 1}}$ such that $I_\lambda(u) \geq a > 0$. □

Lemma 58 *Assume that $\lambda \in (0, \lambda^*)$, where λ^* is given by (5.62). Then there exists $\psi \in X$ such that $\psi \geq 0$, $\psi \neq 0$ and $I_\lambda(t\psi) < 0$ for all small enough $t > 0$.*

Proof. By the conditions of Theorem 42, we have $q^- < p_m^-$. Let $\epsilon_0 > 0$ be such that $q^- + \epsilon_0 < p_m^-$. Since $q \in C(\overline{\Omega})$, there exists an open set $\Omega_0 \subset \Omega$ such that $|q(x) - q^-| < \epsilon_0$ for all $x \in \Omega_0$. It follows that $q(x) < q^- + \epsilon_0 < p_m^-$ for all $x \in \Omega_0$.

Let $\psi \in X$ be such that $\operatorname{supp}(\psi) \supset \overline{\Omega_0}$, $\psi(x) = 1$ for all $x \in \overline{\Omega_0}$ and $0 \leq \psi \leq 1$ in Ω. Then by (5.47) for any $t \in (0,1)$, we have

$$\begin{aligned}
I_\lambda(t\psi) &= \int_\Omega \Big\{ \sum_{i=1}^N A_i(x, \partial_{x_i}(t\psi)) + \frac{b(x)}{p_M(x)}|t\psi|^{p_M(x)} - \frac{\lambda}{q(x)}|t\psi|^{q(x)} \Big\} dx \\
&\leq c_6 \sum_{i=1}^N \int_\Omega \Big(|\partial_{x_i}(t\psi)| + \frac{|\partial_{x_i}(t\psi)|^{p_i(x)}}{p_i(x)} \Big) dx + \frac{1}{p_M^-} \int_\Omega b(x)|t\psi|^{p_M(x)} dx \\
&- \lambda \int_\Omega \frac{1}{q(x)} |t\psi|^{q(x)} dx \\
&\leq c_6 t^{p_m^-} \sum_{i=1}^N \int_\Omega \Big(|\partial_{x_i}\psi| + \frac{1}{p_m^-}|\partial_{x_i}\psi|^{p_i(x)} \Big) dx + \frac{t^{p_M^-}}{p_M^-} \int_\Omega b(x)|\psi|^{p_M(x)} dx \\
&- \frac{\lambda}{q^+} \int_{\Omega_0} t^{q(x)} |\psi|^{q(x)} dx \\
&\leq c_6 t^{p_m^-} \sum_{i=1}^N \int_\Omega \Big(|\partial_{x_i}\psi| + \frac{1}{p_m^-}|\partial_{x_i}\psi|^{p_i(x)} \Big) dx + \frac{t^{p_m^-}}{p_M^-} \int_\Omega b(x)|\psi|^{p_M(x)} dx \\
&- \frac{\lambda t^{q^- + \epsilon_0}}{q^+} \int_{\Omega_0} |\psi|^{q(x)} dx.
\end{aligned}$$

So, $I_\lambda(t\psi) < 0$ for $t < \delta^{\frac{1}{p_m^- - q^- - \epsilon_0}}$, with

$$0 < \delta < \min\left\{ 1, \frac{\lambda}{q^+} \frac{\int_{\Omega_0} |\psi|^{q(x)} dx}{c_6 \sum_{i=1}^N \int_\Omega \Big(|\partial_{x_i}\psi| + \frac{1}{p_m^-}|\partial_{x_i}\psi|^{p_i(x)} \Big) dx + \frac{1}{p_M^-} \int_\Omega b(x)|\psi|^{p_M(x)} dx} \right\}.$$

□

Proof of Theorem 42 (**i**) Let λ^* be defined as in (5.60) and let $\lambda \in (0, \lambda^*)$. By Lemma 57, it follows that on the boundary of the ball centered at the origin and of radius ρ in X, we have

$$\inf_{\partial B_\rho(0)} I_\lambda(u) > 0.$$

On the other hand, by Lemma 58, there exists $\psi \in X$ such that

$$I_\lambda(t\psi) < 0 \quad \text{for small enough } t > 0.$$

Moreover, for $u \in B_\rho(0)$,

$$I_\lambda(u) \geq \frac{\min\{1, b_0\}}{p_M^+ (N+1)^{p_M^+ - 1}} \|u\|^{p_M^+}_{W^{1,\vec{p}(x)}(\Omega)} - \frac{\lambda}{q^-} c_8^{q^-} \|u\|^{q^-}_{W^{1,\vec{p}(x)}(\Omega)}.$$

It follows that

$$-\infty < c_9 = \inf_{B_\rho(0)} I_\lambda(u) < 0.$$

We now let $0 < \varepsilon < \inf_{\partial B_\rho(0)} I_\lambda - \inf_{B_\rho(0)} I_\lambda$. Applying the Ekeland variational principle to the functional $I_\lambda : \overline{B_\rho(0)} \to \mathbb{R}$, we find $u_\varepsilon \in \overline{B_\rho(0)}$ such that

$$\begin{aligned} I_\lambda(u_\varepsilon) &< \inf_{\overline{B_\rho(0)}} I_\lambda + \varepsilon, \\ I_\lambda(u_\varepsilon) &< I_\lambda(u) + \varepsilon \|u - u_\varepsilon\|_{W^{1,\vec{p}(x)}(\Omega)}, \quad u \neq u_\varepsilon. \end{aligned}$$

Since

$$I_\lambda(u_\varepsilon) \leq \inf_{\overline{B_\rho(0)}} I_\lambda + \varepsilon \leq \inf_{B_\rho(0)} I_\lambda + \varepsilon < \inf_{\partial B_\rho(0)} I_\lambda,$$

we deduce that $u_\varepsilon \in B_\rho(0)$. Now, we define $K_\lambda : \overline{B_\rho(0)} \to \mathbb{R}$ by $K_\lambda(u) = I_\lambda(u) + \varepsilon\|u - u_\varepsilon\|_{W^{1,\vec{p}(x)}(\Omega)}$. Since u_ε is a minimum point of K_λ, we deduce that

$$\frac{K_\lambda(u_\varepsilon + tv) - K_\lambda(u_\varepsilon)}{t} \geq 0,$$

for small $t > 0$ and $v \in B_\rho(0)$. The above relation yields

$$\frac{I_\lambda(u_\varepsilon + tv) - I_\lambda(u_\varepsilon)}{t} + \varepsilon\|v\|_{W^{1,\vec{p}(x)}(\Omega)} \geq 0.$$

Letting $t \to 0$ it follows that $\langle I'_\lambda(u_\varepsilon), v\rangle + \varepsilon\|v\|_{W^{1,\vec{p}(x)}(\Omega)} > 0$, hence $\|I'_\lambda(u_\varepsilon)\|_{W^{1,\vec{p}(x)}(\Omega)} \leq \varepsilon$. We deduce that there exists a sequence $(v_n) \subset B_1(0)$ such that

$$I_\lambda(v_n) \to c_9 \quad \text{and} \quad I'_\lambda(v_n) \to 0. \tag{5.63}$$

The sequence (v_n) is bounded in X. Actually, with similar arguments as those used at the end of Theorem 41, we can show that (v_n) strongly converges to u_1 in X. So, by (5.63)

$$I_\lambda(u_1) = c_9 < 0 \quad \text{and} \quad I'_\lambda(u_1) = 0,$$

that is, u_1 is a nontrivial weak solution of the problem (5.44). This completes the proof. □

Next, we intend to construct a global minimizer of the functional I_λ. We start with the following auxiliary result.

Lemma 59 *The functional I_λ is coercive on X.*

Proof. Using the hypothesis $(\mathbf{A_2})$ and relation (5.46) we deduce for all $u \in X$,

$$\begin{aligned} I_\lambda(u) \geq\ & \frac{1}{p_M^+}\sum_{i=1}^N \int_\Omega |\partial_{x_i} u|^{p_i(x)}\,dx + \frac{b_0}{p_M^+}\int_\Omega |u|^{p_M(x)}\,dx \\ & -\frac{\lambda}{q^-}\Big[(c_1\|u\|_{W^{1,\vec{p}(x)}(\Omega)})^{q^-} + (c_2\|u\|_{W^{1,\vec{p}(x)}(\Omega)})^{q^+}\Big]. \end{aligned} \tag{5.64}$$

Now we set $\|u\|_{W^{1,\vec{p}(x)}(\Omega)} > 1$. Using the same techniques as in the proof of (5.51) combined with (5.64) we find that

$$I_\lambda(u) \geq \tilde{c_1}\|u\|^{p_m^-}_{W^{1,\vec{p}(x)}(\Omega)} - \tilde{c_2} - \frac{\lambda}{q^-}\left[(c_1\|u\|_{W^{1,\vec{p}(x)}(\Omega)})^{q^-} + (c_2\|u\|_{W^{1,\vec{p}(x)}(\Omega)})^{q^+}\right],$$

for any $u \in X$ with $\|u\|_{W^{1,\vec{p}(x)}(\Omega)} > 1$. Since $q^- \leq q^+ < p_m^-$, we infer that $I_\lambda(u) \to \infty$ as $\|u\|_{W^{1,\vec{p}(x)}(\Omega)} \to \infty$, that is, I_λ is coercive. □

Proof of Theorem 42 (**ii**) The functional I_λ is weakly lower semicontinuous on X. By Lemma 59, I_λ is also coercive on X. Thus there exists $u_\lambda \in X$, a global minimizer of I_λ and thus weak solution of problem (5.44).
We show that u_λ is nontrivial if λ is large enough. Letting $t_0 > 1$ be a constant and Ω_1 be an open subset of Ω with $|\Omega_1| > 0$, we assume that $v_0 \in X$ is such that $v_0(x) = t_0$ for any $x \in \overline{\Omega_1}$ and $0 \leq v_0(x) \leq t_0$ in $\Omega\backslash\Omega_1$. We have

$$\begin{aligned} I_\lambda(v_0) &= \int_\Omega \Big\{\sum_{i=1}^N A_i(x, \partial_{x_i}(v_0)) + \frac{b(x)}{p_M(x)}|v_0|^{p_M(x)} - \frac{\lambda}{q(x)}|v_0|^{q(x)}\Big\}\,dx \\ &\leq c_6\sum_{i=1}^N \int_\Omega \Big(|\partial_{x_i}(v_0)| + \frac{|\partial_{x_i}(v_0)|^{p_i(x)}}{p_i(x)}\Big)dx + \frac{1}{p_M^-}\int_\Omega b(x)|v_0|^{p_M(x)}\,dx \\ &\quad -\lambda\int_\Omega \frac{1}{q(x)}|v_0|^{q(x)}dx \leq c_{10} - \frac{\lambda}{q^+}t_0^{q^-}\,|\Omega_1|. \end{aligned}$$

So there exists $\lambda^* > 0$ such that $J_\lambda(v_0) < 0$ for any $\lambda \in (\lambda^*, +\infty)$. It follows that for any $\lambda \geq \lambda^*$, u_λ is a nontrivial weak solution of problem (5.44) for large enough λ. □

5.4 Bifurcation for a singular problem modelling the equilibrium of anisotropic continuous media

In this section, we are concerned with the bifurcation analysis of solutions for a class of singular quasilinear elliptic Dirichlet problems on a smooth bounded domain containing the origin. By using a Caffarelli-Kohn-Nirenberg type inequality, a critical point result for differentiable functionals is exploited, in order to prove the existence of a precise open interval of positive eigenvalues for which the treated problem admits at least one nontrivial weak solution. In the case of terms with a sublinear growth near the origin, we deduce the existence of solutions for small positive values of the parameter. Moreover, the corresponding solutions have smaller and smaller energies as the parameter goes to zero.

Consider the following singular elliptic Dirichlet problem

$$\begin{cases} -\mathrm{div}\,(|x|^{-2a}\nabla u) = \lambda|x|^{-2b}f(u) & \text{in } \Omega, \\ u = 0 \quad \text{on } \partial\Omega, \end{cases} \qquad (D_\lambda^{a,b})$$

where λ is a real positive parameter and Ω is a bounded domain in $\mathbb{R}^N$ $(N > 2)$ with smooth boundary $\partial\Omega$ containing the origin, a and b are real numbers such that

$$0 \le a < \frac{N-2}{2} \qquad \text{and} \qquad a \le b < a+1,$$

and $f : \mathbb{R} \to \mathbb{R}$ is a continuous function satisfying the following subcritical growth condition

$$|f(t)| \le a_1 + a_2|t|^{r-1} \qquad \text{for all } t \in \mathbb{R}, \tag{h_∞}$$

for some nonnegative constants a_1, a_2 and $r \in (1, r_\star)$, where

$$r_\star := 2\min\left\{\frac{(N-2b)}{N-2(a+1)}, \frac{N}{N-2}\right\}.$$

If $r_\star = \dfrac{2N}{N-2}$ then some concentration phenomena can occur, due to the action of the noncompact group of dilations in $\mathbb{R}^N$.

The most relevant feature of Problem $(D_\lambda^{a,b})$ is the presence of the singular coefficients $|x|^{-2a}$ and $|x|^{-2b}$, which play a key role in the problem of existence of solutions when $0 \in \overline{\Omega}$.

In 1984, Caffarelli, Kohn and Nirenberg proved in [14], in the context of some more general inequalities, the following result:

Given $1 < p < N$, for every $u \in C_0^\infty(\Omega)$, there exists a positive constant $C_{a,b}$ such that

$$\left(\int_\Omega |x|^{-bq}|u(x)|^q\,dx\right)^{p/q} \le C_{a,b}\int_\Omega |x|^{-ap}|\nabla u(x)|^p\,dx\,, \tag{CKN}$$

where

$$-\infty < a < \frac{N-p}{p}, \quad a \le b \le a+1, \quad q = \frac{Np}{N-p(1+a-b)},$$

and $\Omega \subseteq \mathbb{R}^N$ is an arbitrary open domain.

We observe that the Caffarelli-Kohn-Nirenberg inequality (CKN) reduces to the classical Sobolev inequality (if $a = b = 0$) and to the Hardy inequality (if $a = 0$ and $b = 1$). Relation (2.54) proved to be an important tool in studying degenerate elliptic problems. It is also related with the understanding of some important phenomena such as best constants, existence or nonexistence of extremal functions, symmetry properties of minimizers, compactness of minimizing sequences and concentration phenomena.

In what follows we state, as an example, the following special case of the main results of this section.

Theorem 43 *Let* $0 \leq b < 1$ *and* $f : \mathbb{R} \to \mathbb{R}$ *be a continuous function such that*

$$\lim_{t \to 0^+} \frac{f(t)}{t} = +\infty, \quad \text{and} \quad \lim_{|t| \to \infty} \frac{f(t)}{|t|^s} = 0, \qquad (\mathrm{h}_0^\infty)$$

for some

$$0 \leq s < \frac{N + 2(1-2b)}{N-2}.$$

Then there exists $\lambda^\star > 0$ *such that for all* $\lambda \in (0, \lambda^\star)$*, the following singular Dirichlet problem*

$$\begin{cases} -\Delta u = \lambda |x|^{-2b} f(u) & \text{in } \Omega, \\ u = 0 & \text{on } \partial\Omega, \end{cases} \qquad (D_\lambda^{0,b})$$

admits at least one nontrivial weak solution $u_\lambda \in H_0^1(\Omega)$*. Moreover, we have*

$$\int_\Omega |\nabla u_\lambda(x)|^2 dx \to 0,$$

as $\lambda \to 0^+$ *and the mapping*

$$\lambda \mapsto \frac{1}{2}\Big(\int_\Omega |\nabla u_\lambda(x)|^2 dx\Big) - \lambda \int_\Omega |x|^{-2b} \Big(\int_0^{u_\lambda(x)} f(t)dt\Big) dx,$$

is negative and strictly decreasing in $(0, \lambda^\star)$.

Let $\mathcal{D}_a^{1,2}(\Omega)$ denote the completion of $C_0^\infty(\Omega)$ with respect to the norm

$$\|u\|_a = \Big(\int_\Omega |x|^{-2a} |\nabla u(x)|^2 dx\Big)^{1/2}.$$

Then the embedding $j : \mathcal{D}_a^{1,2}(\Omega) \hookrightarrow L^r(\Omega; |x|^{-q})$ is continuous if

$$1 \leq r \leq 2^* := \frac{2N}{N-2} \quad \text{and} \quad q \leq (1+a)r + N\left(1 - \frac{r}{2}\right),$$

where $L^r(\Omega; |x|^{-q})$ is the weighted Lebesgue space endowed with the norm

$$\|u\|_{r,q} = \Big(\int_\Omega |x|^{-q} |u(x)|^r dx\Big)^{1/r}.$$

Hence there exists a positive constant $K_{r,q}^a$ such that

$$\left(\int_\Omega |x|^{-q}|u(x)|^r dx\right)^{1/r} \le K_{r,q}^a\left(\int_\Omega |x|^{-2a}|\nabla u(x)|^2 dx\right)^{1/2}, \qquad (5.65)$$

for every $u \in \mathcal{D}_a^{1,2}(\Omega)$.

Moreover, the embedding j is compact if

(C) $1 \le r < 2^*$ and $q < (1+a)r + N\left(1 - \frac{r}{2}\right)$.

Set $F(\xi) := \int_0^\xi f(t)dt$, for every $\xi \in \mathbb{R}$. Consider the energy functional $E_\lambda : \mathcal{D}_a^{1,2}(\Omega) \to \mathbb{R}$ associated to $(D_\lambda^{a,b})$ defined by

$$E_\lambda(u) = \Phi(u) - \lambda\Psi(u) \qquad \text{for all } u \in \mathcal{D}_a^{1,2}(\Omega),$$

where

$$\Phi(u) := \frac{1}{2}\int_\Omega |x|^{-2a}|\nabla u(x)|^2 dx, \quad \Psi(u) := \int_\Omega |x|^{-2b}F(u(x))dx.$$

Then Φ is a Gâteaux differentiable functional in $\mathcal{D}_a^{1,2}(\Omega)$ and its derivative is given by

$$\Phi'(u)(v) = \int_\Omega |x|^{-2a}\nabla u(x)\cdot\nabla v(x)dx,$$

for every $v \in \mathcal{D}_a^{1,2}(\Omega)$. Moreover, Φ is weakly lower semicontinuous and coercive.

Lemma 60 *Assume that f satisfies the growth condition* (h_∞). *Then the functional $\Psi : \mathcal{D}_a^{1,2}(\Omega) \to \mathbb{R}$ is well defined, of class C^1, and sequentially weakly continuous.*

Proof. The functional Ψ is well defined since, by using (5.65) and exploiting (C), it follows that

$$\Psi(u) \le \left(a_1 K_{1,2b}^a + \frac{a_2}{r}(K_{r,2b}^a)^r\|u\|_a^{r-1}\right)\|u\|_a < +\infty,$$

for every $u \in \mathcal{D}_a^{1,2}(\Omega)$.

Further, standard arguments show that Ψ is a continuously Gâteaux differentiable functional whose Gâteaux derivative is

$$\Psi'(u)(v) = \int_\Omega |x|^{-2b}f(u(x))v(x)dx \qquad \text{for every } v \in \mathcal{D}_a^{1,2}(\Omega).$$

Next, we prove that Ψ is sequentially weakly continuous arguing by contradiction. Hence, let $\{u_k\} \subset \mathcal{D}_a^{1,2}(\Omega)$ be a sequence such that $u_k \rightharpoonup u$ in

$\mathcal{D}_a^{1,2}(\Omega)$ but $\{\Psi(u_k)\}$ does not converge to $\Psi(u)$ as $n \to \infty$. Next, note that since

$$r \in \left(1, \min\left\{\frac{2(N-2b)}{N-2(a+1)}, 2^*\right\}\right),$$

then r and $2b$ satisfy relation (C). In other words,

$$1 < r < 2^* \qquad \text{and} \qquad 2b < (1+a)r + N\left(1 - \frac{r}{2}\right).$$

Thus, up to a subsequence, there exists $\varepsilon > 0$ such that

$$0 < \varepsilon \leq |\Psi(u_k) - \Psi(u)|,$$

for every $k \in \mathbb{N}$ and $u_k \to u$ in $L^r(\Omega; |x|^{-2b})$. Moreover, for every $k \in \mathbb{N}$, we have $\vartheta_k \in (0,1)$ such that

$$\begin{aligned} |\Psi(u_k) - \Psi(u)| &\leq \int_\Omega |x|^{-2b}|F(u_k(x)) - F(u(x))|dx \\ &= \int_\Omega |x|^{-2b}|f(u(x) + \vartheta_k(u_k(x) - u(x)))||u_k(x) - u(x)|dx. \end{aligned}$$

From the above inequality and (h_∞), it follows that

$$\begin{aligned} |\Psi(u_k) - \Psi(u)| &\leq a_1 \int_\Omega |x|^{-2b}|u_k(x) - u(x)|dx \\ &+ a_2 \int_\Omega |x|^{-2b}|u(x) + \vartheta_k(u_k(x) - u(x))|^{r-1}|u_k(x) - u(x)|dx. \end{aligned}$$

Next, we employ the Hölder inequality (with weight $|x|^{-2b}$) and the following elementary property: for every real number $s > 0$, there exists a constant $c_s > 0$ such that

$$(\alpha + \beta)^s \leq c_s(\alpha^s + \beta^s) \qquad \text{for all } \alpha, \beta \in (0, \infty).$$

Hence (note that $r > 1$) we can write

$$\begin{aligned} &\int_\Omega |x|^{-2b}|u(x) + \vartheta_k(u_k(x) - u(x))|^{r-1}|u_k(x) - u(x)|dx \leq \\ &\int_\Omega |x|^{-2b}(|u(x)| + |u_k(x) - u(x))|)^{r-1}|u_k(x) - u(x)|dx = \\ c_{r-1}&\left(\int_\Omega |x|^{-2b}|u(x)|^{r-1}|u_k(x) - u(x)|dx + \int_\Omega |x|^{-2b}|u_k(x) - u(x)|^r dx\right) \leq \\ c_{r-1}&\left(\int_\Omega |x|^{-2b}|u(x)|^r dx\right)^{(r-1)/r}\left(\int_\Omega |x|^{-2b}|u_k(x) - u(x)|^r dx\right)^{1/r} + \\ &c_{r-1}\int_\Omega |x|^{-2b}|u_k(x) - u(x)|^r dx = \\ &c_{r-1}\left(\|u\|_{r,2b}^{r-1} + \|u_k - u\|_{r,2b}^{r-1}\right)\|u_k - u\|_{r,2b}. \end{aligned}$$

Then

$$\begin{aligned} 0 < \varepsilon \leq |\Psi(u_k) - \Psi(u)| &\leq a_1 \|u_k - u\|_{1,2b} \\ &+ c_{r-1} a_2 \left(\|u\|_{r,2b}^{r-1} + \|u_n - u\|_{r,2b}^{r-1} \right) \|u_k - u\|_{r,2b}. \end{aligned}$$

Since $\{u_k\}$ strongly converges to u in $L^r(\Omega; |x|^{-2b})$ (so $u_k \to u$ in $L^1(\Omega; |x|^{-2b})$), the right-hand side tends to zero as $k \to \infty$, contradicting $\varepsilon > 0$. □

Fix $\lambda > 0$. A function $u : \Omega \to \mathbb{R}$ is said to be a *weak solution* of $(D_\lambda^{a,b})$ if $u \in \mathcal{D}_a^{1,2}(\Omega)$ and

$$\int_\Omega |x|^{-2a} \nabla u(x) \cdot \nabla v(x) dx = \lambda \int_\Omega |x|^{-2b} f(u(x)) v(x) dx \qquad \text{for all } v \in \mathcal{D}_a^{1,2}(\Omega).$$

It follows that the critical points of E_λ are exactly the weak solutions of $(D_\lambda^{a,b})$.

Our main tool to prove the existence of a nontrivial weak solution of problem $(D_\lambda^{a,b})$ is the following critical point theorem due to Ricceri [93].

Theorem 44 *Let X be a reflexive real Banach space and $\Phi, \Psi : X \to \mathbb{R}$ two continuously Gâteaux differentiable functionals such that Φ is sequentially weakly lower semicontinuous and coercive and Ψ is sequentially weakly upper semicontinuous. For every $\rho > \inf_X \Phi$, set*

$$\varphi(\rho) := \inf_{u \in \Phi^{-1}(-\infty,\rho)} \frac{\sup_{v \in \Phi^{-1}(-\infty,\rho)} \Psi(v) - \Psi(u)}{\rho - \Phi(u)}.$$

Then for every $\rho > \inf_{u \in X} \Phi(u)$ and every $\lambda \in \,]0, 1/\varphi(\rho)[$, the restriction of the functional $E_\lambda := \Phi - \lambda \Psi$ to $\Phi^{-1}(-\infty, \rho)$ admits a global minimum, which is a critical point (local minimum) of E_λ in X.

5.4.1 Existence of low-energy solutions

With the previous notations, the main abstract result of this section is the following.

Theorem 45 *Let $f : \mathbb{R} \to \mathbb{R}$ be a continuous function satisfying condition (h_∞). Assume that*

$$\limsup_{\xi \to 0^+} \frac{F(\xi)}{\xi^2} = +\infty, \qquad \text{and} \qquad \liminf_{\xi \to 0^+} \frac{F(\xi)}{\xi^2} > -\infty.$$

Then there exists a positive number $\lambda^\star$ given by

$$\lambda^\star := r \sup_{\gamma > 0} \frac{\gamma}{r a_1 \sqrt{2} K_{1,2b}^a + a_2 2^{r/2} (K_{r,2b}^a)^r \gamma^{r-1}},$$

such that for every $\lambda \in (0, \lambda^\star)$, the following elliptic Dirichlet problem

$$\begin{cases} -div\,(|x|^{-2a}\nabla u) = \lambda |x|^{-2b} f(u) & \text{in } \Omega, \\ u = 0 \quad \text{on } \partial\Omega, \end{cases} \tag{$D_\lambda^{a,b}$}$$

admits at least one nontrivial weak solution $u_\lambda \in \mathcal{D}_a^{1,2}(\Omega)$. Moreover,

$$\lim_{\lambda \to 0^+} \|u_\lambda\|_a = 0,$$

and the function $\lambda \longmapsto E_\lambda(u_\lambda)$ is negative and strictly decreasing in $(0, \lambda^\star)$.

Proof. Fix $\lambda \in (0, \lambda^\star)$. Our aim is to apply Theorem 44 for $X := \mathcal{D}_a^{1,2}(\Omega)$, Φ and Ψ. Note that X is a Hilbert space whose inner product $\langle \cdot, \cdot \rangle : X \times X \to \mathbb{R}$ given by

$$\langle u, v \rangle := \int_\Omega |x|^{-2a} \nabla u(x) \cdot \nabla v(x) dx \qquad \text{for all } u, v \in X$$

induces the norm $\|\cdot\|_a$. Hence X is a reflexive space.

Now, we prove that problem (P_λ) admits a weak solution $u_\lambda \in \Phi^{-1}(-\infty, \bar\rho)$, for a suitable $\bar\rho > 0$.

We first observe that $\Phi : X \to \mathbb{R}$ is a continuously Gâteaux differentiable and sequentially weakly lower semicontinuous functional, and that the map $\Psi : X \to \mathbb{R}$ is continuously Gâteaux differentiable and sequentially weakly upper semicontinuous, by Lemma 60. Moreover, Φ is coercive and $\inf_{u \in X} \Phi(u) = 0$. Due to (h_∞), it follows that

$$F(\xi) \leq a_1 |\xi| + a_2 \frac{|\xi|^r}{r} \qquad \text{for all } \xi \in \mathbb{R}. \tag{5.66}$$

Since $0 < \lambda < \lambda^\star$, there exists $\bar\gamma > 0$ such that

$$\lambda < \frac{r\bar\gamma}{r a_1 \sqrt{2} K_{1,2b}^a + a_2 2^{r/2} (K_{r,2b}^a)^r \bar\gamma^{r-1}} =: \lambda_{\bar\gamma}^\star. \tag{5.67}$$

Let $\rho \in (0, +\infty)$ and consider the function

$$\chi(\rho) := \frac{\sup_{v \in \Phi^{-1}(-\infty, \rho)} \Psi(v)}{\rho}.$$

Taking into account (5.66), it follows that

$$\Psi(v) := \int_\Omega |x|^{-2b} F(v(x)) dx \leq a_1 \|v\|_{1,2b} + \frac{a_2}{r} \|v\|_{r,2b}^r.$$

Thus for every $v \in X : \Phi(v) < \rho$, we obtain

$$\|v\|_a < \sqrt{2\rho}, \tag{5.68}$$

for every $v \in X$.
Now, from (5.65) and (5.68), we get for every $v \in X$ with $\Phi(v) < \rho$,

$$\Psi(v) < a_1 K^a_{1,2b}\sqrt{2\rho} + \frac{a_2 2^{r/2}(K^a_{r,2b})^r}{r}\rho^{r/2}.$$

Hence

$$\sup_{v\in\Phi^{-1}(]-\infty,\rho[)} \Psi(v) \le a_1 K^a_{1,2b}\sqrt{2\rho} + \frac{a_2 2^{r/2}(K^a_{r,2b})^r}{r}\rho^{r/2}.$$

Then

$$\chi(\rho) \le a_1 K^a_{1,2b}\sqrt{\frac{2}{\rho}} + \frac{a_2 2^{r/2}(K^a_{r,2b})^r}{r}\rho^{r/2-1}, \tag{5.69}$$

for every $\rho > 0$. In particular, we deduce that

$$\chi(\bar\gamma^2) \le \frac{a_1\sqrt{2}K^a_{1,2b}}{\bar\gamma} + \frac{a_2 2^{r/2}(K^a_{r,2b})^r}{r}\bar\gamma^{r-2}. \tag{5.70}$$

At this point, observe that

$$\varphi(\bar\gamma^2) := \inf_{u\in\Phi^{-1}(-\infty,\bar\gamma^2)} \frac{\sup_{v\in\Phi^{-1}(-\infty,\bar\gamma^2)} \Psi(v) - \Psi(u)}{\bar\gamma^2 - \Phi(u)} \le \chi(\bar\gamma^2),$$

taking into account that $0_X \in \Phi^{-1}(-\infty,\bar\gamma^2)$ and $\Phi(0_X) = \Psi(0_X) = 0$.

In conclusion, bearing in mind (5.67), the above inequality together with (5.70) yields

$$\varphi(\bar\gamma^2) \le \chi(\bar\gamma^2) \le \frac{a_1\sqrt{2}K^a_{1,2b}}{\bar\gamma} + \frac{a_2 2^{r/2}(K^a_{r,2b})^r}{r}\bar\gamma^{r-2} < \frac{1}{\lambda}.$$

In other words,

$$\lambda \in \left(0, \frac{r\bar\gamma}{r a_1\sqrt{2}K^a_{1,2b} + a_2 2^{r/2}(K^a_{r,2b})^r\bar\gamma^{r-1}}\right) \subseteq (0, 1/\varphi(\bar\gamma^2)).$$

Using Theorem 44, we deduce that there exists a function $u_\lambda \in \Phi^{-1}(-\infty,\bar\gamma^2)$ such that

$$E'_\lambda(u_\lambda) = \Phi'(u_\lambda) - \lambda\Psi'(u_\lambda) = 0,$$

and, in particular, u_λ is a global minimum of the restriction of E_λ to $\Phi^{-1}(-\infty,\bar\gamma^2)$.

Next, we prove that the function u_λ is not trivial. To this end, let us prove that

$$\limsup_{\|u\|\to 0}\frac{\Psi(u)}{\Phi(u)} = \limsup_{\|u\|\to 0}\frac{\displaystyle\int_\Omega |x|^{-2b}F(u(x))dx}{\|u\|_a^2/2} = +\infty. \tag{5.71}$$

Due to our assumptions at zero, we can fix a sequence $\{\xi_k\} \subset \mathbb{R}^+$ converging to zero and constants σ, and κ (with $\sigma > 0$) such that

$$\lim_{k\to\infty} \frac{\displaystyle\int_0^{\xi_k} f(t)dt}{\xi_k^2} = +\infty,$$

and

$$\int_0^{\xi} f(t)dt \geq \kappa\xi^2,$$

for every $\xi \in [0,\sigma]$.

Now, let us consider a closed ball $\bar{B}(x_0,\tau) \subset \Omega$ (not containing the origin) for some $x_0 \in \Omega$ and sufficiently small $\tau > 0$. Let $v \in C_0^\infty(\Omega) \subset X$ such that

i) $v(x) \in [0,1]$, for every $x \in \Omega$;

ii) $v(x) = 1$, for every $x \in \bar{B}(x_0,\tau/2)$;

iii) $v(x) = 0$, for every $x \in \Omega \setminus B(x_0,\tau)$.

Finally, fix $\varepsilon > 0$ and consider a real positive number μ with

$$\varepsilon < 2\frac{\mu\displaystyle\int_{B(x_0,\tau/2)} |x|^{-2b}dx + \kappa\int_C |x|^{-2b}v(x)^2dx}{\|v\|_a^2},$$

where $C := B(x_0,\tau) \setminus B(x_0,\tau/2)$. Note that the right-hand side of the above inequality is finite. Indeed, since $0 \notin B(x_0,\tau/2)$, we get

$$\int_{B(x_0,\tau/2)} |x|^{-2b}dx < +\infty.$$

Moreover, since the injection $j : X \hookrightarrow L^2(\Omega;|x|^{-2b})$ is continuous, taking into account that $1 < 2 < 2^*$ and $2b < 2(a+1)$, we deduce that there exists a constant $K_{2,2b}^a > 0$ (according to our notations) such that

$$\int_C |x|^{-2b}v(x)^2dx < \int_\Omega |x|^{-2b}v(x)^2dx \leq (K_{2,2b}^a)^2\Big(\int_\Omega |x|^{-2a}|\nabla u(x)|^2dx\Big) < +\infty.$$

Thus there is some $\nu \in \mathbb{N}$ such that $\xi_k < \delta$ and

$$\int_0^{\xi_k} f(t)dt \geq \mu\xi_k^2,$$

for every $k > \nu$.

At this point, for every $k > \nu$, and bearing in mind the properties of the function v, we obtain

$$\frac{\Psi(\xi_k v)}{\Phi(\xi_k v)} = 2\frac{\displaystyle\int_{B(x_0,\tau/2)} |x|^{-2b}\left(\int_0^{\xi_k} f(t)dt\right)dx + \int_C |x|^{-2b}F(\xi_k v(x))dx}{\|\xi_k v\|_a^2}$$

$$\geq 2\frac{\displaystyle\mu\int_{B(x_0,\tau/2)} |x|^{-2b}dx + \kappa\int_C |x|^{-2b}v(x)^2dx}{\|v\|_a^2} > \varepsilon.$$

Since ε can be chosen arbitrarily large, it follows that

$$\lim_{k\to\infty}\frac{\Psi(\xi_k v)}{\Phi(\xi_k v)} = +\infty,$$

from which (5.71) clearly follows. Hence, there exists a sequence $\{w_k\} \subset X$ strongly converging to zero, such that for every sufficiently large k, $w_k \in \Phi^{-1}(-\infty, \bar{\gamma}^2)$, and

$$E(w_k) := \Phi(w_k) - \lambda\Psi(w_k) < 0.$$

Since u_λ is a global minimum of the restriction of E_λ to $\Phi^{-1}(-\infty, \bar{\gamma}^2)$, we conclude that

$$\Phi(u_\lambda) - \lambda\Psi(u_\lambda) < 0 = E_\lambda(0_X),$$

hence u_λ is not trivial.

Now, we prove that $\lim\limits_{\lambda\to 0^+} \|u_\lambda\|_a = 0$. Bearing in mind that Φ is a coercive functional and that $u_\lambda \in \Phi^{-1}(-\infty, \bar{\gamma}^2)$, for every $\lambda \in (0, \lambda^\star_{\bar{\gamma}})$, we obtain

$$\|u_\lambda\|_a < \sqrt{2}\bar{\gamma}.$$

Therefore, taking into account (5.65) and the above inequality, it follows from (h_∞) that there exists $M_{\bar{\gamma}} > 0$ such that

$$\left|\int_\Omega |x|^{-2b}f(u_\lambda(x))u_\lambda(x)dx\right| \leq M_{\bar{\gamma}}, \tag{5.72}$$

for every $\lambda \in]0, \lambda^\star_{\bar{\gamma}}[$.

Now, $E'_\lambda(u_\lambda) = 0$, for every $\lambda \in (0, \lambda^\star_{\bar{\gamma}})$ and in particular $E'_\lambda(u_\lambda)(u_\lambda) = 0$, that is,

$$\|u_\lambda\|_a^2 = \lambda\int_\Omega |x|^{-2b}f(u_\lambda(x))u_\lambda(x)dx,$$

for every $\lambda \in (0, \lambda^\star_{\bar{\gamma}})$.

Then it follows from (5.72) that

$$\lim_{\lambda\to 0^+}\|u_\lambda\|_a^2 = \lim_{\lambda\to 0^+}\lambda\Psi'(u_\lambda)(u_\lambda) = 0.$$

Hence, as claimed, $\lim_{\lambda \to 0^+} \|u_\lambda\|_a = 0$.

Next, we observe that

$$E_\lambda(u) = \lambda \left(\frac{\Phi(u)}{\lambda} - \Psi(u) \right),$$

for every $u \in X$. Fix $0 < \lambda_1 < \lambda_2$ (with $\lambda_2 < \lambda_{\bar{\gamma}}^\star$). Set

$$m_{\lambda_1} := \left(\frac{\Phi(u_{\lambda_1})}{\lambda_1} - \Psi(u_{\lambda_1}) \right) = \inf_{u \in \Phi^{-1}(0,\bar{\gamma}^2)} \left(\frac{\Phi(u)}{\lambda_1} - \Psi(u) \right),$$

and

$$m_{\lambda_2} := \left(\frac{\Phi(u_{\lambda_2})}{\lambda_2} - \Psi(u_{\lambda_2}) \right) = \inf_{u \in \Phi^{-1}(0,\bar{\gamma}^2)} \left(\frac{\Phi(u)}{\lambda_2} - \Psi(u) \right).$$

Thus, as claimed before, $m_{\lambda_i} < 0$ (for $i = 1, 2$), and $m_{\lambda_2} \leq m_{\lambda_1}$ thanks to $\lambda_1 < \lambda_2$. Then the mapping $\lambda \mapsto E_\lambda(u_\lambda)$ is strictly decreasing in $(0, \lambda^\star)$ owing to

$$E_{\lambda_2}(u_{\lambda_2}) = \lambda_2 m_{\lambda_2} \leq \lambda_2 m_{\lambda_1} < \lambda_1 m_{\lambda_1} = E_{\lambda_1}(u_{\lambda_1}).$$

This concludes the proof. □

5.4.2 Additional properties and examples

We conclude this section with several remarks concerning the previous abstract results.

Remark 12 *We first point out that since*

$$0 \leq a < \frac{N-2}{2}, \quad \text{and} \quad a \leq b < a+1,$$

it follows that

$$2 < \frac{2(N-2b)}{N-2(a+1)}.$$

Furthermore, we observe that if, instead of $0 \leq a < \dfrac{N-2}{2}$, *we require the more restrictive condition* $0 \leq a < \dfrac{b}{N}(N-2)$, *then*

$$\frac{2(N-2b)}{N-2(a+1)} \in (2, 2^*).$$

Thus in this special case, the above relation yields

$$r_\star = \frac{2(N-2b)}{N-2(a+1)}.$$

On the other hand if $a = b$, it follows that $r_\star = 2^*$, since

$$\frac{N-2a}{N-2(a+1)} \geq \frac{N}{N-2}.$$

We also point out that the parameter $\lambda^\star$ in Theorem 45 has the following expression

$$\lambda^\star = \begin{cases} +\infty \quad \text{if} \quad 1 < r < 2 \\ \dfrac{1}{a_2(K^a_{2,2b})^2} \quad \text{if} \quad r = 2 \\ \dfrac{r\widetilde{\gamma}_{\max}}{ra_1\sqrt{2}K^a_{1,2b} + a_2 2^{r/2}(K^a_{r,2b})^r \widetilde{\gamma}^{r-1}_{\max}} \quad \text{if} \quad r \in (2, r_\star), \end{cases}$$

where

$$\widetilde{\gamma}_{\max} := \frac{1}{\sqrt{2}} \left(\frac{a_1 r K^a_{1,2b}}{a_2(K^a_{r,2b})^r(r-2)} \right)^{1/(r-1)}.$$

Remark 13 *It follows from the above expressions that if f is sublinear at infinity, then Theorem 45 ensures that for all $\lambda > 0$, the problem $(D^{a,b}_\lambda)$ admits at least one nontrivial weak solution. We explicitly observe that, in this case, also the classical direct methods theorem (Tonelli's result) ensures the existence of at least one weak solution that can possibly be trivial.*

Remark 14 *We observe that if $f(0) = 0$, Theorem 45 is a bifurcation result. Indeed, in this setting, it follows that the trivial solution solves problem $(D^{a,b}_\lambda)$ for every parameter λ. Hence, $\lambda = 0$ is a bifurcation point of problem $(D^{a,b}_\lambda)$, in the sense that the point $(0,0)$ belongs to the closure of the set*

$$\mathfrak{F} := \{(u,\lambda) \in \mathcal{D}^{1,2}_a(\Omega)\times]0,+\infty[: \ u \text{ is a weak solution of } (D^{a,b}_\lambda),\ u \neq 0\}$$

in the space $\mathcal{D}^{1,2}_a(\Omega) \times \mathbb{R}$. Otherwise, if $f(0) \neq 0$ we evidently have

$$\lim_{\xi\to 0^+} \frac{F(\xi)}{\xi^2} = +\infty. \tag{h$_0'$}$$

Hence, all the assumptions of Theorem 45 are automatically verified and the conclusions follow.

Remark 15 *Theorem 43 easily follows from Theorem 45 taking into account that the following s-sublinear assumption at infinity*

$$\lim_{|t|\to\infty} \frac{f(t)}{|t|^s} = 0,$$

in which $s \in [0, s_\star)$, where

$$s_\star := \min\left\{ \frac{N+2(a-2b+1)}{N-2(a+1)}, \frac{N+2}{N-2} \right\},$$

implies the growth condition (h_∞). *In addition, if*

$$\lim_{t\to 0^+}\frac{f(t)}{t}=+\infty,$$

condition (h_0') *also holds. Finally, bearing in mind Remark 12, the parameter* $\lambda^\star$ *has the form*

$$\lambda^\star=\begin{cases}+\infty & \text{if}\quad 0\le s<1\\ \dfrac{1}{\bar a_2(K^a_{2,2b})^2} & \text{if}\quad s=1\\ \dfrac{s-1}{2K^a_{1,2b}s}\left(\dfrac{\bar a_1(s+1)K^a_{1,2b}}{\bar a_2(K^a_{s+1,2b})^{s+1}(s-1)}\right)^{1/s} & \text{if}\quad s\in(1,s_\star),\end{cases}$$

for suitable nonnegative real constants $\bar a_1,\bar a_2$.

The next result is an immediate consequence of Remark 12.

Corollary 2 *Let* $f:\mathbb{R}\to\mathbb{R}$ *be a continuous function satisfying the following subcritical growth condition*

$$|f(t)|\le a_1+a_2|t|^{r-1}\qquad \text{for all } t\in\mathbb{R}, \tag{h_∞'}$$

where a_1,a_2 *are nonnegative constants and* $r\in(2,r_\star)$. *Furthermore, assume that condition* (h_0') *holds. Then for every*

$$\lambda\in\left(0,\frac{r-2}{2K^a_{1,2b}(r-1)}\left(\frac{a_1rK^a_{1,2b}}{a_2(K^a_{r,2b})^r(r-2)}\right)^{\frac{1}{r-1}}\right),$$

problem ($D^{a,b}_\lambda$) *admits at least one nontrivial weak solution* $u_\lambda\in\mathcal{D}^{1,2}_a(\Omega)$. *Moreover,*

$$\lim_{\lambda\to 0^+}\|u_\lambda\|_a=0,$$

and the function $\lambda\longmapsto E_\lambda(u_\lambda)$ *is negative and strictly decreasing in* $(0,\lambda^\star)$.

The next example deals with a nonlinearity f that vanishes at zero. The existence of one nontrivial solution for the Dirichlet problem involving the map f is achieved by using Corollary 2.

Example 10 *Let* Ω *be a bounded open subset of* $\mathbb{R}^N$ *(containing the origin) with* $N\ge 3$, *and consider the following problem*

$$\begin{cases}-div(|x|^{-2a}\nabla u)=\lambda|x|^{-2b}f(u) & in\ \Omega,\\ u=0\quad on\ \partial\Omega,\end{cases}\tag{$\widetilde{D}^{a,b}_\lambda$}$$

where $f(u):=|u|^{h-2}u+|u|^{s-2}u$, $1<h<2$ *and* $2<s<r_\star$. *Then for every*

$$\lambda\in\left(0,\frac{s-2}{2K^a_{1,2b}(s-1)}\left(\frac{sK^a_{1,2b}}{(K^a_{s,2b})^s(s-2)}\right)^{\frac{1}{s-1}}\right),$$

problem $(\widetilde{D}_{\lambda}^{a,b})$ *admits at least one nontrivial weak solution* $u_\lambda \in \mathcal{D}_a^{1,2}(\Omega)$ *such that*

$$\lim_{\lambda \to 0^+} \|u_\lambda\|_a = 0.$$

To prove this assertion, we apply Corollary 2 with

$$f(t) := |t|^{h-2}t + |t|^{s-2}t,$$

for every $t \in \mathbb{R}$. *Indeed, clearly* $f(0) = 0$ *and*

$$|f(t)| \leq 2(1 + |t|^{s-1}) \qquad \text{for all } t \in \mathbb{R}.$$

Hence condition (h'_∞) *holds. Moreover, a direct computation shows that*

$$\lim_{\xi \to 0^+} \frac{F(\xi)}{\xi^2} \geq \frac{1}{h}\left(\lim_{\xi \to 0^+} \frac{1}{\xi^{2-h}}\right) = +\infty.$$

Therefore assumption (h'_0) *is verified and the conclusion follows.*

Remark 16 *We point out that the energy functional* E_λ *associated to problem* $(\widetilde{D}_{\lambda}^{a,b})$ *is unbounded from below. Hence, since the functional* E_λ *is not coercive, the classical direct method result cannot be applied to the case treated in Example 10.*

Remark 17 *If* f *is subcritical (hence* (h_∞) *holds), under our assumptions at zero, problem* $(D_{\lambda}^{0,0})$ *admits at least one nontrivial weak solution* $u_{0,\lambda} \in H_0^1(\Omega)$ *for every* $\lambda \in (0, \lambda^\star)$, *in which*

$$\lambda^\star := \begin{cases} +\infty & \text{if} \quad 1 < r < 2 \\ \dfrac{1}{a_2 c_2^2} & \text{if} \quad r = 2 \\ \dfrac{r-2}{2c_1(r-1)}\left(\dfrac{r a_1 c_1}{a_2 c_r^r (r-2)}\right)^{\frac{1}{r-1}} & \text{if} \quad r \in (2, 2^*), \end{cases}$$

where the numbers c_r $(r \in [2, 2^*))$ *are the best Sobolev constants of the continuous inclusion* $H_0^1(\Omega) \hookrightarrow L^r(\Omega)$. *Moreover, we have also in this case*

$$\int_\Omega |\nabla u_{0,\lambda}(x)|^2 dx \to 0,$$

as $\lambda \to 0^+$ *and the mapping*

$$\lambda \mapsto \frac{1}{2}\Big(\int_\Omega |\nabla u_{0,\lambda}(x)|^2 dx\Big) - \lambda \int_\Omega \Big(\int_0^{u_{0,\lambda}(x)} f(t)dt\Big)dx,$$

is negative and strictly decreasing in $(0, \lambda^\star)$.

5.5 Glossary

Anisotropy: This is the property of being directionally dependent, as opposed to isotropy, which implies identical properties in all directions.

Bifurcation Problem: This is a mathematical problem dealing with the study of changes in the qualitative or topological structure of a given family, such as the solutions of a class of partial differential equations.

Indefinite Potential: This notion refers to sign-changing potentials and appears frequently in the qualitative analysis of partial differential equations.

Chapter 6

Difference Equations with Variable Exponents

6.1 Eigenvalue problems associated to anisotropic difference operators 250
6.1.1 Location properties for the spectrum 253
6.2 Homoclinic solutions of difference equations with variable exponents 258
6.3 Low-energy solutions for discrete anisotropic equations 268
6.4 Glossary 278

Often the route to solving problems is to bring new concepts out of the dark.

Alexander Grothendieck (1928–2014)

Partial difference equations usually describe the evolution of certain phenomena over the course of time. Elementary but relevant examples of partial difference equations are concerned with heat diffusion, heat control, temperature distribution, population growth, cellular neural networks, etc. The interest for problems of this type is motivated by major applications of differential and difference operators to various applied fields, such as electrorheological (smart) fluids, space technology, robotics, image processing, etc. In this chapter, we describe several problems described by partial difference equations involving nonhomogeneous operators. The results contained in this chapter are based on the original contributions included in the papers [44, 85, 60, 61, 68, 73].

6.1 Eigenvalue problems associated to anisotropic difference operators

In this section, we study the discrete eigenvalue problem

$$\begin{cases} -\Delta(|\Delta u(k-1)|^{p(k-1)-2}\Delta u(k-1)) = \lambda |u(k)|^{q(k)-2}u(k), & k \in \mathbb{Z}[1,T], \\ u(0) = u(T+1) = 0\,, \end{cases} \tag{6.1}$$

where $T \geq 2$ is a positive integer and $\Delta u(k) = u(k+1) - u(k)$ is the forward difference operator.

We denote by $\mathbb{Z}[a,b]$ the discrete interval $\{a, a+1, ..., b\}$ where a and b are integers and $a < b$. Throughout this section we assume that $p : \mathbb{Z}[0,T] \to [2,\infty)$ and $q : \mathbb{Z}[1,T] \to [2,\infty)$ are bounded functions and λ is a positive constant.

We define the function space

$$H = \{u : \mathbb{Z}[0,T+1] \to \mathbb{R};\ \text{such that } u(0) = u(T+1) = 0\}.$$

Then H is a T-dimensional Hilbert space with the inner product

$$(u,v) = \sum_{k=1}^{T+1} \Delta u(k-1)\Delta v(k-1), \quad \forall\, u,v \in H\,.$$

This associated norm is defined by

$$\|u\| = \left(\sum_{k=1}^{T+1} |\Delta u(k-1)|^2\right)^{1/2}.$$

By a *weak solution* of problem (6.1) we understand a function $u \in H$ such that

$$\sum_{k=1}^{T+1} |\Delta u(k-1)|^{p(k-1)-2}\Delta u(k-1)\Delta v(k-1) - \lambda \sum_{k=1}^{T} |u(k)|^{q(k)-2}u(k)v(k) = 0\,,$$

for any $v \in H$.

Theorem 46 *Assume that functions p and q satisfy the hypothesis*

$$\max_{\mathbb{Z}[0,T]} p < \min_{\mathbb{Z}[1,T]} q\,. \tag{6.2}$$

Then for any $\lambda > 0$ problem (6.1) has a nontrivial weak solution.

Theorem 47 *Assume that functions p and q satisfy the hypothesis*

$$\max_{\mathbb{Z}[1,T]} q < \min_{\mathbb{Z}[0,T]} p\,. \tag{6.3}$$

Then there exists $\lambda^{\star\star} > 0$ such that for any $\lambda > \lambda^{\star\star}$ problem (6.1) has a nontrivial weak solution.

Theorem 48 *Assume that functions p and q satisfy the hypothesis*

$$\min_{\mathbb{Z}[1,T]} q < \min_{\mathbb{Z}[0,T]} p\,. \tag{6.4}$$

Then there exists $\lambda^\star > 0$ such that for any $\lambda \in (0, \lambda^\star)$ problem (6.1) has a nontrivial weak solution.

We point out that if relation (6.4) is satisfied then relation (6.3) is fulfilled, too. Consequently, the result of Theorem 47 can be completed with the conclusion of Theorem 48. More precisely, we deduce the following additional property.

Corollary 3 *Assume that functions p and q satisfy the hypothesis*

$$\min_{\mathbb{Z}[1,T]} q < \min_{\mathbb{Z}[0,T]} p\,.$$

Then there exist $\lambda^\star > 0$ and $\lambda^{\star\star} > 0$ such that for any $\lambda \in (0, \lambda^\star) \cup (\lambda^{\star\star}, \infty)$ problem (6.1) has a nontrivial weak solution.

We observe that the result of Theorem 48 remains true in situations that extend relation (6.3) since in relation (6.4) it can happen that

$$\min_{\mathbb{Z}[1,T]} q < \min_{\mathbb{Z}[0,T]} p < \max_{\mathbb{Z}[1,T]} q\,.$$

We use the following notations:

$$p^- = \min_{\mathbb{Z}[0,T]} p, \quad p^+ = \max_{\mathbb{Z}[0,T]} p, \quad q^- = \min_{\mathbb{Z}[1,T]} q, \quad q^+ = \max_{\mathbb{Z}[1,T]} q\,.$$

On the other hand, it is useful to introduce other norms on H, namely

$$|u|_m = \left(\sum_{k=1}^{T} |u(k)|^m \right)^{1/m}, \quad \forall\, u \in H \text{ and } m \geq 2\,.$$

Then

$$T^{(2-m)/(2m)} \cdot |u|_2 \leq |u|_m \leq T^{1/m} \cdot |u|_2, \quad \forall\, u \in H \text{ and } m \geq 2\,. \tag{6.5}$$

Lemma 61 *(a) There exist positive constants C_1 and C_2 such that*

$$\sum_{k=1}^{T+1} |\Delta u(k-1)|^{p(k-1)} \geq C_1 \cdot \|u\|^{p^-} - C_2, \quad \forall\, u \in H \quad \textit{with } \|u\| > 1\,.$$

(b) There exists a positive constant C_3 such that

$$\sum_{k=1}^{T+1} |\Delta u(k-1)|^{p(k-1)} \geq C_3 \cdot \|u\|^{p^+}, \quad \forall\, u \in H \quad \textit{with } \|u\| < 1\,.$$

(c) For any $m \geq 2$ there exists a positive constant c_m such that

$$\sum_{k=1}^{T} |u(k)|^m \leq c_m \cdot \sum_{k=1}^{T+1} |\Delta u(k-1)|^m, \quad \forall\, u \in H\,.$$

Proof. (a) Fix $u \in H$ with $\|u\| > 1$. We define

$$\alpha_k = \begin{cases} p^+, & \text{if} \quad |\Delta u(k)| < 1 \\ p^-, & \text{if} \quad |\Delta u(k)| > 1\,, \end{cases}$$

for each $k \in \mathbb{Z}[0, T]$.

We deduce that

$$\begin{aligned} \sum_{k=1}^{T+1} |\Delta u(k-1)|^{p(k-1)} &\geq \sum_{k=1}^{T+1} |\Delta u(k-1)|^{\alpha_{k-1}} \\ &\geq \sum_{k=1}^{T+1} |\Delta u(k-1)|^{p^-} \\ &- \sum_{\{k \in \mathbb{Z}[0,T];\ \alpha_{k-1}=p^+\}} \left(|\Delta u(k-1)|^{p^-} - |\Delta u(k-1)|^{p^+}\right) \\ &\geq \sum_{k=1}^{T+1} |\Delta u(k-1)|^{p^-} - T\,. \end{aligned}$$

The above inequality in combination with relation (6.5) yields

$$\sum_{k=1}^{T+1} |\Delta u(k-1)|^{p(k-1)} \geq T^{(2-p^-)/2} \cdot \|u\|^{p^-} - T, \quad \forall\, u \in H \text{ with } \|u\| > 1\,.$$

(b) Assume that $u \in H$ with $\|u\| < 1$. It follows that $|\Delta u(k)| < 1$ for each $k \in \mathbb{Z}[0, T]$. So by (6.5) we deduce that

$$\begin{aligned} \sum_{k=1}^{T+1} |\Delta u(k-1)|^{p(k-1)} &\geq \sum_{k=1}^{T+1} |\Delta u(k-1)|^{p^+} \\ &\geq 1/T^{(2-p^+)/2} \cdot \|u\|^{p^+}\,. \end{aligned}$$

(c) Since

$$|u(k)| \leq \sum_{i=0}^{k-1} |\Delta u(i)|, \quad \forall\, u \in H \text{ and } k \in \mathbb{Z}[0, T]\,,$$

we deduce that for any positive real number $m \geq 2$ there exists a positive constant $c_{m,k}$ such that

$$|u(k)|^m \leq c_{m,k} \cdot \sum_{i=0}^{k-1} |\Delta u(i)|^m, \quad \forall\, u \in H \text{ and } k \in \mathbb{Z}[0, T]\,.$$

Thus there exists a positive constant c_m such that

$$\sum_{k=1}^{T} |u(k)|^m \leq c_m \cdot \sum_{k=1}^{T+1} |\Delta u(k-1)|^m, \quad \forall\, u \in H\,. \tag{6.6}$$

The proof of Lemma 61 is complete. □

6.1.1 Location properties for the spectrum

For any $\lambda > 0$ the energy functional corresponding to problem (6.1) is defined as $J_\lambda : H \to \mathbb{R}$,

$$J_\lambda(u) = \sum_{k=1}^{T+1} \frac{1}{p(k-1)} |\Delta u(k-1)|^{p(k-1)} - \lambda \cdot \sum_{k=1}^{T} \frac{1}{q(k)} |u(k)|^{q(k)}\,.$$

Then $J_\lambda \in C^1(H, \mathbb{R})$ and

$$\begin{aligned} \langle J'_\lambda(u), v\rangle \; &= \sum_{k=1}^{T+1} |\Delta u(k-1)|^{p(k-1)-2} \Delta u(k-1) \Delta v(k-1) \\ &\quad -\lambda \sum_{k=1}^{T} |u(k)|^{q(k)-2} u(k) v(k)\,, \end{aligned}$$

for all $u,\ v \in H$. Thus the weak solutions of problem (6.1) coincide with the critical points of J_λ.

In the sequel, we are concerned with the proof of Theorem 46. We first establish that the energy functional J_λ satisfies the geometric hypotheses of the mountain pass theorem.

Lemma 62 *There exist $\eta > 0$ and $\alpha > 0$ such that $J_\lambda(u) \geq \alpha > 0$ for any $u \in H$ with $\|u\| = \eta$.*

Proof. We observe that

$$|u(k)|^{q^-} + |u(k)|^{q^+} \geq |u(k)|^{q(k)}, \quad \forall\, k \in \mathbb{Z}[1, T] \text{ and } u \in H\,. \tag{6.7}$$

Therefore

$$J_\lambda(u) \geq \frac{1}{p^+} \sum_{k=1}^{T+1} \frac{1}{p(k-1)} |\Delta u(k-1)|^{p(k-1)} - \frac{\lambda}{q^-} (|u|_{q^-}^{q^-} + |u|_{q^+}^{q^+}), \quad \forall\, u \in H\,. \tag{6.8}$$

Next, we focus on the case when $u \in H$ with $\|u\| < 1$. Thus, $|\Delta u(k)| < 1$ for any $k \in \mathbb{Z}[0, T+1]$. Then using Lemma 61 (c) and relation (6.5) we obtain

$$\begin{aligned} |u|_{q^-}^{q^-} + |u|_{q^+}^{q^+} \; &\leq c_{q^-} \sum_{k=1}^{T+1} |\Delta u(k-1)|^{q^-} + c_{q^+} \sum_{k=1}^{T+1} |\Delta u(k-1)|^{q^+} \\ &\leq c_{q^-} T \|u\|^{q^-} + c_{q^+} T \|u\|^{q^+}\,. \end{aligned} \tag{6.9}$$

For $u \in H$ with $\|u\| < 1$ the above inequalities combined with relation (4.27), Lemma 61 (b) and relation (6.5) imply

$$\begin{aligned} J_\lambda(u) &\geq \frac{C_3}{p^+}\|u\|^{p^+} - \frac{\lambda}{q^-}(c_{q^-}T \cdot \|u\|^{q^-} + c_{q^+}T \cdot \|u\|^{q^+}) \\ &= (d_1 - d_2 \cdot \|u\|^{q^- - p^+} - d_3 \cdot \|u\|^{q^+ - p^+}) \cdot \|u\|^{p^+}, \end{aligned}$$

where d_1, d_2 and d_3 are positive constants.

We remark that by continuity arguments, the function $g : [0,1] \to \mathbb{R}$ defined by

$$g(t) = d_1 - d_2 \cdot t^{q^+ - p^+} - d_3 \cdot t^{q^- - p^+}$$

is positive in a neighborhood of the origin. We conclude that Lemma 62 is true. □

Lemma 63 *There exists $e \in H$ with $\|e\| > \eta$ (where η is given by Lemma 62) such that $J_\lambda(e) < 0$.*

Proof. Consider the function $\psi : \mathbb{Z}[0, T+1] \to \mathbb{R}$ such that there exists an integer k_0 satisfying $0 < k_0 < T+1$ for which $\psi(k_0) = 1$ and $\psi(k) = 0$ for any $k \in \mathbb{Z}[0, T+1] \setminus \{k_0\}$. Thus we deduce that $\psi \in H$. For each $t > 1$ we have

$$J_\lambda(t\psi) = \frac{t^{p(k_0)}}{p(k_0)} + \frac{t^{p(k_0-1)}}{p(k_0-1)} - \lambda \cdot \frac{t^{q(k_0)}}{q(k_0)} \leq \frac{2 \cdot t^{p^+}}{p^-} - \lambda \cdot \frac{t^{q^-}}{q^+}.$$

Since $q^- > p^+$ it is clear that $\lim_{t\to\infty} J_\lambda(t\psi) = -\infty$. Then for large enough $t > 1$ we can take $e = t\psi$ such that $\|e\| > \eta$ and $J_\lambda(e) < 0$.

The proof of Lemma 63 is complete. □

Proof of Theorem 46. By Lemmas 62 and 63 and the mountain-pass theorem of Ambrosetti and Rabinowitz, we deduce the existence of a sequence $\{u_n\} \subset H$ such that

$$J_\lambda(u_n) \to \bar{c} > 0 \quad \text{and} \quad J'_\lambda(u_n) \to 0 \quad \text{as } n \to \infty. \tag{6.10}$$

We prove that $\{u_n\}$ is bounded in H. Arguing by contradiction, we assume that passing eventually to a subsequence, still denoted by $\{u_n\}$, we have $\|u_n\| \to \infty$. Thus, we can assume that for large enough n we have $\|u_n\| > 1$.

Relation (6.10) and the above considerations imply that for large enough n we have

$$\begin{aligned} 1 + \bar{c} + \|u_n\| &\geq J_\lambda(u_n) - \frac{1}{q^-}\langle J'_\lambda(u_n), u_n\rangle \\ &\geq \left(\frac{1}{p^+} - \frac{1}{q^-}\right)\sum_{k=1}^{T+1} |\Delta u_n(k-1)|^{p(k-1)}. \end{aligned}$$

By Lemma 61 (a) and the above inequality, we deduce that there exist positive constants D_1 and D_2 such that

$$1 + \bar{c} + \|u_n\| \geq D_1 \cdot \|u_n\|^{p^-} - D_2,$$

for large enough n. Dividing by $\|u_n\|^{p^-}$ in the above inequality and passing to the limit as $n \to \infty$ we obtain a contradiction. It follows that $\{u_n\}$ is bounded in H. Since H is a finite dimensional Hilbert space, there exist a subsequence, still denoted by $\{u_n\}$, and $u_0 \in H$ such that $\{u_n\}$ converges to u_0 in H.

Then, we have by relation (6.10)

$$J_\lambda(u_0) = \overline{c} > 0 \quad \text{and} \quad J'_\lambda(u_0) = 0.$$

We conclude that u_0 is a nontrivial weak solution of problem (6.1). □

Proof of Theorem 47. The main idea is to show that J_λ has a nontrivial global minimum point in H. Lemma 61 (a) implies that J_λ is coercive on H. On the other hand, it is weakly lower semicontinuous on the finite dimensional Hilbert space H. Thus, there exists a global minimizer $u_\lambda \in H$ of J_λ, hence a weak solution of problem (6.1).

We show that u_λ is not trivial for large enough λ. Indeed, letting $t_0 > 1$ be a fixed real and defining the function $v_0 : \mathbb{Z}[0, T+1] \to \mathbb{R}$ such that there exists an integer k_0 with $0 < k_0 < T+1$ for which $v_0(k_0) = t_0$ and $v_0(k) = 0$ for any $k \in \mathbb{Z}[0, T+1] \setminus \{k_0\}$ we deduce that $v_0 \in H$ and

$$J_\lambda(v_0) = \frac{t_0^{p(k_0-1)}}{p(k_0-1)} + \frac{t_0^{p(k_0)}}{p(k_0)} - \frac{\lambda \cdot t_0^{q(k_0)}}{q(k_0)} \le L_1 - L_2 \cdot \lambda,$$

where L_1 and L_2 are positive constants. Thus there exists $\lambda^{**} > 0$ such that $J_\lambda(v_0) < 0$ for any $\lambda \in [\lambda^{**}, \infty)$. It follows that $J_\lambda(u_\lambda) < 0$ for any $\lambda \ge \lambda^{**}$ and thus u_λ is a nontrivial weak solution of problem (6.1) for large enough λ. The proof of Theorem 47 is complete. □

Proof of Theorem 48. We start with two auxiliary results.

Lemma 64 *There exists $\lambda^\star > 0$ such that for any $\lambda \in (0, \lambda^\star)$ there exist ρ and $a > 0$ such that $J_\lambda(u) \ge a > 0$ for any $u \in H$ with $\|u\| = \rho$.*

Proof. For all $u \in H$, Lemma 61 (c) implies that

$$c_2 \cdot \|u\| \ge |u|_2 \,.$$

Using now (6.5), we deduce that

$$c_2 \cdot T^{1/q^-} \cdot \|u\| \ge |u|_{q^-}, \quad \forall\, u \in H\,. \tag{6.11}$$

We fix $\rho \in (0,1)$ such that $\rho < \min\{1, 1/(c_2 \cdot T^{1/q^-})\}$. Thus, for any $u \in H$ with $\|u\| = \rho$ we have $|u|_{q^-} < 1$. It follows that $|u(k)| < 1$ for all $k \in \mathbb{Z}[0, T+1]$. Therefore

$$\sum_{k=1}^{T} |u(k)|^{q(k)} \le |u|_{q^-}^{q^-}, \quad \forall\, u \in H \text{ with } \|u\| = \rho\,. \tag{6.12}$$

By relations (6.11) and (6.12) we obtain

$$\sum_{k=1}^{T} |u(k)|^{q(k)} \le c_2^{q^-} \cdot T \cdot \|u\|^{q^-}, \quad \forall\, u \in H \text{ with } \|u\| = \rho\,.$$

By Lemma 61 (b) and the above relation we deduce that for any $u \in H$ with $\|u\| = \rho$ the following inequalities hold:

$$\begin{aligned} J_\lambda(u) &\geq \frac{C_3}{p^+} \cdot \|u\|^{p^+} - \frac{\lambda \cdot c_2^{q^-} \cdot T}{q^-} \cdot \|u\|^{q^-} \\ &= (C_4 \cdot \rho^{p^+ - q^-} - \lambda \cdot C_5) \cdot \rho^{q^-}, \end{aligned}$$

where C_4 and C_5 are positive constants. Using the above inequality and the fact that $q^- < p^- \leq p^+$, and setting

$$\lambda^\star = \frac{C_4 \cdot \rho^{p^+ - q^-}}{2 \cdot C_5}, \tag{6.13}$$

we get that for any $\lambda \in (0, \lambda^\star)$ and any $u \in H$ with $\|u\| = \rho$ there exists $a = \frac{C_4 \cdot \rho^{p^+}}{2}$ such that

$$J_\lambda(u) \geq a > 0.$$

The proof of Lemma 64 is complete. □

Lemma 65 *There exists $\phi \in H$ such that $\phi \geq 0$, $\phi \neq 0$, and $J_\lambda(t\phi) < 0$, for small enough $t > 0$.*

Proof. Since $q^- < p^-$, it follows that there exists an integer k_0 such that $0 < k_0 < T + 1$ and $q^- = q(k_0) < p^- \leq p(k_0)$. We define the function $\phi : \mathbb{Z}[0, T+1] \to \mathbb{R}$ such that $\phi(k_0) = 1$ and $\phi(k) = 0$ for any $k \in \mathbb{Z}[0, T+1] \setminus \{k_0\}$. We deduce that $\phi \in H$ and for any $t \in (0, 1)$ we have

$$J_\lambda(t\phi) = \frac{t^{p(k_0 - 1)}}{p(k_0 - 1)} + \frac{t^{p(k_0)}}{p(k_0)} - \lambda \cdot \frac{t^{q(k_0)}}{q(k_0)} \leq \frac{2 \cdot t^{p^-}}{p^-} - \frac{\lambda \cdot t^{q^-}}{q^+}.$$

The above inequality implies

$$J_\lambda(t\phi) < 0$$

for any $t < \delta^{1/(p^- - q^-)}$ where

$$0 < \delta < \frac{p^- \cdot \lambda}{2 \cdot q^+}.$$

The proof of Lemma 65 is complete. □

Now we return to the proof of Theorem 48. Let $\lambda^\star > 0$ be defined as in (6.13) and $\lambda \in (0, \lambda^\star)$. It follows by Lemma 64 that on the boundary of the ball centered at the origin and of radius ρ in H, denoted by $B_\rho(0)$, we have

$$\inf_{\partial B_\rho(0)} J_\lambda > 0. \tag{6.14}$$

On the other hand, by Lemma 65, there exists $\phi \in H$ such that $J_\lambda(t\phi) < 0$

for all $t > 0$ small enough. Moreover, relation (6.5) and Lemma 61 (b) imply that for any $u \in B_\rho(0)$ we have

$$J_\lambda(u) \geq \frac{C_3}{p^+}\|u\|^{p^+} - \frac{\lambda}{q^-}(c_{q^-}T \cdot \|u\|^{q^-} + c_{q^+}T \cdot \|u\|^{q^+}).$$

It follows that

$$-\infty < \underline{c} := \inf_{\overline{B_\rho(0)}} J_\lambda < 0.$$

Fix $0 < \epsilon < \inf_{\partial B_\rho(0)} J_\lambda - \inf_{B_\rho(0)} J_\lambda$. Applying the Ekeland variational principle to the functional $J_\lambda : \overline{B_\rho(0)} \to \mathbb{R}$, we find $u_\epsilon \in \overline{B_\rho(0)}$ such that

$$\begin{aligned} J_\lambda(u_\epsilon) &< \inf_{\overline{B_\rho(0)}} J_\lambda + \epsilon \\ J_\lambda(u_\epsilon) &< J_\lambda(u) + \epsilon \cdot \|u - u_\epsilon\|, \quad u \neq u_\epsilon. \end{aligned}$$

Since

$$J_\lambda(u_\epsilon) \leq \inf_{\overline{B_\rho(0)}} J_\lambda + \epsilon \leq \inf_{B_\rho(0)} J_\lambda + \epsilon < \inf_{\partial B_\rho(0)} J_\lambda\,,$$

we deduce that $u_\epsilon \in B_\rho(0)$. Now, we define $I_\lambda : \overline{B_\rho(0)} \to \mathbb{R}$ by $I_\lambda(u) = J_\lambda(u) + \epsilon \cdot \|u - u_\epsilon\|$. Then u_ϵ is a minimum point of I_λ and hence

$$\frac{I_\lambda(u_\epsilon + t \cdot v) - I_\lambda(u_\epsilon)}{t} \geq 0$$

for small $t > 0$ and any $v \in B_1(0)$. This relation yields

$$\frac{J_\lambda(u_\epsilon + t \cdot v) - J_\lambda(u_\epsilon)}{t} + \epsilon \cdot \|v\| \geq 0.$$

Letting $t \to 0$ it follows that $\langle J'_\lambda(u_\epsilon), v\rangle + \epsilon \cdot \|v\| > 0$ and we infer that $\|J'_\lambda(u_\epsilon)\| \leq \epsilon$.

We deduce that there exists a sequence $\{w_n\} \subset B_\rho(0)$ such that

$$J_\lambda(w_n) \to \underline{c} \quad \text{and} \quad J'_\lambda(w_n) \to 0. \tag{6.15}$$

Since the sequence $\{w_n\}$ is bounded in H, there exists $w \in H$ such that, up to a subsequence, $\{w_n\}$ converges to w in H. Therefore, by (6.15),

$$J_\lambda(w) = \underline{c} < 0 \quad \text{and} \quad J'_\lambda(w) = 0.$$

We conclude that w is a nontrivial weak solution of problem (6.1). The proof of Theorem 48 is complete. □

6.2 Homoclinic solutions of difference equations with variable exponents

Let $T > 0$ be a natural number and let $p, q : \mathbb{Z} \to [2, \infty)$, $V(\cdot) : \mathbb{Z} \to \mathbb{R}$ be three T-periodic functions and $f(k,t) : \mathbb{Z} \times \mathbb{R} \to \mathbb{R}$ be a continuous function in $t \in \mathbb{R}$ and T-periodic in k.

This section is devoted to the study of the difference nonhomogeneous equations of type

$$\begin{cases} \Delta^2_{p(k-1)} u(k-1) - V(k)|u(k)|^{q(k)-2}u(k) + f(k, u(k)) = 0 & \text{for} \quad k \in \mathbb{Z}, \\ u(k) \to 0 & \text{as} \quad |k| \to \infty\,, \end{cases} \tag{6.16}$$

where $\Delta^2_{p(\cdot)}$ stands for the $p(\cdot)$-Laplace difference operator, that is,

$$\Delta^2_{p(k-1)} u(k-1) = |\Delta u(k)|^{p(k)-2} \Delta u(k) - |\Delta u(k-1)|^{p(k-1)-2} \Delta u(k-1)\,, \tag{6.17}$$

for each $k \in \mathbb{Z}$. We have denoted by Δ the difference operator, which is defined by

$$\Delta u(k) = u(k+1) - u(k)\,,$$

for each $k \in \mathbb{Z}$.

Our goal is to establish the existence of nontrivial *homoclinic solutions* for problem (6.16). In order to explain the notion of *homoclinic solution* we go back to the definition of *homoclinic orbit*, which was introduced by Poincaré [81] in the context of Hamiltonian systems. More exactly, Poincaré called a trajectory $x(t)$ a *homoclinic orbit* (or *doubly asymptotic* trajectory) if it is asymptotic to a constant as $|t| \to \infty$. Since we are seeking solutions $u(k)$ for problem (6.16) satisfying $\lim_{|k|\to\infty} u(k) = 0$, in accordance with the above discussion, we are interested in finding nontrivial homoclinic solutions for problem (6.16).

Set

$$p^+ := \sup_{k\in\mathbb{Z}} p(k) \qquad p^- := \inf_{k\in\mathbb{Z}} p(k)$$

$$q^+ := \sup_{k\in\mathbb{Z}} q(k) \qquad q^- := \inf_{k\in\mathbb{Z}} q(k)\,,$$

and we assume that

$$1 < q^- \le q^+ < p^- \le p^+\,. \tag{6.18}$$

We assume that the T-periodic function V satisfies the following hypotheses:

(V1) $0 < V_0 := \min\{V(0), ..., V(T-1)\}$;

(V2) $V_0 < q^+$.

The continuous function $f = f(k,t) : \mathbb{Z} \times \mathbb{R} \to \mathbb{R}$ is assumed to be T-periodic in k and satisfying

(F1) there exist $\alpha > p^+$ and $r > 0$ such that

$$\alpha F(k,t) := \alpha \int_0^t f(k,s)\,ds \le t f(k,t), \quad \forall\, k \in \mathbb{Z},\ t \neq 0\,,$$

and

$$F(k,t) > 0, \quad \forall\, k \in \mathbb{Z},\ t \ge r\,;$$

(F2) $f(k,t) = o(|t|^{q^+-1})$ as $|t| \to 0$.

If $p(\cdot) : \mathbb{Z} \to (1,\infty)$ then we consider the function space

$$\ell^{p(\cdot)} := \left\{ u : \mathbb{Z} \to \mathbb{R};\ \rho_{p(\cdot)}(u) := \sum_{k\in\mathbb{Z}} |u(k)|^{p(k)} < \infty \right\}.$$

On $\ell^{p(\cdot)}$ we introduce the discrete Luxemburg norm

$$|u|_{p(\cdot)} := \inf \left\{ \mu > 0;\ \sum_{k\in\mathbb{Z}} \left| \frac{u(k)}{\mu} \right|^{p(k)} \le 1 \right\}.$$

Then the following relations hold:

$$|u|_{p(\cdot)} < 1 \Rightarrow |u|_{p(\cdot)}^{p^+} \le \rho_{p(\cdot)}(u) \le |u|_{p(\cdot)}^{p^-}, \tag{6.19}$$

$$|u|_{p(\cdot)} > 1 \Rightarrow |u|_{p(\cdot)}^{p^-} \le \rho_{p(\cdot)}(u) \le |u|_{p(\cdot)}^{p^+}, \tag{6.20}$$

$$|u|_{p(\cdot)} \to 0 \Leftrightarrow \rho_{p(\cdot)}(u) \to 0\,. \tag{6.21}$$

We also consider the space

$$\ell^\infty = \left\{ u : \mathbb{Z} \to \mathbb{R};\ |u|_\infty := \sup_{k\in\mathbb{Z}} |u(k)| < \infty \right\}.$$

We start with the following embedding property.

Proposition 14 *Assume that condition* (6.18) *is fulfilled. Then* $\ell^{q(\cdot)} \subset \ell^{p(\cdot)}$.

Proof. If $\sum_{k\in\mathbb{Z}} |u(k)|^{p(k)} < \infty$, then there exists $S > 0$ such that

$$|u(k)|^{q(k)} \le 1, \quad \forall\, |k| > S\,.$$

It follows that

$$|u(k)| \le 1, \quad \forall\, |k| > S\,.$$

By relation (6.18) we infer that $q(k) < p(k)$ for all $k \in \mathbb{Z}$. That fact and the above inequality assure that

$$|u(k)|^{p(k)} \leq |u(k)|^{q(k)}, \quad \forall\, |k| > S\,,$$

which concludes the proof. □

By Proposition 14, relation (6.18) and the hypotheses on functions V and f we infer that the natural space where we should seek homoclinic solutions for (6.16) is $\ell^{q(\cdot)}$. Thus, we say that $u \in \ell^{q(\cdot)}$ is a *homoclinic solution* for (6.16) if

$$\sum_{k\in\mathbb{Z}} |\Delta u(k-1)|^{p(k-1)-2}\Delta u(k-1)\Delta v(k-1) \quad + \sum_{k\in\mathbb{Z}} V(k)|u(k)|^{q(k)-2}u(k)v(k) - \sum_{k\in\mathbb{Z}} f(k,u(k))v(k) = 0\,,$$

for all $v \in \ell^{q(\cdot)}$ and

$$\lim_{|k|\to\infty} u(k) = 0\,.$$

The main result of this section is given by the following theorem.

Theorem 49 *Assume that hypotheses* (6.18)*, (V1)-(V2) and (F1)-(F2) are fulfilled. Then problem* (6.16) *has at least one nontrivial homoclinic solution. Moreover, given a nontrivial homoclinic solution u of problem* (6.16)*, there exist integers S_1 and S_2 with $S_1 \leq S_2$ such that for all $k > S_2$ and all $k < S_1$ the sequence $u(k)$ is strictly monotone.*

We first introduce the following notations:

$$\phi_{p(t)}(t) := |t|^{p(t)-2}t \qquad \Phi_{p(t)}(t) := \frac{|t|^{p(t)}}{p(t)}\,.$$

Note that

$$\Delta^2_{p(k-1)}u(k-1) = \Delta(\phi_{p(k-1)}(\Delta u(k-1)))\,.$$

Next, we introduce the functional $A : \ell^{q(\cdot)} \to \mathbb{R}$ defined by

$$A(u) := \sum_{k\in\mathbb{Z}} \Phi_{p(k-1)}(\Delta u(k-1)) + \sum_{k\in\mathbb{Z}} V(k)\Phi_{q(k)}(u(k))\,.$$

The energy functional $J : \ell^{q(\cdot)} \to \mathbb{R}$ associated to problem (6.16) is defined by

$$J(u) := A(u) - \sum_{k\in\mathbb{Z}} F(k,u(k))\,.$$

Then $J \in C^1(\ell^{q(\cdot)},\mathbb{R})$ and

$$\begin{aligned}\langle J'(u),v\rangle \;&= \sum_{k\in\mathbb{Z}} [|\Delta u(k-1)|^{p(k-1)-2}\Delta u(k-1)\Delta v(k-1) \\ &\quad + V(k)|u(k)|^{q(k)-2}u(k)v(k) - f(k,u(k))v(k)]\,,\end{aligned}$$

for all $u, v \in \ell^{q(\cdot)}$. Thus, critical points of J correspond to the weak solutions of problem (6.16).

In order to facilitate further computations we point out that on $\ell^{q(\cdot)}$ we can introduce an equivalent norm to $|\cdot|_{q(\cdot)}$, namely

$$\|u\|_{q(\cdot)} := \inf\left\{\mu > 0;\ \sum_{k\in\mathbb{Z}} \frac{V(k)}{q(k)}\left|\frac{u(k)}{\mu}\right|^{q(k)} \le 1\right\}.$$

It follows that

$$\|u\|_{q(\cdot)} < 1 \Rightarrow \|u\|_{q(\cdot)}^{q^+} \le \sum_{k\in\mathbb{Z}} \frac{V(k)}{q(k)}|u(k)|^{q(k)} \le \|u\|_{q(\cdot)}^{q^-}, \tag{6.22}$$

$$\|u\|_{q(\cdot)} > 1 \Rightarrow \|u\|_{q(\cdot)}^{q^-} \le \sum_{k\in\mathbb{Z}} \frac{V(k)}{q(k)}|u(k)|^{q(k)} \le \|u\|_{q(\cdot)}^{q^+}, \tag{6.23}$$

$$\|u\|_{q(\cdot)} \to 0 \Leftrightarrow \sum_{k\in\mathbb{Z}} \frac{V(k)}{q(k)}|u(k)|^{q(k)} \to 0. \tag{6.24}$$

The next property shows that J has the mountain-pass geometry.

Lemma 66 *Assume that hypotheses of Theorem 49 are fulfilled. Then there exist $\varrho > 0$ and $\nu > 0$ and $e \in \ell^{q(\cdot)}$ with $\|e\|_{q(\cdot)} > \varrho$ such that*

(i) $J(u) \ge \nu$ for all $u \in \ell^{q(\cdot)}$ with $\|u\|_{q(\cdot)} > \varrho$;

(ii) $J(e) < 0$.

Proof. (i) Condition (F2) implies that there exists $\delta \in (0,1)$ such that

$$F(k,t) \le \frac{V_0}{2q^+}|t|^{q^+} \le \frac{V_0}{2q^+}|t|^{q(k)}, \tag{6.25}$$

for all $|t| \le \delta$ and all $k \in \mathbb{Z}$. Define

$$\varrho := \left(\frac{V_0}{q^+}\right)^{1/q^-} \delta^{q^+/q^-}.$$

By condition (V2) we deduce that $\varrho \in (0,1)$. Then for any $\|u\|_{q(\cdot)} = \varrho$ relation (6.22) yields

$$\varrho^{q^-} = \frac{V_0}{q^+}\delta^{q^+} = \|u\|_{q(\cdot)}^{q^-} \ge \sum_{i\in\mathbb{Z}} \frac{V(i)}{q(i)}|u(i)|^{q(i)} \ge \frac{V_0}{q^+}|u(k)|^{q(k)},$$

for all $k \in \mathbb{Z}$. It follows that

$$1 > \delta^{q^+} \ge |u(k)|^{q(k)},$$

for all $k \in \mathbb{Z}$. Consequently, $|u(k)| < 1$ for every $k \in \mathbb{Z}$ and thus

$$|u(k)|^{q(k)} \geq |u(k)|^{q^+},$$

for all $k \in \mathbb{Z}$. The above inequalities show that actually

$$\delta \geq |u(k)|,$$

for all $k \in \mathbb{Z}$. Next, by (6.25) we obtain

$$\sum_{k\in\mathbb{Z}} F(k,u(k)) \leq \frac{V_0}{2q^+}\sum_{k\in\mathbb{Z}} |u(k)|^{q(k)} \leq \frac{1}{2}\sum_{k\in\mathbb{Z}} \frac{V(k)}{q(k)}|u(k)|^{q(k)},$$

provided that $\|u\|_{q(\cdot)} = \varrho$. Define

$$\nu := \frac{\varrho^{q^+}}{2}.$$

Then by (6.22), for each u with $\|u\|_{q(\cdot)} = \varrho$ we have

$$\begin{aligned} J(u) &= A(u) - \sum_{k\in\mathbb{Z}} F(k,u(k)) \\ &\geq \sum_{k\in\mathbb{Z}} \frac{V(k)}{q(k)}|u(k)|^{q(k)} - \frac{1}{2}\sum_{k\in\mathbb{Z}} \frac{V(k)}{q(k)}|u(k)|^{q(k)} \\ &\geq \frac{1}{2}\sum_{k\in\mathbb{Z}} \frac{V(k)}{q(k)}|u(k)|^{q(k)} \\ &\geq \frac{1}{2}\|u\|_{q(\cdot)}^{q^+} = \nu. \end{aligned}$$

(ii) We infer by condition (F1) that there exist constants $c_1 > 0$ and $c_2 > 0$ such that

$$F(k,t) \geq c_1|t|^\alpha - c_2,$$

for all $k \in \mathbb{Z}$ and all $t \in \mathbb{R}$. Define $v \in \ell^{q(\cdot)}$ by $v(0) = a > 0$ and $v(k) = 0$ if $k \neq 0$. We find

$$\begin{aligned} J(\eta v) &= A(\eta v) - \sum_{k\in\mathbb{Z}} F(k,\eta v(k)) \\ &= \frac{2}{p(0)}\eta^{p(0)}a^{p(0)} + V(0)\frac{\eta^{q(0)}a^{q(0)}}{q(0)} - c_1\eta^\alpha a^\alpha + c_2, \end{aligned}$$

for each $\eta > 0$. Since by relation (6.18) we have $\alpha > p^+ \geq p(0) \geq p^- > q^+ \geq q(0)$, the above relation shows that for any sufficiently large $\eta > 0$ we have

$$J(\eta v) < 0.$$

The proof of Lemma 66 is complete. □

We recall that given $c \in \mathbb{R}$, we say that a sequence $(u(k)) \subset \ell^{q(\cdot)}$ is said to satisfy the $(PS)_c$ *condition* if

$$J(u(k)) \to c \text{ and } J'(u(k)) \to 0.$$

Lemma 67 *Assume that hypotheses of Theorem 49 are fulfilled. Then there exist $c > 0$ and a bounded $(PS)_c$ sequence for J in $\ell^{q(\cdot)}$.*

Proof. Lemma 66 and the mountain-pass theorem imply that there exists a sequence $\{u_n\} \subset \ell^{q(\cdot)}$ such that

$$J(u_n) \to c, \qquad J'(u_n) \to 0\,, \tag{6.26}$$

where

$$c := \inf_{\gamma\in\Gamma} \max_{t\in[0,1]} J(\gamma(t))$$

and

$$\Gamma := \{\gamma \in C([0,1], \ell^{q(\cdot)});\ \ \gamma(0) = 0,\ \gamma(1) = e\}\,,$$

with e is given by Lemma 66 (ii).

We verify that $\{u_n\}$ is bounded in $\ell^{q(\cdot)}$. Indeed, assuming that $\|u_n\|_{q(\cdot)} > 1$ for each n we deduce by condition (F1) and relation (6.23) that

$$\begin{aligned} \alpha J(u_n) - \langle J'(u_n), u_n\rangle &= \sum_{k\in\mathbb{Z}} \left(\frac{\alpha}{p(k)} - 1\right) |\Delta u_n(k-1)|^{p(k-1)} \\ &+ \sum_{k\in\mathbb{Z}} \left(\frac{\alpha}{q(k)} - 1\right) V(k)|u_n(k)|^{q(k)} \\ &- \sum_{k\in\mathbb{Z}} [\alpha F(k, u_n(k)) - f(k, u_n(k))u_n(k)] \\ &\geq (\alpha - q^+)\|u_n\|_{q(\cdot)}^{q^-}\,, \end{aligned}$$

for all n. The above estimates and condition (6.26) show that $\{u_n\}$ is bounded in $\ell^{q(\cdot)}$. The proof of Lemma 67 is complete. □

We now return to the proof of Theorem 49. Assume that $\{u_n\}$ is the sequence given by Lemma 67. Then for each $n \in \mathbb{N}$ the sequence $\{|u_n(k)|;\ k \in \mathbb{Z}\} \subset \ell^{q(\cdot)}$ is bounded and $|u_n(k)| \to 0$ as $|k| \to \infty$.

Assume that $\{|u_n(k)|\}_{k\in\mathbb{Z}}$ achieves its maximum in $k_n \in \mathbb{Z}$. There exists $j_n \in \mathbb{Z}$ such that

$$j_n T \leq k_n < (j_n + 1)T\,.$$

Set

$$w_n(k) := u_n(k - j_n T)\,.$$

Then $\{|w_n(k)|\}_{k\in\mathbb{Z}}$ attains its maximum in

$$i_n := k_n - j_n T \in [0, T]\,.$$

The T-periodicity of $p(\cdot)$, $q(\cdot)$ and $V(\cdot)$ implies

$$\sum_{k\in\mathbb{Z}} \frac{V(k)}{q(k)}|u_n(k)|^{q(k)} = \sum_{k\in\mathbb{Z}} \frac{V(k)}{q(k)}|w_n(k)|^{q(k)}\,,$$

and

$$J(u_n) = J(w_n)\,.$$

Since $\{u_n\}_{n\in\mathbb{N}}$ is bounded in $\ell^{q(\cdot)}$ the above estimates and relations (6.22) and (6.23) yield that $\{w_n\}_{n\in\mathbb{N}}$ is bounded in $\ell^{q(\cdot)}$, too. Then there exists $w \in \ell^{q(\cdot)}$ such that w_n weakly converges to w in $\ell^{q(\cdot)}$ as $n \to \infty$.

We claim that $w_n(k) \to w(k)$ as $n \to \infty$ for each $k \in \mathbb{Z}$. Indeed, defining the test function $v_m \in \ell^{q(\cdot)}$ by

$$v_m(j) := \begin{cases} 1 & \text{if} \quad j = m \\ 0 & \text{if} \quad j \neq m\,, \end{cases}$$

and taking into account the weak convergence of w_n to w in $\ell^{q(\cdot)}$ we find

$$\lim_{n\to\infty} w_n(k) = \lim_{n\to\infty} \langle w_n, v_k\rangle = \langle w, v_k\rangle = w(k)\,,$$

for all $k \in \mathbb{Z}$. This proves our claim.

Next, we point out that for each $v \in \ell^{q(\cdot)}$ we have

$$\begin{aligned} |\langle J'(w_n), v\rangle| &= |\langle J'(u_n), v(\cdot + j_n T)\rangle| \\ &\leq \|J'(u_n)\|_\star \|v\|_{q(\cdot)}\,. \end{aligned}$$

Thus, using relation (6.26) from Lemma 67 we obtain $J'(w_n) \to 0$ as $n \to \infty$. It follows that for all $v \in \ell^{q(\cdot)}$, we have as $n \to \infty$

$$\sum_{k\in\mathbb{Z}} [\phi_{p(k-1)}(\Delta w_n(k-1))\Delta v(k-1) + V(k)\phi_{q(k)}(w_n(k))v(k) \\ - f(k, w_n(k))v(k)] \to 0\,. \tag{6.27}$$

Consider that $v \in \ell^{q(\cdot)}$ has compact support, hence there exist $a, b \in \mathbb{Z}$, $a < b$ such that $v\,(k) = 0$ if $k \in \mathbb{Z}\backslash[a, b]$ and $v\,(k) \neq 0$ if $k \in \{a+1, b-1\}$. The set of compact support functions, denoted by $\ell_0^{q(\cdot)}$ is dense in $\ell^{q(\cdot)}$. This fact can be easily explained since for each $v \in \ell^{q(\cdot)}$ we can define $v_n \in \ell_0^{q(\cdot)}$ by $v_n(j) = 0$ if $|j| \geq n+1$ and $v_n(j) = v(j)$ if $|j| \neq n$ and we have $\sum_{j\in\mathbb{Z}} \frac{V(j)}{q(j)} |v(j) - v_n(j)| \to 0$ as $n \to \infty$, or, by relation (6.24), $\|v - v_n\|_{q(\cdot)} \to 0$ as $n \to \infty$.

Now, for each $v \in \ell_0^{q(\cdot)}$ in (6.27) taking into account the finite sums and the continuity of $f(k, \cdot)$ we obtain by passing to the limit as $n \to \infty$ that

$$\sum_{k\in\mathbb{Z}} [\phi_{p(k-1)}(\Delta w(k-1))\Delta v(k-1) + V(k)\phi_{q(k)}(w(k))v(k) - f(k, w(k))v(k)] \to 0\,.$$

This shows that w is a critical point of J and hence a solution of problem (6.16).

We prove that $w \neq 0$. Assume to the contrary, that $w = 0$. Then we have

$$|u_n|_\infty = |w_n|_\infty = \max\{|w_n(k)|;\ k \in \mathbb{Z}\} \to 0\,,$$

as $n \to \infty$. On the other hand, condition (F2) implies that for a given $\epsilon > 0$ there exists $\delta \in (0,1)$ such that

$$\left\{ \begin{array}{l} |F(k,t)| \leq \epsilon |t|^{q^+} \\ |f(x,t)t| \leq \epsilon |t|^{q^+} , \end{array} \right. \tag{6.28}$$

for all $k \in \{0,1,...,T-1\}$ and all $|t| < \delta$. The above inequalities show that for every $k \in \{0,1,...,T-1\}$ there exists M_k such that for $n > M_k$ we have

$$|w_n(k)| < \delta .$$

Since $i_n \in \{0,1,...,T-1\}$ it follows that for $n > M := \max\{M_n;\ k \in \{0,1,...,T-1\}\}$ and every $k \in \mathbb{Z}$ we have

$$|w_n(k)| \leq |w_n(i_n)| < \delta < 1 .$$

This fact and relation (6.28) imply that

$$|F(k, w_n(k))| \leq \epsilon |w_n(k)|^{q^+} \leq \epsilon |w_n(k)|^{q(k)} ,$$

and

$$|f(k, w_n(k))w_n(k)| \leq \epsilon |w_n(k)|^{q^+} \leq \epsilon |w_n(k)|^{q(k)} .$$

We infer that for each $n > M$ and every $k \in \mathbb{Z}$ the following estimates hold:

$$\begin{aligned}
0 \quad < \quad & q^- J(w_n) = q^- \sum_{k\in\mathbb{Z}} \frac{1}{p(k)} \left|\Delta w_n\,(k-1)\right|^{p(k-1)} + q^- \sum_{k\in\mathbb{Z}} \frac{V\,(k)}{q(k)} \left|w_n\,(k)\right|^{q(k)} \\
& -q^- \sum_{k\in\mathbb{Z}} F\,(k, w_n\,(k)) \\
\leq \quad & \sum_{k\in\mathbb{Z}} \left|\Delta w_n\,(k-1)\right|^{p(k-1)} + \sum_{k\in\mathbb{Z}} V\,(k) \left|w_n\,(k)\right|^{q(k)} - \sum_{k\in\mathbb{Z}} f\,(k, w_n\,(k))\, w_n\,(k) \\
& - \sum_{k\in\mathbb{Z}} \left(q^- F\,(k, w_n\,(k)) - f\,(k, w_n\,(k))\, w_n\,(k)\right) \\
\leq \quad & \left\langle J'\,(w_n)\,, w_n \right\rangle + q^- \sum_{k\in\mathbb{Z}} F(k, w_n\,(k)) + \sum_{k\in\mathbb{Z}} \left|f\,(k, w_n\,(k))\, w_n\,(k)\right| \\
\leq \quad & \left\langle J'\,(w_n)\,, w_n \right\rangle + q^- \epsilon \sum_{k\in\mathbb{Z}} |w_n(k)|^{q(k)} + \epsilon \sum_{k\in\mathbb{Z}} |w_n(k)|^{q(k)} \\
\leq \quad & \left\langle J'\,(w_n)\,, w_n \right\rangle + \left(q^- \epsilon \frac{q^+}{V_0} + \epsilon \frac{q^+}{V_0}\right) \sum_{k\in\mathbb{Z}} \frac{V(k)}{q(k)} |w_n(k)|^{q(k)} \\
\leq \quad & \|J'\,(w_n)\,\|_* \|w_n\|_{q(\cdot)} + \epsilon \frac{q^+(q^- + 1)}{V_0} \left[\|w_n\|^{q^+}_{q(\cdot)} + \|w_n\|^{q^-}_{q(\cdot)}\right] .
\end{aligned}$$

Taking into account that $\|u_n\|_{q(\cdot)}$ is bounded, $J'(u_n) \to 0$ as $n \to \infty$ and $\epsilon > 0$ is arbitrary we find by the above estimates a contradiction with $J(w_n) \to c > 0$ as $n \to \infty$. Thus, w is a nontrivial solution of problem (6.16).

Next, let u be a nonzero homoclinic solution of problem (6.16). Assume that it attains positive local maximums and negative local minimums at infinitely many points k_n. In particular, we can assume that $\{|k_n|\} \to \infty$. Consequently,

$$\Delta^2_{p(k_n-1)} u\,(k_n - 1)\; u(k_n) \le 0\,,$$

and

$$u(k_n) \to 0\,,$$

as $n \to \infty$. Using this fact and multiplying in (6.16) by $u(k_n)/|u(k_n)|^{q(k_n)}$, we have

$$\begin{aligned} \frac{f\,(k_n, u\,(k_n))\; u(k_n)}{|u(k_n)|^{q(k_n)}} &\ge \frac{\Delta^2_{p(k_n-1)} u\,(k_n - 1)\; u(k_n)}{|u(k_n)|^{q(k_n)}} + \frac{f\,(k_n, u\,(k_n))\; u(k_n)}{|u(k_n)|^{q(k_n)}} \\ &= V\,(k_n) \ge V_0 > 0. \end{aligned} \tag{6.29}$$

Using (6.29) and condition (F2), we deduce

$$0 = \lim_{n\to\infty} \frac{f\,(k_n, u\,(k_n))\; u(k_n)}{|u(k_n)|^{q(k_n)}} \ge V_0 > 0\,,$$

which is a contradiction. Consequently, u does not attain positive local maxima and negative local minima at infinitely many points.

Assume now that u vanishes at infinitely many points l_n. By condition (F2) we find that $\Delta^2_{p(l_n-1)} u\,(l_n - 1) = 0$ and, consequently, $u(l_n - 1)\, u(l_n + 1) < 0$. Therefore it has an unbounded sequence of positive local maxima and negative local minima, in contradiction with the previous assertion.

In conclusion, for large enough $|k|$, the function u has constant sign and it is strictly monotone. The proof of Theorem 49 is complete. □

We notice that the "homogeneous" problem

$$\begin{cases} \Delta^2_{p(k-1)} u(k-1) - V(k)|u(k)|^{q(k)-2}u(k) = 0 & \text{for} \quad k \in \mathbb{Z}, \\ u(k) \to 0 & \text{as} \quad |k| \to \infty\,, \end{cases} \tag{6.30}$$

has only the trivial solution. Indeed, if u is positive or negative, let k_0 be the point of his positive maximum or negative minimum. Then

$$\Delta^2_{p(k_0-1)} u(k_0 - 1)u(k_0) \le 0$$

and

$$0 = \Delta^2_{p(k_0-1)} u(k_0 - 1)u(k_0) - V(k_0)|u(k_0)|^{q(k_0)} < 0,$$

which is a contradiction. The same conclusion can be made if u is sign-changing.

Note that under assumptions of Theorem 49, one can prove with the same approach, that for all $\lambda > 0$ there exists a nontrivial solution of the eigenvalue problem

$$\begin{cases} \Delta^2_{p(k-1)} u(k-1) - V(k)|u(k)|^{q(k)-2}u(k) + \lambda f(k, u(k)) = 0 & \text{for} \quad k \in \mathbb{Z}, \\ u(k) \to 0 & \text{as} \quad |k| \to \infty\,. \end{cases} \tag{6.31}$$

The above remark shows that $\lambda = 0$ is a bifurcation point of problem (6.31).

Let us now assume that in addition to conditions (F1) and (F2), the following condition holds:

(F3) $f(k,t) \geq 0$ for any $t < 0$ and all $k \in \mathbb{Z}$.

Then the homoclinic solution of problem (6.16) is positive.

Indeed, let u be a homoclinic solution of the problem and assume that (F3) holds. Suppose that there exists k_0 such that $u(k_0) < 0$ and let k_1 be such that $u(k_1) = \min\{u(k),\, k \in \mathbb{Z}\} < 0$. It follows that $\Delta^2_{p(k_1-1)} u(k_1 - 1) \geq 0$, which yields

$$f(k_1, u(k_1)) = -\Delta^2_{p(k_1-1)} u(k_1 - 1) + V(k_1) |u(k_1)|^{q(k_1)-2} u(k_1) < 0,$$

in contradiction with (F3). Then $u(k) \geq 0$ for every $k \in \mathbb{Z}$.

Suppose that $u(k_2) = 0$ for some $k_2 \in \mathbb{Z}$. Note, that if $u(k_2 + 1) = 0$ or $u(k_2 - 1) = 0$, the solution is identically zero by a recursion and $f(k, 0) = 0$. Hence $\Delta^2_{p(k_2-1)} u(k_2 - 1) > 0$ and we arrive to a contradiction. Hence, the solution u is positive.

We now summarize the above remarks.

Theorem 50 *Suppose that the functions $V : \mathbb{Z} \to \mathbb{R}$ and $f : \mathbb{Z} \to \mathbb{R}$ satisfy assumptions of Theorem 49 and (F3). Then problem* (6.16) *has a positive homoclinic solution.*

In the case $q(k) = 2$, we can estimate the maximum of the positive solution u, provided that the following additional assumption holds:

(F4) Let $f(k,t)$ have the form $f(k,t) = tg(k,t)$, where $g(k,t)$ is T-periodic in k, $g(k,0) = 0$ and for each k, $g(k,t)$ is increasing in t for $t > 0$.

Let $g^{-1}(k,t)$ be the inverse function of $g(k,t)$ for $t > 0$. Then $g^{-1}(k,t)$ is increasing in t for $t > 0$. Let u be a positive homoclinic solution of the problem (6.16) and $u(k_0) > 0$ be its maximum. Note that, in view of the periodicity of coefficients, if $u(.)$ is a solution of problem (6.16), then $u(. + jT), j \in \mathbb{Z}$ is also a solution of (6.16). Hence, we may assume that $k_0 \in [0, T-1]$. We have $\Delta^2_{p(k_0-1)} u(k_0 - 1) \leq 0$ and

$$u(k_0) g(k_0, u(k_0)) - V(k_0) u(k_0) \geq 0,$$

and by properties of g and V

$$u(k_0) \geq g^{-1}(k_0, V_0).$$

Therefore

$$\max\{u(k) : k \in [0, T-1]\} \geq \min\{g^{-1}(k, V_0) : k \in [0, T-1]\}. \tag{6.32}$$

Thus, we have the following result.

Corollary 4 *Suppose that the functions $V : \mathbb{Z} \to \mathbb{R}$ and $f : \mathbb{Z} \to \mathbb{R}$ satisfy assumptions of Theorem 50 and (F4). Then the positive homoclinic solution of the equation*

$$\Delta^2_{p(k-1)} u\,(k-1) - V(k)u(k) + u(k)g(k, u(k)) = 0$$

satisfies relation (6.32).

We illustrate with an example the abstract result established in Theorem 50. Consider the equations

$$\Delta^2_{p(k-1)} u(k-1) - V(k)u(k) + au^2(k) + bu^3(k) = 0, \quad k \in \mathbb{Z}, \tag{6.33}$$

and

$$\Delta^2_{p(k-1)} u(k-1) - V(k)u(k) + au^2(k) + bu^3_+(k) = 0, \quad k \in \mathbb{Z}, \tag{6.34}$$

where $u_+(k) = \max\{u(k), 0\}$, $2 < p(k) < 3$, $p(k)$ and $V(k)$ are T-periodic, $V_0 < 2$, $a > 0$ and $b > 0$. All hypotheses of Theorem 50 are satisfied for $f(k,t) = at^2 + bt^5_+$ and there exists a positive homoclinic solution u of equation (6.34), which is a homoclinic solution of problem (6.33). Moreover, we can estimate $\max\{u(k) : k \in \mathbb{Z}\}$. Let $u(k_0) > 0$ be the maximum of u. Then $\Delta^2_{p(k_0-1)} u(k_0 - 1) \le 0$ and by equation (6.33) we have

$$-V(k_0)u(k_0) + au^2(k_0) + bu^3(k_0) \ge 0,$$

which implies

$$au(k_0) + bu^2(k_0) - V_0 \ge 0.$$

Hence

$$\max\{u(k) : k \in \mathbb{Z}\} \ge \frac{\sqrt{a^2 + 4bV_0} - a}{2b}.$$

6.3 Low-energy solutions for discrete anisotropic equations

For every $a, b \in \mathbb{Z}$, such that $a < b$, set $\mathbb{Z}[a,b] := \{a, a+1, ..., b\}$. This section is concerned with sufficient conditions for the existence of solutions to the following anisotropic difference equation

$$\begin{cases} -\Delta(|\Delta u(k-1)|^{p(k-1)-2}\Delta u(k-1)) = \lambda f_k(u(k)), \ k \in \mathbb{Z}[1,T] \\ u(0) = u(T+1) = 0, \end{cases} \tag{A^f_λ}$$

where λ is a positive parameter, $f_k : \mathbb{R} \to \mathbb{R}$ is a continuous function for every $k \in \mathbb{Z}[1,T]$ (with $T \ge 2$), and $\Delta u(k-1) := u(k) - u(k-1)$ is the forward difference operator. Further, we assume that the map $p : \mathbb{Z}[0,T] \to \mathbb{R}$ satisfies $p^- := \min_{\mathbb{Z}[0,T]} p(k) > 1$ as well as $p^+ := \max_{\mathbb{Z}[0,T]} p(k) > 1$.

The main idea here is to transfer the problem of existence of solutions for problem (A^f_λ) to the problem of existence of critical points for a suitable associated energy functional, namely J_λ. More precisely, the main purpose of this section is to investigate the existence of infinitely many solutions of problem (A^f_λ) by using a critical point theorem obtained due to Ricceri.

In our approach we do not require any symmetry hypothesis. A special case of the main result in this section reads as follows.

Theorem 51 *Let $g : \mathbb{R} \to \mathbb{R}$ be a nonnegative and continuous function. Assume that*

$$\liminf_{t\to 0^+} \frac{\int_0^t g(s)ds}{t^{p^+}} = 0 \quad \text{and} \quad \limsup_{t\to 0^+} \frac{\int_0^t g(s)ds}{t^{p^-}} = +\infty.$$

Then for each $\lambda > 0$, the problem

$$\begin{cases} -\Delta(|\Delta u(k-1)|^{p(k-1)-2}\Delta u(k-1)) = \lambda g(u(k)), \ k \in \mathbb{Z}[1,T] \\ u(0) = u(T+1) = 0, \end{cases} \qquad (A^g_\lambda)$$

admits a sequence of nonzero solutions which converges to zero.

Let $(X, \|\cdot\|)$ be a finite dimensional Banach space and let $J_\lambda : X \to \mathbb{R}$ be a function satisfying the following structure hypothesis:

(Λ) $J_\lambda(u) := \Phi(u) - \lambda\Psi(u)$ *for all $u \in X$, where $\Phi, \Psi : X \to \mathbb{R}$ are functions of class C^1 on X with Φ coercive, that is $\lim_{\|u\|\to\infty} \Phi(u) = +\infty$, and λ is a real positive parameter.*

Moreover, provided that $r > \inf_X \Phi$, put

$$\varphi(r) := \inf_{u\in\Phi^{-1}((-\infty,r))} \frac{\left(\sup_{v\in\Phi^{-1}((-\infty,r))} \Psi(v)\right) - \Psi(u)}{r - \Phi(u)},$$

and

$$\delta := \liminf_{r\to(\inf_X \Phi)^+} \varphi(r).$$

We observe that $\delta \geq 0$. Furthermore, when $\delta = 0$, we agree in the sequel that $1/\delta$ is $+\infty$.

Theorem 52 *Assume that the condition (Λ) is satisfied. If $\delta < +\infty$ then for each $\lambda \in (0, 1/\delta)$, one of the following properties holds:*

(b_1) *either there is a global minimum of Φ which is a local minimum of J_λ,*

(b_2) *or there is a sequence* $\{u_m\}$ *of pairwise distinct critical points (local minima) of* J_λ*, with* $\lim_{m\to\infty}\Phi(u_m) = \inf_X \Phi$*, which converges to a global minimum of* Φ.

Theorem 52 is a special form of the quoted variational principle of Ricceri contained in [93].

By a *strong solution* (briefly called a "solution") of (A_λ^f) we mean a function $u : \mathbb{Z}[0, T+1] \to \mathbb{R}$ which satisfies the given equation and the associated boundary conditions.

We point out that solutions will be investigated in the space

$$H = \{u : \mathbb{Z}[0, T+1] \to \mathbb{R};\ u(0) = u(T+1) = 0\}.$$

Then H is a T-dimensional Hilbert space with the inner product

$$(u, v) := \sum_{k=1}^{T+1} \Delta u(k-1)\Delta v(k-1), \quad \forall\, u, v \in H.$$

The associated norm is defined by

$$\|u\| = \left(\sum_{k=1}^{T+1} |\Delta u(k-1)|^2\right)^{1/2}.$$

For every $u \in H$, we set

$$\Phi(u) := \sum_{k=1}^{T+1} \frac{1}{p(k-1)}|\Delta u(k-1)|^{p(k-1)}, \quad \text{and} \quad \Psi(u) := \sum_{k=1}^{T} F_k(u(k)),$$

where $F_k(t) := \int_0^t f_k(s)ds$, for every $t \in \mathbb{R}$ and $k \in \mathbb{Z}[1, T]$. Furthermore, we denote

$$J_\lambda(u) := \Phi(u) - \lambda\Psi(u),$$

for every $u \in H$.

Then $J_\lambda \in C^1(H; \mathbb{R})$ and

$$\langle J_\lambda'(u), v\rangle = \sum_{k=1}^{T+1} |\Delta u(k-1)|^{p(k-1)-2}\Delta u(k-1)\Delta v(k-1) - \lambda\sum_{k=1}^{T} f_k(u(k))v(k),$$

for all $u, v \in H$.

The next result will be crucial in the sequel.

Lemma 68 *The functional* $\Phi : H \to \mathbb{R}$ *is coercive, that is,*

$$\lim_{\|u\|\to\infty} \sum_{k=1}^{T+1} \frac{1}{p(k-1)}|\Delta u(k-1)|^{p(k-1)} = +\infty.$$

Proof. There exist positive constants C_1 and C_2 such that

$$\sum_{k=1}^{T+1} |\Delta u(k-1)|^{p(k-1)} \geq C_1 \|u\|^{p^-} - C_2,$$

for every $u \in H$ with $\|u\| > 1$. Hence the conclusion clearly holds, taking into account that

$$\begin{aligned} \Phi(u) &:= \sum_{k=1}^{T+1} \frac{1}{p(k-1)} |\Delta u(k-1)|^{p(k-1)} \\ &\geq \frac{1}{p^+} \left(\sum_{k=1}^{T+1} |\Delta u(k-1)|^{p(k-1)} \right) \\ &\geq \frac{C_1 \|u\|^{p^-} - C_2}{p^+} \to +\infty, \end{aligned}$$

as $\|u\| \to \infty$. □

A *critical point* for J_λ, that is, an element $u \in H$ such that

$$\langle J'_\lambda(u), v \rangle = 0,$$

for every $v \in H$, is said to be a *weak solution* of problem (A^f_λ).

Integrating by parts we see that any weak solution of problem (A^f_λ) is in fact a strong one. Hence, in order to solve (A^f_λ) we check to find critical points for J_λ and investigate their multiplicity.

Finally, let us fix a constant $p > 1$. Then

$$\|u\|_\infty := \max_{k \in \mathbb{Z}[1,T]} |u(k)| \leq \frac{(T+1)^{(p-1)/p}}{2} \left(\sum_{k=1}^{T+1} |\Delta u(k-1)|^p \right)^{1/p}, \tag{6.35}$$

for every $u \in H$.

Set

$$A_0 := \liminf_{t \to 0^+} \frac{\sum_{k=1}^{T} \max_{|\xi| \leq t} F_k(\xi)}{t^{p^+}}, \qquad B^0 := \limsup_{t \to 0^+} \frac{\sum_{k=1}^{T} F_k(t)}{t^{p^-}},$$

and

$$\kappa := \frac{2^{p^+-1} p^-}{p^+ (T+1)^{p^+-1}}.$$

The main result in this section establishes the following multiplicity property.

Theorem 53 *Assume that the following inequality holds*

(h_0) $\qquad A_0 < \kappa B^0.$

Then for each $\lambda \in (\frac{2}{p^- B^0}, \frac{2^{p^+}}{p^+(T+1)^{p^+-1}A_0})$, *problem* (A_λ^f) *admits a sequence of nonzero solutions which converges to zero.*

Proof. Fix $\lambda \in (\frac{2}{p^- B^0}, \frac{2^{p^+}}{p^+(T+1)^{p^+-1}A_0})$. Our aim is to apply Theorem 52 to function J_λ. Now, by standard arguments and bearing in mind Lemma 68, condition (Λ) is fulfilled. Therefore, our conclusion follows provided that $\delta < +\infty$ as well as that 0_H is not a local minimum point for the functional J_λ. To this end, let $\{c_m\} \subset]0, +\infty[$ be a sequence such that $\lim_{m\to\infty} c_m = 0$ and

$$\lim_{m\to\infty} \frac{\sum_{k=1}^{n} \max_{|\xi|\le c_m} F_k(\xi)}{c_m^2} = A_0.$$

Set

$$r_m := \left(\frac{2^{p^+}}{p^+(T+1)^{p^+-1}}\right) c_m^{p^+},$$

for every $m \in \mathbb{N}$.

For sufficiently large $m > 0$, it follows that

$$\Phi^{-1}(]-\infty, r_m[) \subset \{v \in X : |v(k)| \le c_m, \ \forall\, k \in \mathbb{Z}[0, T+1]\}. \tag{6.36}$$

Indeed, if $v \in X$ and $\Phi(v) < r_m$, we have

$$\sum_{k=1}^{T+1} \frac{1}{p(k-1)} |\Delta v(k-1)|^{p(k-1)} < r_m.$$

Therefore

$$|\Delta v(k-1)| < (p(k-1)r_m)^{1/p(k-1)},$$

for every $k \in \mathbb{Z}[1, T+1]$. Consequently, since $r_m < 1/p^+$ for every $m \ge \bar{m}$, one immediately has

$$(p^+ r_m)^{1/p(k-1)} \le (p^+ r_m)^{1/p^+},$$

hence

$$|\Delta v(k-1)| < (p^+ r_m)^{1/p^+} < 1,$$

for every $k \in \mathbb{Z}[1, T+1]$. So

$$\sum_{k=1}^{T+1} \frac{1}{p^+} |\Delta v(k-1)|^{p^+} \le \sum_{k=1}^{T+1} \frac{1}{p(k-1)} |\Delta v(k-1)|^{p(k-1)} < r_m,$$

for every $m \geq \bar{m}$. At this point, by (6.35), the following inequality

$$\max_{k\in\mathbb{Z}[1,T]} |v(k)| \leq \left(p^+ \frac{(T+1)^{p^+-1}}{2^{p^+}} r_m\right)^{1/p^+} = c_m,$$

is satisfied for every $m \geq \bar{m}$. Thus the above computations ensure that (6.36) holds. The algebraic inclusion (6.36) implies that

$$\begin{aligned}\varphi(r_m) &\leq \frac{\sup\limits_{\Phi(v)<r_m} \sum\limits_{k=1}^{T} F_k(v(k))}{r_m} \leq \frac{\sum\limits_{k=1}^{T} \max\limits_{|t|\leq c_m} F_k(t)}{r_m} \\ &= \frac{p^+(T+1)^{p^+-1}}{2^{p^+}} \frac{\sum\limits_{k=1}^{T} \max\limits_{|t|\leq c_m} F_k(t)}{c_m^{p^+}},\end{aligned}$$

for every sufficiently large m. Hence it follows that

$$\delta \leq \lim_{m\to\infty} \varphi(r_m) \leq \frac{p^+(T+1)^{p^+-1}}{2^{p^+}} A_0 < \frac{1}{\lambda} < +\infty.$$

In the next step we show that 0_H is not a local minimum point for the functional J_λ. First, assume that $B^0 = +\infty$. Accordingly, fix M such that $M > \frac{2}{p^-\lambda}$ and let $\{b_m\}$ be a sequence of positive numbers, with $\lim\limits_{m\to\infty} b_m = 0$, such that

$$\sum_{k=1}^{T} F_k(b_m) > M b_m^{p^-} \qquad (\forall\, m \in \mathbb{N}).$$

Thus, take in H a sequence $\{s_m\}$ such that, for each $m \in \mathbb{N}$, $s_m(k) := b_m$ for every $k \in \mathbb{Z}[1,T]$. Observe that $\|s_m\| \to 0$ as $m \to \infty$. So it follows that

$$\sum_{k=1}^{T+1} \frac{1}{p(k-1)} |\Delta s_m(k-1)|^{p(k-1)} \leq \frac{2b_m^{p^-}}{p^-},$$

taking into account that

$$\sum_{k=1}^{T+1} |\Delta s_m(k-1)|^{p^-} = 2b_m^{p^-}.$$

Therefore

$$J_\lambda(s_m) \leq \frac{2b_m^{p^-}}{p^-} - \lambda \sum_{k=1}^{T} F_k(b_m) < \left(\frac{2}{p^-} - \lambda M\right) b_m^{p^-},$$

that is, $J_\lambda(s_m) < 0$ for every sufficiently large m.

Next, assume that $B^0 < +\infty$. Since $\lambda > \dfrac{2}{p^- B^0}$, we can fix $\varepsilon > 0$ such that $\varepsilon < B^0 - \dfrac{2}{p^-\lambda}$. Therefore, also taking $\{b_m\}$ a sequence of positive numbers such that $\lim\limits_{m\to\infty} b_m = 0$ and

$$(B^0 - \varepsilon)b_m^{p^-} < \sum_{k=1}^{T} F_k(b_m) < (B^0 + \varepsilon)b_m^{p^-}, \qquad (\forall\, m \in \mathbb{N})$$

arguing as before and by choosing $\{s_m\}$ in H as above, we have

$$J_\lambda(s_m) < \left(\frac{2}{p^-} - \lambda(B^0 - \varepsilon)\right) b_m^{p^-}.$$

So also in this case, $J_\lambda(s_m) < 0$ for every sufficiently large m. Finally, since $J_\lambda(0_H) = 0$, the above fact means that 0_H is not a local minimum of J_λ. Therefore, the unique global minimum of Φ is not a local minimum of the functional J_λ. Hence, by Theorem 52 we obtain a sequence $\{u_m\} \subset H$ of critical points of J_λ such that

$$\lim_{m\to\infty} \sum_{k=1}^{T+1} \frac{1}{p(k-1)} |\Delta u_m(k-1)|^{p(k-1)} = \lim_{m\to\infty} \|u_m\| = 0.$$

Thus it follows that $\|u_m\|_\infty \to 0$ as $m \to \infty$. The proof is complete. □

Theorem 53 has been obtained by exploiting Theorem 52. Via the approach developed in this section, we are able to determine an open subinterval of $(0, 1/\delta)$, where $\delta := \liminf\limits_{r\to 0^+} \varphi(r)$, such that problem (A_λ^f) admits infinitely many solutions. Indeed, the main condition (h_0) implies that the real interval of parameter

$$\left(\frac{2}{p^- B^0}, \frac{2^{p^+}}{p^+(T+1)^{p^+-1} A_0}\right),$$

is well-defined and nonempty. Furthermore, since $B_0 > 0$ and

$$\delta \le \lim_{m\to\infty} \varphi(r_m) \le \frac{p^+(T+1)^{p^+-1}}{2^{p^+}} A_0,$$

the following inclusion

$$\left(\frac{2}{p^- B^0}, \frac{2^{p^+}}{p^+(T+1)^{p^+-1} A_0}\right) \subseteq \left(0, \frac{1}{\delta}\right),$$

is verified.

We note that, if f_k is a nonnegative continuous function, for every $k \in$

$\mathbb{Z}[1,T]$, condition (h$_0$) assumes the form

$$\liminf_{t\to 0^+}\frac{\sum_{k=1}^{n}F_k(t)}{t^{p^+}}<\kappa\limsup_{t\to 0^+}\frac{\sum_{k=1}^{n}F_k(t)}{t^{p^-}}.$$

Consequently, Theorem 51 immediately follows from Theorem 53.

Set

$$\widehat{A}_0:=\liminf_{t\to 0^+}\frac{\sum_{k=1}^{T}\max_{|\xi|\le t}F_k(\xi)}{t^p},\qquad\text{and}\qquad \widehat{B}^0:=\limsup_{t\to 0^+}\frac{\sum_{k=1}^{T}F_k(t)}{t^p}.$$

With the previous notations, taking the map $p:\mathbb{Z}[0,T]\to\mathbb{R}$ such that $p(k)=p$, for every $k\in\mathbb{Z}[0,T]$, we have the following immediate consequence of Theorem 53.

Corollary 5 *Assume that*

$(\widehat{h}_0)$ $$\widehat{A}_0<\frac{2^{p-1}}{(T+1)^{p-1}}\widehat{B}^0.$$

Then for each $\lambda\in(\frac{2}{p\widehat{B}^0},\frac{2^p}{p(T+1)^{p-1}\widehat{A}_0})$, *the following problem*

$$\begin{cases}-\Delta(\phi_p(\Delta u(k-1)))=\lambda f_k(u(k)), & k\in\mathbb{Z}[1,T]\\ u(0)=u(T+1)=0,\end{cases}\qquad (D_\lambda^f)$$

admits a sequence of nonzero solutions which converges to zero.

A more technical version of Theorem 53 can be stated as follows.

Theorem 54 *Assume that there exist nonnegative real sequences* $\{a_m\}$ *and* $\{b_m\}$, *with* $\lim_{m\to\infty}b_m=0$, *such that*

(k_1) $$a_m^{p^-}<\left(\frac{2^{p^+-1}p^-}{p^+(T+1)^{p^+-1}}\right)b_m^{p^+},\ \textit{for each } m\in\mathbb{N};$$

(k_2) $$G_0<\frac{B^0}{p^+(T+1)^{p^+-1}},\ \textit{where}$$

$$G_0:=\lim_{m\to\infty}\frac{\sum_{k=1}^{T}\max_{|t|\le b_m}F_k(t)-\sum_{k=1}^{T}F_k(a_m)}{2^{p^+-1}p^-b_m^{p^+}-p^+(T+1)^{p^+-1}a_m^{p^-}}.$$

Then for each $\lambda\in(\frac{2}{p^-B^0},\frac{2}{p^-p^+(T+1)^{p^+-1}G_0})$, *problem* (A_λ^f) *admits a sequence of nonzero solutions which converges to zero.*

Proof. We first observe that

$$\varphi(r_m) \leq \inf_{w \in \Phi^{-1}((-\infty, r_m))} \frac{\sum_{k=1}^{T} \max_{|t| \leq b_m} F_k(t) - \sum_{k=1}^{T} F_k(w(k))}{\left(\frac{2^{p^+}}{p^+(T+1)^{p^+-1}}\right) b_m^{p^+} - \Phi(u)}. \tag{6.37}$$

Now, for each $m \in \mathbb{N}$, let $w_m \in H$ be defined by $w_m(k) := a_m$ for every $k \in \mathbb{Z}[1, T]$. Since $\|w_m\| \to 0$ as $m \to \infty$, it follows that $|\Delta w_m(k-1)| < 1$ for every $k \in \mathbb{Z}[1, T+1]$ and sufficiently large m. Then there exists $\bar{m} \in \mathbb{N}$ such that

$$\Phi(w_m) := \sum_{k=1}^{T+1} \frac{1}{p(k-1)} |\Delta w_m(k-1)|^{p(k-1)} \leq \frac{1}{p^-} \left(\sum_{k=1}^{T+1} |\Delta w_m(k-1)|^{p^-} \right),$$

for every $m \geq \bar{m}$.

Now, by condition (k_1) and taking into account that $\sum_{k=1}^{T+1} |\Delta w_m(k-1)|^{p^-} = 2a_m^{p^-}$, the above inequality implies that

$$0 < \left(\frac{2^{p^+}}{p^+(T+1)^{p^+-1}}\right) b_m^{p^+} - \frac{2a_m^{p^-}}{p^-} \leq \left(\frac{2^{p^+}}{p^+(T+1)^{p^+-1}}\right) b_m^{p^+} - \Phi(w_m),$$

for every $m \geq \bar{m}$. We have $w_m \in \Phi^{-1}((-\infty, r_m))$ and inequality (6.37) yields

$$\varphi(r_m) \leq \left(\frac{p^- p^+(T+1)^{p^+-1}}{2}\right) \frac{\sum_{k=1}^{T} \max_{|t| \leq b_m} F_k(t) - \sum_{k=1}^{T} F_k(a_m)}{p^- 2^{p^+-1} b_m^{p^+} - p^+(T+1)^{p^+-1} a_m^{p^-}},$$

for every large enough m.
Further, by hypothesis (k_2), we obtain

$$\delta \leq \lim_{m \to \infty} \varphi(r_m) \leq \frac{p^- p^+(T+1)^{p^+-1}}{2} G_0 < \frac{1}{\lambda} < +\infty.$$

From now on, arguing exactly as in the proof of Theorem 53 we obtain the assertion. □

We now provide an application of Theorem 51. Let $\gamma > 2$ be a real positive constant and $p : \mathbb{Z}[0, T] \to \mathbb{R}^+$ a map such that $p^+ := \max_{\mathbb{Z}[0,T]} p(k) = \gamma$ and $p^- := \min_{\mathbb{Z}[0,T]} p(k) = \gamma - 1$, with $T \geq 2$. Let $\{s_m\}$, $\{t_m\}$ and $\{\delta_m\}$ be sequences defined by

$$s_m := 2^{-\frac{m!}{2}}, \qquad t_m := 2^{-2m!}, \qquad \delta_m := 2^{-(m!)^2},$$

and consider $\nu \in \mathbb{N}$ such that

$$s_{m+1} < t_m < s_m - \delta_m, \quad \forall\, m \geq \nu.$$

Moreover, let $g : \mathbb{R} \to \mathbb{R}$ be the nonnegative continuous function given by

$$g(s) := \begin{cases} 2^{-(\gamma-1)\nu!} & \text{if} \quad s \in]s_\nu - \delta_\nu, +\infty[\\ y_m(s) & \text{if} \quad s \in \bigcup_{m\geq\nu}]s_{m+1} - \delta_{m+1}, s_{m+1}[\\ 2^{-(\gamma-1)m!} & \text{if} \quad s \in \bigcup_{m\geq\nu} [s_{m+1}, s_m - \delta_m] \\ 0 & \text{if} \quad s \leq 0, \end{cases}$$

where

$$y_m(s) := \left(2^{-(\gamma-1)m!} - 2^{-(\gamma-1)(m+1)!}\right)\left(\frac{s - s_{m+1} + \delta_{m+1}}{\delta_{m+1}}\right) + 2^{-(\gamma-1)(m+1)!}.$$

Set $G(t) := \displaystyle\int_0^t g(s)\, ds$ for every $t \in \mathbb{R}$. Then

$$\frac{G(s_m)}{s_m^\gamma} \leq \frac{g(s_{m+1})s_m + g(s_m)\delta_m}{s_m^\gamma},$$

and

$$\frac{G(t_m)}{t_m^{\gamma-1}} \geq \frac{g(s_{m+1})(t_m - s_{m+1})}{t_m^{\gamma-1}},$$

for every large enough m.

Since

$$\lim_{m\to\infty} \frac{g(s_{m+1})s_m + g(s_m)\delta_m}{s_m^\gamma} = 0, \quad \text{and} \quad \lim_{m\to\infty} \frac{g(s_{m+1})(t_m - s_{m+1})}{t_m^{\gamma-1}} = +\infty,$$

it follows that

$$\lim_{m\to\infty} \frac{G(s_m)}{s_m^\gamma} = 0, \quad \text{and} \quad \lim_{m\to\infty} \frac{G(t_m)}{t_m^{\gamma-1}} = +\infty.$$

Thus

$$\liminf_{t\to 0^+} \frac{\displaystyle\int_0^t g(s)ds}{t^\gamma} = 0, \quad \text{and} \quad \limsup_{t\to 0^+} \frac{\displaystyle\int_0^t g(s)ds}{t^{\gamma-1}} = +\infty.$$

Then by Theorem 51, for each $\lambda > 0$, the following anisotropic discrete Dirichlet problem

$$\begin{cases} -\Delta(|\Delta u(k-1)|^{p(k-1)-2}\Delta u(k-1)) = \lambda g(u(k)), \ k \in \mathbb{Z}[1,T] \\ u(0) = u(T+1) = 0, \end{cases} \qquad (A_\lambda^g)$$

admits a sequence of nonzero solutions which converges to zero.

6.4 Glossary

Discrete Laplace Operator: This is an analog of the continuous Laplace operator, which is defined so that it has meaning on a graph or a discrete grid.

Homoclinic Orbit: This is a trajectory of a Hamiltonian system, which is asymptotic to a constant at infinity.

Partial Difference Equation: This is a functional relation with recursive structure.

Appendix A. Ekeland Variational Principle

> We are servants rather than masters in mathematics.
>
> Charles Hermite (1822–1901)

The Ekeland variational principle [26] was established in 1974 and is the nonlinear version of the Bishop–Phelps theorem [79], with its main feature how to use the norm completeness and a partial ordering to obtain a point where a linear functional achieves its supremum on a closed bounded convex set. For any bounded from below, lower semi-continuous functional f, the Ekeland variational principle provides a minimizing sequence whose elements minimize an appropriate sequence of perturbations of f which converges locally uniformly to f. Roughly speaking, the Ekeland variational principle states that there exist points which are almost points of minima and where the "gradient" is small. In particular, it is not always possible to minimize a nonnegative continuous function on a complete metric space. A major consequence of the Ekeland variational principle is that even if it is not always possible to minimize a nonnegative C^1 functional f on a Banach space, there is always a minimizing sequence $(u_n)_{n\geq 1}$ such that $f'(u_n) \to 0$ as $n \to \infty$. The Ekeland variational principle is a fundamental tool that is effective in numerous situations, which led to many new results and strengthened a series of known results in various fields of analysis, geometry, the Hamilton-Jacobi theory, extremal problems, the Ljusternik-Schnirelmann theory, etc.

In what follows, we state the original version of the Ekeland variational principle, which is valid in the general framework of complete metric spaces.

Theorem 55 *Let (X, d) be a complete metric space and let $f : X \to \mathbb{R} \cup \{\infty\}$ be a lower semi-continuous, bounded from below functional with $D(f) = \{u \in X : f(u) < \infty\} \neq \emptyset$. Then for every $\varepsilon > 0$, $\lambda > 0$, and $u \in X$ such that*

$$f(u) \leq \inf_X f + \varepsilon$$

there exists an element $v \in X$ such that

(a) $f(v) \leq f(u)$;

(b) $d(v, u) \leq \frac{1}{\lambda}$;

(c) $f(w) > f(v) - \varepsilon\lambda d(w, v)$ *for each* $w \in X \setminus \{v\}$.

Proof. It suffices to prove our assertion for $\lambda = 1$. The general case is then obtained by replacing d with an equivalent metric λd. We define the relation on X by:

$$w \leq v \Longleftrightarrow f(w) + \varepsilon d(v, w) \leq f(v).$$

It is easy to see that this relation defines a partial ordering on X. We now inductively construct a sequence $\{u_n\} \subset X$ as follows: $u_0 = u$, and assuming that u_n has been defined, we set

$$S_n = \{w \in X : w \leq u_n\}$$

and choose $u_{n+1} \in S_n$ so that

$$f(u_{n+1}) \leq \inf_{S_n} f + \frac{1}{n+1}.$$

Since $u_{n+1} \leq u_n$ then $S_{n+1} \subset S_n$ and by the lower semi-continuity of f, S_n is closed. We now show that $\operatorname{diam} S_n \to 0$. Indeed, if $w \in S_{n+1}$, then $w \leq u_{n+1} \leq u_n$ and consequently,

$$\varepsilon d(w, u_{n+1}) \leq f(u_{n+1}) - f(w) \leq \inf_{S_n} f + \frac{1}{n+1} - \inf_{S_n} f = \frac{1}{n+1}.$$

This estimate implies that

$$\operatorname{diam} S_{n+1} \leq \frac{2}{\varepsilon(n+1)}$$

and our claim follows. The fact that X is complete implies that $\bigcap_{n \geq 0} S_n = \{v\}$ for some $v \in X$. In particular, $v \in S_0$, that is, $v \leq u_0 = u$ and hence

$$f(v) \leq f(u) - \varepsilon d(u, v) \leq f(u)$$

and moreover,

$$d(u, v) \leq \frac{1}{\varepsilon}(f(u) - f(v)) \leq \frac{1}{\varepsilon}(\inf_X f + \varepsilon - \inf_X f) = 1.$$

Now, let $w \neq v$. To complete the proof we must show that $w \leq v$ implies $w = v$. If $w \leq v$, then $w \leq u_n$ for each integer $n \geq 0$, that is $w \in \cap_{n \geq 0} S_n = \{v\}$. So, $w \not\leq v$, which is actually (c). □

In $\mathbb{R}^N$ equipped with the Euclidean metric, properties (a) and (c) in the statement of the Ekeland variational principle are completely intuitive as Figure 6.4 shows. Indeed, assuming that $\lambda = 1$, let us consider a cone lying below

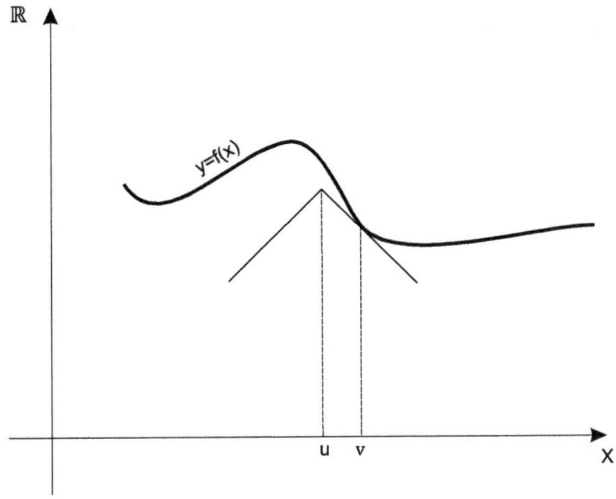

FIGURE A.1: Geometric illustration of the Ekeland variational principle

the graph of f, with slope $+1$, and vertex projecting onto u. We move up this cone until it first touches the graph of f at some point $(v, f(v))$. Then the point v satisfies both (a) and (c).

In the case when $X = \mathbb{R}^N$ we can give the following simple alternative proof of the Ekeland variational principle, which is due to Hiriart-Urruty [39]. Indeed, consider the perturbed functional

$$g(w) := f(w) + \varepsilon\lambda \|w - u\|, \qquad w \in \mathbb{R}^N.$$

Since f is lower semi-continuous and bounded from below, then g is lower semi-continuous and $\lim_{\|w\|\to\infty} g(w) = +\infty$. Therefore there exists $v \in \mathbb{R}^N$ minimizing g on $\mathbb{R}^N$ such that for all $w \in \mathbb{R}^N$

$$f(v) + \varepsilon\lambda \|v - u\| \le f(w) + \varepsilon\lambda \|w - u\|. \tag{6.38}$$

By letting $w = u$ we find

$$f(v) + \varepsilon\lambda \|v - u\| \le f(u)$$

and (a) follows. Now, since $f(u) \le \inf_{\mathbb{R}^N} f + \varepsilon$, we also deduce that $\|v - u\| \le 1/\lambda$.

We infer from relation (6.38) that for any w,

$$f(v) \le f(w) + \varepsilon\lambda \left[\|w - u\| - \|v - u\|\right] \le f(w) + \varepsilon\lambda \|w - u\|,$$

which is the desired inequality (c).

Taking $\lambda = \dfrac{1}{\sqrt{\varepsilon}}$ in the above theorem we obtain the following property.

Corollary 6 *Let (X, d) be a complete metric space and let $f : X \to \mathbb{R} \cup \{\infty\}$ be a lower semi-continuous, bounded from below and $D(f) = \{u \in X : f(u) < \infty\} \neq \emptyset$. Then for every $\varepsilon > 0$ and every $u \in X$ such that*

$$f(u) \leq \inf_X f + \varepsilon$$

there exists an element $u_\varepsilon \in X$ such that

a) $f(u_\varepsilon) \leq f(u)$;

b) $d(u_\varepsilon, u) \leq \sqrt{\varepsilon}$;

c) $f(w) > f(u_\varepsilon) - \sqrt{\varepsilon} d(w, u_\varepsilon)$ *for each* $w \in X \setminus \{u_\varepsilon\}$.

Let $(X, \|\cdot\|)$ be a real Banach space, $X^\star$ its topological dual endowed with its natural norm, denoted for simplicity also by $\|\cdot\|$. We denote by $\langle \cdot, \cdot \rangle$ the duality mapping between X and $X^\star$, that is, $\langle x^\star, u \rangle = x^\star(u)$ for every $x^\star \in X^\star, u \in X$. Theorem 55 readily implies the following property, which asserts the existence of *almost critical points.* In other words, the Ekeland variational principle can be viewed as a generalization of the Fermat theorem which establishes that interior extrema points of a smooth functional are necessarily critical points of this functional.

Corollary 7 *Let X be a Banach space and let $f : X \to \mathbb{R}$ be a lower semi-continuous functional which is bounded from below. Assume that f is Gâteaux differentiable at every point of X. Then for every $\varepsilon > 0$ there exists an element $u_\varepsilon \in X$ such that*

(i) $f(u_\varepsilon) \leq \inf_X f + \varepsilon$;

(ii) $\|f'(u_\varepsilon)\| \leq \varepsilon$.

Letting $\varepsilon = 1/n$, $n \in \mathbb{N}$, Corollary 7 gives rise to a minimizing sequence for the infimum of a given function which is bounded from below. Note however that such a sequence need not converge to any point. Indeed, let $f : \mathbb{R} \to \mathbb{R}$ be defined by $f(s) = e^{-s}$. Then $\inf_{\mathbb{R}} f = 0$, and any minimizing sequence fulfilling (a) and (b) from Corollary 7 tends to $+\infty$.

The following consequence of the Ekeland variational principle has been of particular interest in our arguments. Roughly speaking, this property establishes the existence of *almost critical points* for bounded from below C^1 functionals defined on Banach spaces.

Sullivan [101] observed that the Ekeland variational principle characterizes complete metric spaces in the following sense.

Theorem 56 *Let (M, d) be a metric space. Then M is complete if and only if the following holds: for every mapping $f : M \to (-\infty, \infty]$, $f \not\equiv \infty$, which is bounded from below, and for every $\varepsilon > 0$, there exists $z_\varepsilon \in M$ such that*

(i) $f(z_\varepsilon) \leq \inf_M f + \varepsilon$;

(ii) $f(x) > f(z_\varepsilon) - \varepsilon\, d(x, z_\varepsilon)$, *for any* $x \in M \setminus \{z_\varepsilon\}$.

Appendix B. Mountain Pass Theorem

It is very simple to be happy, but
it is very difficult to be simple.

Rabindranath Tagore
(1861–1941)

The mountain pass theorem was established by Ambrosetti and Rabinowitz [4]. Their original proof relies on some deep deformation techniques developed by Palais and Smale [77], [78], who put the main ideas of the Morse theory into the framework of differential topology on infinite dimensional manifolds. In this way, Palais and Smale replaced the finite dimensionality assumption with an appropriate compactness hypothesis, which will be stated in the sequel. As pointed out by Brezis and Browder [11], the mountain pass theorem "*extends ideas already presented by Poincaré and Birkhoff*".

The mountain pass theorem is a powerful tool for proving the existence of critical points of energy functionals, hence of weak solutions of wide classes of nonlinear problems. In this result, we consider a function $f : X \to \mathbb{R}$ of class C^1, where X is a real Banach space. It is assumed that f satisfies the following geometric conditions:
(H1) *there exist numbers* $R > 0$ *and* $c_0 \in \mathbb{R}$ *such that* $f(u) \geq c_0$ *for every* $u \in S_R := \{v \in X;\ \|v\| = R\}$;
(H2) $f(0) < c_0$ *and* $f(e) < c_0$ *for some* $e \in X$ *with* $\|e\| > R$.

With an additional compactness condition of Palais-Smale type it then follows that the function f has a critical point $u_0 \in X \setminus \{0, e\}$ with critical value $c \geq c_0$. In essence, this critical value occurs because 0 and e are low points on both sides of the *mountain* S_R, so that between 0 and e there must be a lowest critical point, or a *mountain pass*. Condition (H2) signifies that the mountain should have *positive altitude*. Pucci and Serrin [83, 84] proved that the mountain pass theorem continues to hold for a mountain of *zero altitude*, provided it also has nonzero thickness. In addition, if $c = c_0$, then the "pass" itself occurs precisely on the mountain. Roughly speaking, Pucci and Serrin showed that the mountain pass theorem still remains true if (H1) is strengthened a bit, namely to the form
(H1)′ *there exist real numbers* c_0, R, r *such that* $0 < r < R$ *and* $f(u) \geq c_0$ *for every* $u \in A := \{v \in X;\ r < \|v\| < R\}$,

while hypothesis (H2) is weakened and replaced with
(H2)′ $f(0) \le c_0$ *and* $f(e) \le c_0$ *for some* $e \in X$ *with* $\|e\| > R$.

The geometric interpretation of the mountain pass theorem is described in the sequel. Denote by f the function which measures the altitude of a mountain terrain and assume that there are two points in the horizontal plane L_1 and L_2, representing the coordinates of two locations such that $f(L_1)$ and $f(L_2)$ are the deepest points of two separated valleys. Roughly speaking, our aim is to walk along an optimal path on the mountain from the point $(L_1, f(L_1))$ to the point $(L_2, f(L_2))$, spending the least amount of energy by passing the mountain ridge between the two valleys. Walking on a path $(\gamma, f(\gamma))$ from $(L_1, f(L_1))$ to $(L_2, f(L_2))$ such that the maximal altitude along γ is the smallest among all such continuous paths connecting $(L_1, f(L_1))$ and $(L_2, f(L_2))$, we reach a point L on γ passing the ridge of the mountain which is called a *mountain pass point*.

Let $f : X \to \mathbb{R}$ be a function of class C^1 satisfying the following assumption: there exists an open neighborhood $\mathcal{N}$ of some $e_0 \in X$ and there are $e_1 \notin \bar{\mathcal{N}}$ and $c_0 \in \mathbb{R}$ such that

$$\max\{f(e_0),\, f(e_1)\} < c_0 \le f(u) \qquad \text{for all } u \in \partial\mathcal{N}. \tag{6.39}$$

Next, we consider the family $\mathcal{P}$ of all continuous paths $p : [0,1] \to X$ joining e_0 and e_1, that is, $p(0) = e_0$ and $p(1) = e_1$. Denote

$$c := \inf_{p\in\mathcal{P}} \max_{t\in[0,1]} f(p(t))\,. \tag{6.40}$$

Since each path $p \in \mathcal{P}$ crosses the boundary of $\mathcal{N}$, we have $\max_{t\in[0,1]} f(p(t)) \ge c_0$, hence $c \ge c_0$. In fact, in the original version of the mountain pass theorem it was assumed that the mountain has *positive altitude*, that is,

$$c > \max\{f(e_0),\, f(e_1)\}\,. \tag{6.41}$$

The number c defined by relation (6.40) is an "approximate critical value" of the functional f. The reason for this notion appears in the following version of the mountain pass theorem.

Theorem 57 *Assume that* $f \in C^1(X, \mathbb{R})$ *satisfies condition* (6.41). *Then there exists a sequence* (u_n) *in* X *such that*

$$f(u_n) \to c \quad \textit{and} \quad \|f'(u_n)\|_{X^*} \to 0 \qquad \textit{as } n \to \infty. \tag{6.42}$$

We refer to Figure 6.4 for a geometric illustration of Theorem 57.

In order to assert that c is really a critical value of f it has become quite standard to assume the following *compactness condition*, originally introduced by Palais and Smale [78]: the function f is said to satisfy the Palais-Smale condition $(\mathrm{PS})_a$ at level $a \in \mathbb{R}$ provided that

$$\left\{ \begin{array}{l} \text{any sequence } (u_n) \text{ in } X \text{ such that } f(u_n) \to a \text{ and } \|f'(u_n)\|_{X^*} \to 0 \\ \text{has a convergent subsequence.} \end{array} \right.$$

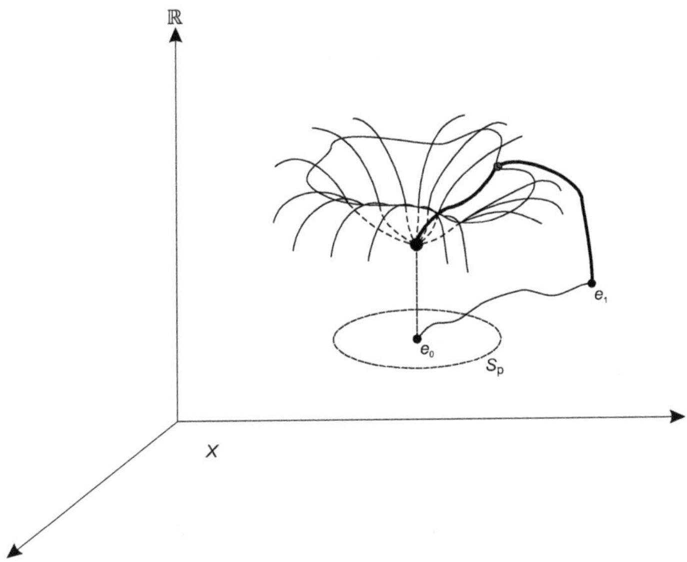

FIGURE B.1: Mountain pass landscape between "villages" e_0 and e_1

If we now assume that Palais-Smale condition $(\mathrm{PS})_c$ is fulfilled with c defined in relation (6.40), we obtain the following version of the mountain pass theorem with compactness assumption.

Theorem 58 *Assume that $f \in C^1(X, \mathbb{R})$ satisfies conditions* (6.41) *and $(PS)_c$. Then the number c defined in* (6.40) *is a critical value of f.*

The proof of Theorem 57 relies on a version of the deformation lemma, which is obtained by applying the following *pseudo-gradient* lemma.

Lemma 69 *Let M be a metric space and assume that $\Phi : M \to X^* \setminus \{0\}$ is a continuous function. Then for any $\varepsilon > 0$ there exists a continuous function $v : M \to X$ such that for all $x \in M$,*

$$\|v(x)\| \leq (1+\varepsilon)\|\Phi(x)\| \tag{6.43}$$

and

$$\langle \Phi(x), v(x) \rangle \geq \|\Phi(x)\|^2 . \tag{6.44}$$

Proof. Fix $x \in M$. Then there exists $z \in M$ such that

$$\|z\| < (1+\varepsilon)\|\Phi(x)\|$$

and

$$\langle \Phi(x), z \rangle > \|\Phi(x)\|^2 .$$

Now, for a fixed z and using the continuity of Φ, we can deduce that these relations hold for the entire open neighborhood $\mathcal{N}_x$ of x. However $M = \cup_{x\in M}\mathcal{N}_x$. Thus by Theorem 5.3 from Dugundji [22], there exists a locally finite subcovering U_j of M with associated z_j. Set $\rho_j := \text{dist}\,(x, M\setminus U_j)$. Then the mapping ρ_j is continuous, $\rho_j \equiv 0$ on $M \setminus U_j$, and (ϕ_j) is a partition of unity associated to the covering (U_j), where

$$\phi_j(x) := \frac{\rho_j(x)}{\sum_k \rho_k(x)}\,.$$

Then the mapping $v : M \to X$ defined by

$$v(x) = \sum_j \phi_j(x) z_j$$

satisfies relations (6.43) and (6.44). □

If $\Phi : X \to \mathbb{R}$ is an arbitrary function and $a \in \mathbb{R}$, we set

$$\Phi_a := \{u \in X;\ \Phi(u) \le a\}\,.$$

The key point of the proof of Theorem 57 is the following deformation lemma.

Lemma 70 *Let X be a real Banach space and assume that $f : X \to \mathbb{R}$ is a function of class C^1. Assume that there exist $c \in \mathbb{R}$, $\varepsilon > 0$, and $\delta > 0$ such that*

$$\|f'(u)\| \ge \delta \qquad \text{for all } u \in X \text{ with } f(u) \in [c-\varepsilon, c+\varepsilon]. \tag{6.45}$$

Then there exists a continuous deformation $\eta : [0,1] \times X \to X$ such that

$$\eta(0,u) = 0 \qquad \text{for every } u \in X; \tag{6.46}$$

$$\eta(t,u) = u \text{ for every } (t,u) \in [0,1] \times X \text{ with } f(u) \notin [c-\varepsilon, c+\varepsilon]; \tag{6.47}$$

$$\eta(1, f_{c+\varepsilon/2}) \subset f_{c-\varepsilon/2}\,. \tag{6.48}$$

Proof. Set

$$M := \{u \in X;\ c-\varepsilon < f(u) < c+\varepsilon\}$$

and

$$M_0 := \{u \in X;\ c-\varepsilon/2 \le f(u) \le c+\varepsilon/2\}\,.$$

The continuous function $h : X \to [0,1]$ defined by

$$h(u) := \frac{\text{dist}\,(u, X\setminus M)}{\text{dist}\,(u, X\setminus M) + \text{dist}\,(u, M_0)}$$

satisfies $h \equiv 1$ on M_0 and $h \equiv 0$ on $X \setminus M$.

According to Lemma 69, there is a pseudo-gradient $v : \{u \in X;\ f'(u) \neq 0\} \to X$. Define the vector field $V : X \to X$ by

$$V(u) := \begin{cases} -h(u)\, \dfrac{v(u)}{\|v(u)\|^2} & \text{if } u \in M \\ 0 & \text{if } u \notin M. \end{cases}$$

Then V is locally Lipschitz on X and, for any $u \in X$, $\|V(u)\| \leq 1/\delta$. Thus for any fixed $u \in X$, the problem

$$\begin{cases} \eta'(t) = V(\eta(t)) & \text{if } t > 0 \\ \eta(0) = u \end{cases}$$

has a unique solution $\eta(t) = \eta(t,u)$, defined for all $0 \leq t < \infty$. Moreover, $\eta(t,u) = u$ for all $t \geq 0$ and every $u \in X \setminus M$.

Without loss of generality we can assume that $\varepsilon \in (0, 1/4)$. Then for all $(t,u) \in [0,\infty) \times X$,

$$f'(\eta(t)) = \langle f'(\eta(t)), V(\eta(t)) \rangle \leq -\frac{1}{4}\, h(\eta(t))\,.$$

The mapping η clearly has all the required properties. □

We now have all ingredients to prove Theorem 57. Arguing by contradiction, we assume that there is no sequence (u_n) in X satisfying (6.42). Thus there exist $\varepsilon > 0$ and $\delta > 0$ such that

$$\|f'(u)\| \geq \delta \quad \text{for all } u \in X \text{ with } c - \varepsilon < f(u) < c + \varepsilon.$$

We can assume without loss of generality that $\varepsilon \leq 1/4$, $f(e_0) < c - \varepsilon$, and $f(e_1) < c - \varepsilon$. Using the definition of c given in relation (70), we deduce that there is a path $p \in \mathcal{P}$ such that

$$f(p(t)) \leq c + \varepsilon/2 \qquad \text{for all } t \in [0,1].$$

Thus, by Lemma 70, there is a continuous deformation η satisfying relations (6.46)–(6.48). Define the path $q(t) = \eta(1, p(t))$ for all $t \in [0,1]$. Since $q(0) = \eta(1, e_0) = e_0$, $q(1) = \eta(1, e_1) = e_1$, we deduce that $q \in \mathcal{P}$. On the other hand, by Lemma 70, $q(t) \in f_{c-\varepsilon/2}$ for all $t \in [0,1]$, which contradicts our basic assumption (6.41). This concludes the proof of Theorem 57. □

A straightforward argument shows that the conclusions of Lemma 70 still remain true provided that assumption (6.45) is replaced with the following weaker hypothesis: there exist $\varepsilon > 0$ and $\delta > 0$ such that

$$(1 + \|u\|)\, \|f'(u)\| \geq \delta \qquad \text{for all } u \in X \text{ with } f(u) \in [c - \varepsilon, c + \varepsilon]. \tag{6.49}$$

This enables us to show that the conclusion (6.42) of Theorem 57 can be strengthened to the following *Cerami compactness condition* (see [16])

$$f(u_n) \to c \quad \text{and} \quad (1 + \|u_n\|)\, \|f'(u_n)\|_{X^*} \to 0 \qquad \text{as } n \to \infty. \tag{6.50}$$

More generally, the same conclusions remain true if $(1 + \|u\|)$ is replaced in relations (6.49) and (6.50) with $\psi(\|u\|)$, where $\psi : [0, \infty) \to [1, \infty)$ is a continuous function satisfying

$$\int_0^\infty \frac{\mathrm{dt}}{\psi(t)} = \infty \,.$$

In our arguments developed in this book we have also used the following $\mathbb{Z}_2$-symmetric version of the mountain pass theorem.

Theorem 59 *Let X be an infinite dimensional real Banach space and let $I \in C^1(X, \mathbb{R})$ be even, satisfying the Palais-Smale condition and $I(0) = 0$. Suppose that*

(I1) there exist constants ρ, $b > 0$ such that $I(x) \geq b$ if $\|x\| = \rho$;

(I2) for each finite dimensional subspace $X_1 \subset X$, the set $\{x \in X_1;\ I(x) \geq 0\}$ is bounded.

Then I has an unbounded sequence of critical values.

For a proof of Theorem 59 we refer to Willem [104].

Bibliography

[1] Acerbi E. and Mingione G. Regularity results for a class of functionals with nonstandard growth. *Arch. Ration. Mech. Anal.*, 156:121–140, 2001.

[2] Adams R. *Sobolev spaces.* Academic Press, New York, 1975.

[3] Ambrosetti A., Brezis H., and Cerami G. Combined effects of concave and convex nonlinearities in some elliptic problems. *J. Funct. Anal.*, 122:519–543, 1994.

[4] Ambrosetti A. and Rabinowitz P. Dual variational methods in critical point theory and applications. *J. Funct. Anal.*, 14:349–381, 1973.

[5] Ashbaugh M.S. and Harrell E.M. Maximal and minimal eigenvalues and their associated nonlinear equations. *J. Math. Phys.*, 28:1770–1786, 1987.

[6] Berestycki H. and Brezis H. On a free boundary problem arising in plasma physics. *Nonlinear Anal.*, 4:415–436, 1980.

[7] Bocher M. The smallest characteristic numbers in a certain exceptional case. *Bull. Amer. Math. Soc.*, 21:6–9, 1914.

[8] Boureanu M.M., Pucci P., and Rădulescu V. Multiplicity of solutions for a class of anisotropic elliptic equations with variable exponent. *Complex Var. Elliptic Equ.*, 56:755–767, 2011.

[9] Boureanu M.M. and Rădulescu V. Eigenvalue problems for anisotropic elliptic equations: an Orlicz-Sobolev setting. *Nonlinear Anal.*, 75:4471–4482, 2012.

[10] Brezis H. *Analyse fonctionnelle: théorie et applications.* Masson, Paris, 1992.

[11] Brezis H. and Browder F. Partial differential equations in the 20th century. *Adv. Math.*, 135:76–144, 1998.

[12] Brezis H. and Oswald L. Remarks on sublinear elliptic equations. *Nonlinear Anal.*, 10:55–64, 1986.

[13] Brezis H. and Kamin S. Sublinear elliptic equations in $\mathbb{R}^n$. *Manuscripta Math.*, 74:87–106, 1992.

[14] Caffarelli L., Kohn R., and Nirenberg L. First order interpolation inequalities with weights. *Compositio Math.*, 53:259–275, 1984.

[15] Cencelj M., Repovš D., and Žiga V. Multiple perturbations of a singular eigenvalue problem. *Nonlinear Anal.*, 119:37–45, 2015.

[16] Cerami G. Un criterio di esistenza per i punti critici su variet illimitate. *Istit. Lombardo Accad. Sci. Lett. Rend. A*, 112:332–336, 1978.

[17] Chen Y., Levine S., and Rao M. Variable exponent, linear growth functionals in image restoration. *SIAM J. Appl. Math.*, 66:1383–1406, 2006.

[18] Ciarlet Ph.G. *Linear and nonlinear functional analysis with applications.* Society for Industrial and Applied Mathematics, Philadelphia, 2013.

[19] Clément Ph., de Pagter B., Sweers G., and de Thélin F. Existence of solutions to a semilinear elliptic system through Orlicz-Sobolev spaces. *Mediterr. J. Math.*, 1:241–267, 2004.

[20] DiBenedetto E. $C^{1+\alpha}$–local regularity of weak solutions of degenerate elliptic equations. *Nonlinear Anal.*, 7:827–850, 1983.

[21] Diening L., Harjulehto P., Hästö P., and Ruzicka M. *Lebesgue and Sobolev spaces with variable exponents.* Lecture Notes in Mathematics, vol. 2017, Springer-Verlag, Heidelberg, 2011.

[22] Dugundji J. *Topology.* Allyn and Bacon, Inc., Boston, Mass.-London-Sydney, 1966.

[23] Edmunds D.E. and Rakosnik J. Density of smooth functions in $W^{k,p(x)}(\Omega)$. *Proc. Royal Soc. London A*, 437:229–236, 1992.

[24] Edmunds D.E. and Rakosnik J. Sobolev embedding with variable exponent. *Studia Math.*, 143:267–293, 2000.

[25] Egnell H. Extremal properties of the first eigenvalue of a class of elliptic eigenvalue problems. *Ann. Sc. Norm. Sup. Pisa*, 14:1–48, 1987.

[26] Ekeland I. On the variational principle. *J. Math. Anal. Appl.*, 47:324–353, 1974.

[27] Filippucci R., Pucci P., and Rădulescu V. Existence and non-existence results for quasilinear elliptic exterior problems with nonlinear boundary conditions. *Comm. Partial Differential Equations*, 33:706–717, 2008.

[28] Fragalà I., Gazzola F., and Kawohl B. Existence and nonexistence results for anisotropic quasilinear equations. *Ann. Inst. H. Poincaré, Analyse Non Linéaire*, 21:715–734, 2004.

[29] Fan X.L. Solutions for $p(x)$-Laplacian Dirichlet problems with singular coefficients. *J. Math. Anal. Appl.*, 312:464–477, 2005.

[30] Fan X.L. and Zhao D. On the spaces $L^{p(x)}$ and $W^{m,p(x)}$. *J. Math. Anal. Appl.*, 263:424–446, 2001.

[31] Fan X.L. and Zhang Q. Existence of solutions for $p(x)$-Laplacian Dirichlet problems. *Nonlinear Anal.*, 52:1843–1852, 2003.

[32] Fan X.L., Zhang Q., and Zhao D. Eigenvalues of $p(x)$-Laplacian Dirichlet problem. *J. Math. Anal. Appl.*, 302:306–317, 2005.

[33] Fan X.L. and Han X.Y. Existence and multiplicity of solutions for $p(x)$-Laplacian equations in $\mathbb{R}^N$. *Nonlinear Anal.*, 59:173–188, 2004.

[34] Gierer A. and Meinhardt H. A theory of biological pattern formation. *Kybernetik*, 12:30–39, 1972.

[35] Gilbarg D. and Trudinger N.S. *Elliptic partial differential equations of second order.* Springer, Berlin, 1998.

[36] Halsey T.C. Electrorheological fluids. *Science*, 258:761–766, 1992.

[37] Harjulehto P., Hästö P., Le U.V., and Nuortio M. Overview of differential equations with non-standard growth. *Nonlinear Anal.*, 72:4551–4574, 2010.

[38] Hess P. and Kato T. On some linear and nonlinear eigenvalue problems with an indefinite weight function. *Comm. Partial Differential Equations*, 5:999–1030, 1980.

[39] Hiriart-Urruty J.-B. A short proof of the variational principle for approximate solutions of a minimization problem. *Amer. Math. Monthly*, 90:206–207, 1983.

[40] Kajikia R. A critical point theorem related to the symmetric mountain pass lemma and its applications to elliptic equations. *J. Funct. Anal.*, 225:352–370, 2005.

[41] Keller E.F. and Segel L.A. Initiation of slime mold aggregation viewed as an instability. *J. Theor. Biol.*, 26:399–415, 1970.

[42] Kovacik O. and Rakosnik J. On spaces $L^{p(x)}$ and $W^{1,p(x)}$. *Czechoslovak Math. J.*, 41:592–618, 1991.

[43] Krasnoselskii M.A. *The operator of translation along the trajectories of differential equations.* Nauka, Moscow, 1963.

[44] Kristaly A., Mihăilescu M., Rădulescu V., and Tersian S. Spectral estimates for a nonhomogeneous difference problem. *Commun. Contemp. Math.*, 12:1015–1029, 2010.

[45] Lavrentiev M. Sur quelques problèmes du calcul des variations. *Ann. Mat. Pura Appl.*, 4:7–28, 1926.

[46] Lindqvist P. On the equation $\mathrm{div}(|\nabla u|^{p-2}\nabla u) + \lambda|u|^{p-2}u = 0$. *Proc. Amer. Math. Soc.*, 109:157–164, 1990.

[47] Liu J.Q. and Su J.B. Remarks on multiple nontrivial solutions for quasilinear resonant problems. *J. Math. Anal. Appl.*, 258:209–222, 2001.

[48] Marcellini P. Regularity and existence of solutions of elliptic equations with (p,q)-growth conditions. *J. Differential Equations*, 90:1–30, 1991.

[49] Mihăilescu M. and Rădulescu V. A multiplicity result for a nonlinear degenerate problem arising in the theory of electrorheological fluids. *Proc. R. Soc. Lond. Ser. A Math. Phys. Eng. Sci.*, 462:2625–2641, 2006.

[50] Mihăilescu M. and Rădulescu V. Existence and multiplicity of solutions for quasilinear nonhomogeneous problems: an Orlicz-Sobolev space setting. *J. Math. Anal. Appl.*, 330:416–432, 2007.

[51] Mihăilescu M. and Rădulescu V. On a nonhomogeneous quasilinear eigenvalue problem in Sobolev spaces with variable exponent. *Proc. Amer. Math. Soc.*, 135:2929–2937, 2007.

[52] Mihăilescu M. and Rădulescu V. Continuous spectrum for a class of nonhomogeneous differential operators. *Manuscripta Math.*, 125:157–167, 2008.

[53] Mihăilescu M. and Rădulescu V. Neumann problems associated to nonhomogeneous differential operators in Orlicz-Sobolev spaces. *Ann. Inst. Fourier*, 58:2087–2111, 2008.

[54] Mihăilescu M. and Rădulescu V. Spectrum in an unbounded interval for a class of nonhomogeneous differential operators. *Bull. London Math. Soc.*, 40:972–984, 2008.

[55] Mihăilescu M. and Rădulescu V. Continuous spectrum for a class of nonhomogeneous differential operators. *Mathematica Scand.*, 104:132–146, 2009.

[56] Mihăilescu M. and Rădulescu V. Concentration phenomena in nonlinear eigenvalue problems with variable exponents and sign-changing potential. *J. d'Anal. Math.*, 111:267–287, 2010.

[57] Mihăilescu M. and Rădulescu V. Sublinear eigenvalue problems associated to the Laplace operator revisited. *Israel J. Math.*, 181:317–326, 2011.

[58] Mihăilescu M., Rădulescu V., and Repovš D. On a non-homogeneous eigenvalue problem involving a potential: an Orlicz-Sobolev space setting. *J. Math. Pures Appl.*, 93:132–148, 2010.

[59] Mihăilescu M., Rădulescu V., and Stancu D. A Caffarelli-Kohn-Nirenberg type inequality with variable exponent and applications to PDE's. *Complex Var. Elliptic Equ.*, 56:659–669, 2011.

[60] Mihăilescu M., Rădulescu V., and Tersian S. Eigenvalue problems for anisotropic discrete boundary value problems. *J. Difference Equ. Appl.*, 15:557–567, 2009.

[61] Mihăilescu M., Rădulescu V., and Tersian S. Homoclinic solutions of difference equations with variable exponents. *Topol. Methods Nonlinear Anal.*, 38:277–289, 2011.

[62] Mihăilescu M., Pucci P., and Rădulescu V. Eigenvalue problems for anisotropic quasilinear elliptic equations with variable exponent. *J. Math. Anal. Appl.*, 340:687–698, 2008.

[63] Mihăilescu M., Moroşanu G., and Rădulescu V. Anisotropic Neumann problems in Sobolev spaces with variable exponent. *Nonlinear Anal.*, 73:3239–3253, 2010.

[64] Minakshisundaram S. and Pleijel A. Some properties of the eigenfunctions of the Laplace-operator on Riemannian manifolds. *Canadian J. Math.*, 1:242–256, 1949.

[65] Molica Bisci G. and Repovš D. Nonlinear algebraic systems with discontinuous terms. *J. Math. Anal. Appl.*, 398:846–856, 2013.

[66] Molica Bisci G. and Repovš D. On some variational algebraic problems. *Adv. Nonlinear Anal.*, 2:127–146, 2013.

[67] Molica Bisci G. and Repovš D. Existence and localization of solutions for nonlocal fractional equations. *Asymptotic Anal.*, 90:367–378, 2014.

[68] Molica Bisci G. and Repovš D. Existence of solutions for p-Laplacian discrete equations. *Appl. Math. Comput.*, 242:454–461, 2014.

[69] Molica Bisci G. and Repovš D. Higher nonlocal problems with bounded potential. *J. Math. Anal. Appl.*, 420:167–176, 2014.

[70] Molica Bisci G. and Repovš D. Multiple solutions for elliptic equations involving a general operator in divergence form. *Ann. Acad. Sci. Fenn. Math.*, 39:259–273, 2014.

[71] Molica Bisci G. and Repovš D. Multiple solutions of p-biharmonic equations with Navier boundary conditions. *Complex Var. Elliptic Equ.*, 59:271–284, 2014.

[72] Molica Bisci G. and Repovš D. Nonlinear Neumann problems driven by a nonhomogeneous differential operator. *Bull. Math. Soc. Sci. Math. Roumanie*, 57:13–25, 2014.

[73] Molica Bisci G. and Repovš D. On sequences of solutions for discrete anisotropic equations. *Expo. Math.*, 32:284–295, 2014.

[74] Musielak J. *Orlicz spaces and modular spaces.* Lecture Notes in Mathematics, Vol. 1034, Springer, Berlin, 1983.

[75] Nakano H. *Modulared semi-ordered linear spaces.* Maruzen Co., Ltd., Tokyo, 1950.

[76] Orlicz W. Über konjugierte Exponentenfolgen. *Studia Math.*, 3:200–212, 1931.

[77] Palais R. Ljusternik–Schnirelmann theory on Banach manifolds. *Topology*, 5:115–132, 1966.

[78] Palais R. and Smale S. A generalized Morse theory. *Bull. Amer. Math. Soc.*, 70:165–171, 1964.

[79] Phelps R. Support cones in Banach spaces and their applications. *Adv. Math.*, 13:1–19, 1974.

[80] Poincaré H. Sur les équations aux dérivées partielles de la physique mathématique. *Amer. J. Math.*, 12:211–294, 1890.

[81] Poincaré H. *Les méthodes nouvelles de la mécanique céleste.* Gauthier-Villars, Paris, 1899.

[82] Pucci P. and Rădulescu V. Remarks on eigenvalue problems for nonlinear polyharmonic equations. *C. R. Acad. Sci. Paris, Ser. I*, 348:161–164, 2010.

[83] Pucci P. and Serrin J. Extensions of the mountain pass theorem. *J. Functional Anal.*, 59:185–210, 1984.

[84] Pucci P. and Serrin J. A mountain pass theorem. *J. Differential Equations*, 60:142–149, 1985.

[85] Rădulescu V. Nonlinear elliptic equations with variable exponent: old and new. *Nonlinear Anal.*, 121:336–369, 2015.

[86] Rădulescu V. and Zhang B. Morse theory and local linking for a nonlinear degenerate problem arising in the theory of electrorheological fluids. *Nonlinear Anal. Real World Appl.*, 17:311–321, 2014.

[87] Rădulescu V. and Repovš D. Perturbation effects in nonlinear eigenvalue problems. *Nonlinear Anal.*, 70:3030–3038, 2009.

[88] Rădulescu V. and Repovš D. Combined effects in nonlinear problems arising in the study of anisotropic continuous media. *Nonlinear Anal.*, 75:1524–1530, 2012.

[89] Rabinowitz P. Variational methods for nonlinear eigenvaalue problems. *Arch. Math. (Basel)*, 59:139–195, 1974.

[90] Rajagopal K.R. and Ruzicka M. Mathematical modelling of electrorheological fluids. *Cont. Mech. Term.*, 13:59–78, 2001.

[91] Repovš D. Asymptotics for singular solutions of quasilinear elliptic equations with an absorption term. *J. Math. Anal. Appl.*, 395:78–85, 2012.

[92] Repovš D. and Semenov P.V. *Continuous selections of multivalued mappings.* Mathematics and its Applications, 455, Kluwer Academic Publishers, Dordrecht, 1998.

[93] Ricceri B. A general variational principle and some of its applications. *J. Comput. Appl. Math.*, 133:401–410, 2000.

[94] Ricceri B. On three critical points theorem. *Arch. Math. (Basel)*, 75:220–226, 2000.

[95] Ruzicka M. *Electrorheological fluids modeling and mathematical theory.* Springer, Berlin, 2002.

[96] Serrin J. Local behavior of solutions of quasi-linear equations. *Acta Math.*, 111:247–302, 1964.

[97] Sharapudinov I. On the topology of the space $L^{p(t)}([0;1])$. *Math. Notes*, 26:796–806, 1979.

[98] Simons J. Régularité de la solution d'une équation non linéaire dans $\mathbb{R}^n$. *Journées d'Analyse Non Linéaire, Lecture Notes in Math., Springer, Berlin*, pages 205–227, 1978.

[99] Stampacchia G. *Équations elliptiques du second ordre à coefficients discontinus.* Presses de l'Université de Montreal, 1966.

[100] Struwe M. *Variational methods: applications to nonlinear partial differential equations and Hamiltonian systems.* Springer-Verlag, Berlin, 1986.

[101] Sullivan F. A characterization of complete metric spaces. *Proc. Amer. Math. Soc.*, 83:345–346, 1981.

[102] Trudinger N. On embeddings into Orlicz spaces and some applications. *J. Math. Mech.*, 17:473–483, 1967.

[103] Tsenov I. Generalization of the problem of best approximation of a function in the space L^s. *Uch. Zap. Dagestan Gos. Univ.*, 7:25–37, 1961.

[104] Willem M. *Minimax theorems.* Progress in Nonlinear Differential Equations and their Applications, 24, Birkhäuser Boston, Inc., Boston, MA, 1996.

[105] Zeidler E. *Nonlinear functional analysis and its applications II/B.* Springer, Berin, 1985.

[106] Zhikov V. Averaging of functionals of the calculus of variations and elasticity theory. *Math. USSR Izv.*, 29:33–66, 1987.

[107] Zou W., Li F., Liu M., and Lv B. Existence of solutions for a nonlocal problem arising in plasma physics. *J. Differential Equations*, 256:1653–1682, 2014.

Index

(S^+), 58
$C_q(\Phi, \infty)$, 62
$C_q(\Phi, u)$, 61
$L^{p(x)}(\Omega)$, 7
$L_{\Phi_i}(\Omega)$, 16
N-function, 16
$W^1 L_\Phi(\Omega)$, 16
$W^{1,(\infty,q(\cdot))}(\Omega)$, 13
$W^{1,p(x)}(\Omega)$, 10
$W_0^1 L_\Phi(\Omega)$, 17
$X^\star$, 282
Δ_2-condition, 17
$\Delta^2_{p(k)}$, 258
Φ^c, 61
ℓ^∞, 259
$\ell^{p(\cdot)}$, 259
$\mathcal{D}_0^{1,p(x)}(\Omega)$, 69
$\mathcal{P}$, 284
AC$[0,1]$, 12
Lip$[0,1]$, 12

Abelian group, 61
absolutely continuous trajectory, 12
activator-inhibitor system, 108
almost critical point, 282
Ambrosetti, A., 30, 47, 74, 283
Ambrosetti-Rabinowitz condition, 55
anisotropic variable exponent, 14, 196, 199, 223
anisotropic variable exponent
    Orlicz-Sobolev space, 199
anisotropic variable exponent
    Sobolev space, 14, 15
approximate critical value, 284
Aristotle, 145

Banach manifold, 192
bifurcation, 244
bifurcation point, 244, 267
Bingham, E., 20
Bingham fluid, 5, 20
biological pattern formation, 115
boundary condition
    Dirichlet, 56, 73
    Neumann, 171
    no-flux, 220, 223
Brezis, H., 25, 30, 33, 221, 283
Browder, F., 283

Caffarelli-Kohn-Nirenberg inequality, 66, 233, 234
Carathéodory function, 48
Cerami compactness condition, 287
chemotactic aggregation model, 108
compact operator, 42
complementary functions, 16
complete metric space, 279, 282
concentration phenomenon, 92
condition
    Ambrosetti-Rabinowitz, 55
    Cerami, 287
    Palais-Smale, 125, 140, 166, 169, 192, 288
cone, 52
conjugate Young function, 172
convolution product, 9
critical
    group, 61
    point, 283

deformation lemma, 285
derivative
    Fréchet, 116
    Gâteaux, 31, 116, 282

Dirichlet boundary condition, 56, 73
Dirichlet principle, 11
Dirichlet, P.G.L., 20
discrete Laplace operator, 258, 278
discrete Luxemburg norm, 259
doubly asymptotic trajectory, 258
Doyle, Sir A.C., 25
duality mapping, 282

effect
  staircase, 6
  Winslow, 21
eigenfunction, 48
eigenvalue, 42
Ekeland variational principle, 43, 46, 73
electromagnetic field, 26
electrorheological fluid, 6, 73
embedding
  compact, 15, 65, 69
  continuous, 8, 11
  Lebesgue, 8
  Sobolev, 20
equation
  Euler-Lagrange, 12
  minimal surface, 28
Euclidean metric, 280
Euler-Lagrange equation, 12
exterior domain, 82

Fermat theorem, 282
Feynman, R., 220
Flemish medium, 5
fluid
  Bingham, 5, 20
  electrorheological, 6, 73
  generalized Newtonian, 122
  non-Newtonian, 26
  smart, 21
flux-divergence theorem, 67, 73
forward difference operator, 250, 269
fountain theorem, 216
Fréchet derivative, 116
Fréchet, M., 116
function
  Carathéodory, 48
  logarithmic Hölder continuous, 13
  Young, 16, 134
  Young conjugate, 172
functional
  continuous differentiable, 282
  locally Lipschitz, 287
  lower semi-continuous, 279
  perturbed, 281
  upper-semicontinuous, 116

Gâteaux derivative, 31, 116
Gâteaux differentiability, 282
Gaussian smoothing, 6
generalized Newtonian fluid, 122
global minimizer, 33, 35, 36, 110, 112, 133, 189
Grothendieck, A., 249

Hamilton-Jacobi theory, 279
Hamiltonian system, 258, 278
Hardy inequality, 66
Hermite, Ch., 279
high-energy solution, 211, 212
Hilbert space, 250, 255, 270
Hölder inequality, 8, 16, 21, 31, 34, 39, 70, 90, 97, 99, 102, 111, 151, 154, 157, 183, 184, 188, 202, 216, 220, 237
Hölder, O., 21
homoclinic orbit, 258, 278
homoclinic solution, 258
homologically nontrivial critical point, 61

image processing, 5
image restoration, 5
indefinite potential, 191
inequality
  Caffarelli-Kohn-Nirenberg, 66, 233, 234
  Hardy, 66
  Hölder, 8, 16, 21, 31, 34, 39, 70, 90, 97, 99, 102, 111, 140, 151, 154, 157, 183, 184, 188, 202, 216, 220, 237

Poincaré, 13, 21, 119, 209
Sobolev, 13, 66
Young, 9, 16, 64, 67, 110, 119, 134, 145, 162

lack of compactness, 82
Lagrange, J.L., 116
Lavrentiev, M., 21
Lavrentiev phenomenon, 12, 21
Lebesgue dominated convergence theorem, 33, 131, 183
Lebesgue, H., 21
Lebesgue measure, 59
lemma
deformation, 285
pseudo-gradient, 285
Lindqvist, P., 91
Lipschitzian trajectory, 12
Ljusternik-Schnirelmann theory, 42, 279
local linking, 54
locally finite subcovering, 286
locally Lipschitz functional, 287
logarithmic Hölder continuous function, 13
low-energy solution, 134, 140, 238
lower semi-continuity, 280
lower semi-continuous functional, 279
Luxemburg norm, 8, 16
discrete, 259

mean continuity property, 8
mean curvature operator, 28
mean value theorem, 31, 74, 182
metric space, 279, 282
minimal surface equation, 28
minimizer, 11
concave, 12
convex, 12
global, 33, 35, 36, 110, 112, 133, 189
linear, 12
minimizing sequence, 279
modular, 9

Morse theory, 54, 62, 74, 283
Moser iteration, 115
mountain of zero altitude, 201
mountain pass theorem, 33, 37, 41, 44, 45, 47, 74, 121, 125, 137, 201, 253, 263, 288

negative part, 51
Neumann boundary condition, 171
Newton, Sir I., 195
no-flux boundary condition, 220, 223
non-Newtonian fluid, 26
nonhomogeneous differential operator, 26, 27
nonlinear elasticity, 122
nonlinear plasticity, 122
norm
equaivalent, 10
Lebesgue, 21
Luxemburg, 8, 16

operator
compact, 42
discrete Laplace, 258, 278
forward difference, 250, 269
homogeneous, 94
mean curvature, 28
nonhomogeneous differential, 26, 27
poly-harmonic, 54
self-adjoint, 42
optimization, 93, 116, 147
orbit
homoclinic, 258, 278
Orlicz space, 16, 17
Orlicz, W., 192
Orlicz-Sobolev space, 17

Palais-Smale condition, 125, 140, 166, 169, 192, 288
path, 284
pattern formation, 108
perturbed functional, 281
phenomenon
concentration, 92
Lavrentiev, 12, 21

Plateau problem, 28, 74
Poincaré inequality, 13, 21, 119, 209
Poincaré, H., 3, 21
poly-harmonic operator, 54
positive part, 51
potential
  indefinite, 191
  sign-changing, 116
  variable exponent, 29
principle
  Dirichlet, 11
  Ekeland variational, 43, 46, 73
problem
  bifurcation, 244
  optimization, 94
  Plateau, 28, 74
property
  compactness, 15
  mean continuity, 8
  multiplicity, 162
  optimization, 156
pseudo-gradient, 287
pseudo-gradient lemma, 285
Pucci, P., 54, 166, 192, 201, 283

Rabinowitz, P., 47, 74, 166, 283
Rayleigh quotient, 34, 43, 48, 74, 77, 78, 83, 93, 94, 147, 148, 156, 158
reflexive space, 8, 10, 14, 17
Rellich-Kondrachov theorem, 96
Ricceri, B., 166, 192, 238, 269, 270
Riesz-Fredholm theory, 42
Roosevelt, E., 117

saddle point, 74, 166
Schopenhauer, A., 75
self-adjoint operator, 42
Serrin, J., 166, 192, 201, 283
set function, 116
sign-changing potential, 116
singular elliptic problem, 233
singular relative homology, 61
smart fluid, 21
Sobolev inequality, 66
Sobolev, S., 21
solution
  high-energy, 211, 212
  homoclinic, 258
  low-energy, 134, 140, 238
  strong, 270
  weak, 29, 58, 71
space
  anisotropic variable exponent Sobolev, 14
  Hilbert, 250, 255, 270
  Orlicz, 16, 17
  Orlicz-Sobolev, 17
  reflexive, 8, 10, 14, 17
  uniformly convex, 8, 10
  variable exponent Lebesgue, 7
  variable exponent Sobolev, 4
  weighted Lebesgue, 109, 235
spectrum, 42
  concentration, 77, 95, 149, 179
  continuous, 42, 49, 134
  discrete, 42, 48
staircase effect, 6
Stampacchia theorem, 32
Struwe, M., 47
symmetric mountain pass theorem, 125, 140, 288

Tagore, R., 283
theorem
  Ambrosetti-Rabinowitz, 283
  Bishop–Phelps, 279
  Dugundji, 286
  Fermat, 282
  flux-divergence, 67, 73
  fountain, 216
  Lagrange multipliers, 116, 147
  Lebesgue dominated convergence, 33, 131, 183
  mean value, 31, 74, 182
  mountain pass, 33, 37, 41, 44, 45, 47, 74, 121, 125, 137, 201, 253, 263, 283, 288
  mountain pass of zero altitude, 283

Pucci-Serrin, 283
Rellich-Kondrachov, 96
Ricceri, 166
Stampacchia, 32
symmetric mountain pass, 125, 140, 288
three critical points, 166, 192
theory
Ljusternik-Schnirelmann, 42
Morse, 54, 62, 74
Riesz-Fredholm, 42
three critical points theorem, 166, 192
topological index, 166
total variation, 6
trajectory
absolutely continuous, 12
doubly asymptotic, 258
Lipschitzian, 12
Trudinger, N., 20

uniformly convex space, 8, 10
upper-semicontinuous functional, 116

Variable exponent Lebesgue space, 4, 7, 8
Variable exponent Sobolev space, 4, 10, 13, 14
vector field, 287
viscosity, 21

weak solution, 29, 58, 71
weighted Lebesgue space, 109, 235
Winslow effect, 21
Winslow, W., 21

Young function, 16, 134
Young inequality, 9, 16, 64, 67, 110, 119, 134, 145, 162

For Product Safety Concerns and Information please contact our EU representative GPSR@taylorandfrancis.com
Taylor & Francis Verlag GmbH, Kaufingerstraße 24, 80331 München, Germany

www.ingramcontent.com/pod-product-compliance
Lightning Source LLC
Chambersburg PA
CBHW060746150926
51501CB00003B/109
9781498703413